Two Manuals Combined into One Book

Marine Operations on Urbanized Terrain

And

Security Force Assistance

Department of Defense

Pentagon
Publishing

Military Operations on Urbanized Terrain

Table of Contents

Chapter 1

Modern Urban Warfare

> **"... the likelihood is high that in the future, the National Command Authorities will again commit Marines to missions in urban areas."**
>
> *A Concept for Future Military Operations on Urbanized Terrain*

1001. Importance of Urban Areas. Throughout history, military planners have viewed cities as centers of gravity. As such, in war, cities are something to be either protected or taken away, depending upon one's perspective (MCDP 1, *Warfighting*). Cities house the population centers, transportation hubs, seats of government, sources of wealth, centers for industry, information networks, and key nodes of communication within a nation. Recent forecasts based on population statistics and the worldwide migration trend from agrarian to industrialized societies predict that 85 percent of the world's population will reside in urbanized areas by the year 2025. As the world trend toward urbanization increases, the military significance of cities is likely to increase proportionally.

Urbanized areas, themselves, may be significant sources of future conflict. Cities historically are where radical ideas ferment, dissenters find allies, mixtures of people cause ethnic friction, and discontented groups receive media attention. Adversaries may focus on the capture of radio and television stations in an attempt to influence public opinion and attain their political goals. Our political leaders may take advantage to neutralize or stabilize some extremely volatile political situations, or to provide assistance to allies in need of support, by deploying U.S. forces into urban environments,

1002. The Marine Corps Role in Urban Warfare. As the Nation's force in readiness, forward deployed with expeditionary forces, Marines must be prepared to fight on urbanized terrain. In the past two decades, MAGTFs ranging in size from MEFs (Saudi Arabia, Desert Shield/Desert Storm; Somalia, Restore Hope) through Marine expeditionary units (MEUs) (Beirut, Lebanon; Grenada, Urgent Fury; Somalia, Eastern Exit and Restore Hope) have participated in MOUT. The task-organization and combined-arms aspect of the MAGTF makes it well suited for combat on urbanized terrain.

The results of geographical studies show that 60 percent of politically significant urban areas outside allied or former Warsaw Pact territory are located along or within 25 miles of a coastline; 75 percent are within 150 miles; 87 percent are within 300 miles; 95 percent are within 600 miles; and all are within 800 miles. U.S. embassies and diplomatic facilities are primarily located in cities where the host country's political and economic leadership is concentrated. The Marine Corps will continue to play a prominent role in future evacuations of U.S. citizens, as well as the conduct of peace, counterinsurgency, and contingency operations centered on urbanized areas.

Today's Marine air-ground task forces (MAGTFs) are deployed as part of naval expeditionary forces (NEFs) that maintain a *global forward presence for rapid crisis response*. These integrated combined-arms forces are part of the Nation's proven contingency and naval power projection force. Therefore, Marines may find themselves rapidly deployed and employed in actions across the spectrum of military operations. Many of these trouble spots will likely be located in or around large urban centers.

In the years since World War II, the United States has employed military force more than 200 times. Of these, four out of five involved naval forces, and the majority of the naval efforts included Marines embarked in amphibious ships. The reasons are straightforward: availability and adaptability. Availability derives from the loiter time of forward deployed forces embarked on amphibious shipping. Adaptability comes from the Marine Corps' MAGTF organization, doctrine, training, and equipment, which prepare us for expeditionary missions from the sea in support of a variety of missions, including forcible entry. Enhancing our adaptability are the maritime prepositioning forces (MPFs). MPFs provide a rapid buildup of combat and logistics equipment that is joined with Marines on a distant shore, creating a substantial combat force. Despite our availability and adaptability, the prospect of urban warfare combined with an amphibious assault is a complex task which requires special preparation.

At the outset of a developing situation, forward-deployed expeditionary forces can move quickly within range of a crisis that threatens the political stability of a country. Urban intervention operations must often be planned and executed in a matter of hours or days (rather than weeks or months) to take advantage of the internal turmoil surrounding a developing crisis. Navy and Marine forces should anticipate deployment to urbanized areas on a day-to-day basis.

1003. Distinguishing Features of Urbanized Terrain. Urbanized terrain is a complex and challenging environment. It possesses all of the characteristics of the natural landscape, coupled with manmade construction, resulting in an incredibly complicated and fluid environment that influences the conduct of military operations in unique ways.

Military operations on urbanized terrain (MOUT) is defined as *all military actions planned and conducted on a topographical complex and its adjacent terrain where manmade construction is the dominant feature. It includes combat in cities, which is that portion of MOUT involving house-to-house and street-by-street fighting in towns and cities* (Marine Corps Reference Publication (MCRP) 5-12A, *Operational Terms and Graphics*). MOUT effects the tactical options available to a commander. A built-up area is *a concentration of structures, facilities, and populations, such as villages, cities, and towns,* that form the economic and cultural focus for the surrounding area. (MCRP 5-12A)

a. Cities. Cities are centers of finance, politics, transportation, communication, industry, and culture. They generally have large population concentrations ranging from tens of thousands to millions of people. Because of their psychological, political, or logistical value, control of cities have often been the scenes of pitched battles.

(1) Operations in built-up areas are normally conducted to capitalize on the operational or tactical significance of a particular city. In developing nations, control of only a few cities is often the key to the control of national resources. The side that controls a major city usually has a psychological advantage, which can be enough to significantly affect the outcome of a countrywide conflict.

(2) The abundance of guerrilla and terrorist operations conducted in built-up areas (e.g., Santo Domingo, Caracas, Belfast, Managua, and Beirut) demonstrates the importance many insurgent groups place on urban warfare.

(3) In the past 40 years, many cities have expanded dramatically, losing their well-defined boundaries as they extended into the countryside. New transportation systems (highways, canals, and railroads) have been built to connect population centers. Industries have grown along those connectors, creating "strip areas." Rural areas, although retaining much of their farmlike character, are connected to the towns by a network of secondary roads.

b. Multiple Avenues of Approach. Urbanized terrain is a unique battlespace that provides both attacker and defender with numerous and varied avenues of approach and fields of fire. The urban battlespace is divided into four basic levels: *building*, *street*, *subterranean,* and *air*. Operations can be conducted from above ground, on ground level, inside buildings, or below the ground. Most operations will include fighting on all levels simultaneously.

(1) Building Level. Buildings provide cover and concealment; limit or increase fields of observation and fire; and canalize, restrict, or block movement of forces, especially mechanized forces. They provide optimum perches for snipers and antiair weapons. Buildings also provide antitank weapons optimum positioning to allow engagement from above, exploiting an inherent weakness found in most armored vehicles.

(2) Street Level. While streets provide the means for rapid advance or withdawal, forces moving along streets are often canalized by buildings and have little space for off-road maneuver. Because they are more difficult to bypass, obstacles on streets in urbanized areas are usually more effective than those on roads in open terrain.

(3) Subterranean Level. Subterranean systems are easily overlooked but can be important to the outcome of operations. These areas may be substantial and include subways, sewers, cellars, and utility systems (Figure 1-1 on page 1-4). The city of Los Angeles alone has more than 200 miles of storm sewers located under the city streets. Both attacker and defender can use subterranean avenues to maneuver to the rear or the flanks of an enemy. These avenues also facilitate the conduct of ambushes, counterattacks, and infiltrations. (See Appendix E)

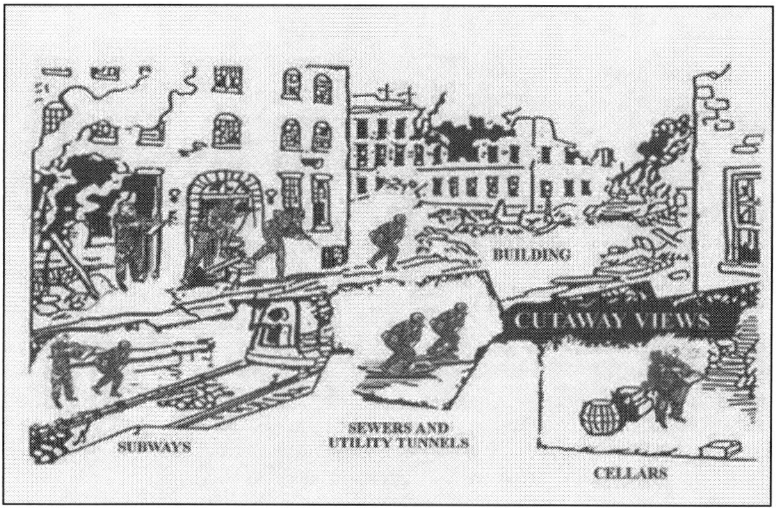

Figure 1-1. Subterranean Systems

(4) Air Level. The air provides another avenue of approach in urbanized areas. Aviation assets can be used for high speed insertion or extraction of troops, supplies, and equipment. While aviation assets are not affected by obstacles on the streets, they are affected by light towers, signs, power lines, and other aerial obstructions. They are also vulnerable to the man-portable surface-to-air missile threat, crew served weapons, and small arms fire.

c. Categories of Built-Up Areas. Built-up areas are generally classified as:

- Villages (populations of 3,000 or less)
- Strip areas (industrialized zones built along roads connecting towns or cities)
- Towns or small cities (populations of up to 100,000 and not part of a major urban complex)
- Large cities with associated urban sprawl (populations in the millions, covering hundreds of square kilometers).

d. Characteristics of Urbanized Areas. A typical urban area consists of combinations of the *city core*, *commercial ribbon*, *core periphery*, *residential sprawl*, *outlying industrial areas*, and *outlying high-rise areas*. Each of the urban area's regions has distinctive characteristics that may weigh heavily in planning for MOUT. Most urbanized areas resemble the generalized model shown in Figure 1-2.

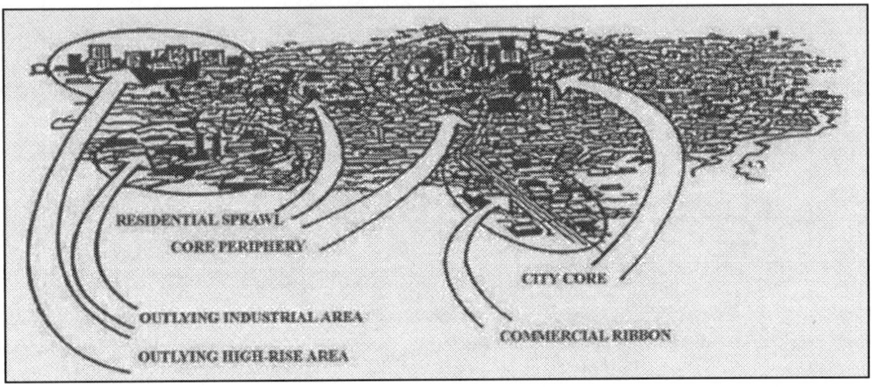

Figure 1-2. Typical Urban Area

(1) City Core. In most cities, the city core has undergone more recent development than the core periphery. As a result, the two regions are often quite different. Typical city cores are made up of high-rise buildings which vary in height. Modern urban planning for built-up areas allows for more open spaces between buildings than in old city cores or in core peripheries (Figure 1-3).

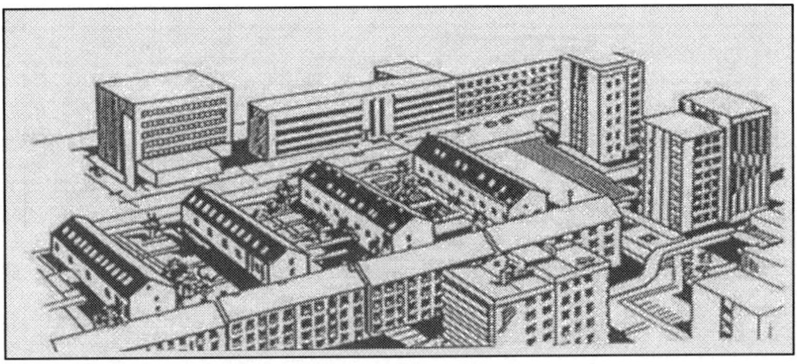

Figure 1-3. City Core

(2) Commercial Ribbon. Commercial ribbons are composed of rows of stores, shops, and restaurants that are built along both sides of major streets through built-up areas.

Typically, such streets are 25 meters wide or more. The buildings in the outer areas are uniformly two to three stories tall—about one story taller than the dwellings on the streets behind them (Figure 1-4).

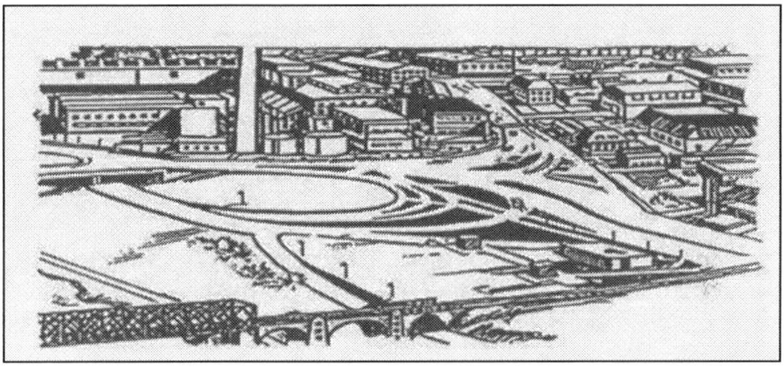

Figure 1-4. Commercial Ribbons

(3) Core Periphery. The core periphery generally consists of streets 12 - 20 meters wide with continuous fronts of brick or concrete buildings. The building heights are fairly uniform—2 or 3 stories in small towns, 5 to 10 or more stories in large cities (Figure 1-5).

Figure 1-5. Core Periphery

(4) Residential Sprawl. Residential sprawl areas consist mainly of low houses or apartments that are one to three stories tall. The area is primarily composed of detached dwellings that are usually arranged in irregular patterns along streets, with many smaller open areas between structures (Figure 1-6).

Figure 1-6. Residential Sprawl

(5) Outlying Industrial Areas. These areas generally consist of clusters of industrial buildings varying from one to five stories in height. Buildings generally vary dramatically in size and composition to match the needs of the particular businesses they house. Industrial parks are good examples of this category (Figure 1-7).

Figure 1-7. Outlying Industrial Areas

(6) Outlying High-Rise Areas. These areas are similar in composition to city core areas, but may be composed of clusters of more modern multistory high-rise buildings in outlying parts of the city. Building height and size may vary dramatically (Figure 1-8). Generally, there is more open space between buildings located in the outlying high-rise areas than is found within the city core area.

Figure 1-8. Outlying High-Rise Area

1004. Modern Battles for Urbanized Terrain. Urban warfare is as old as war itself. Since man began building villages, he has fought battles in and around them. Geography, politics, and economics dictate that cities will continue to be an objective of armies in warfare. From the armies that invaded and liberated Europe twice during the 20th century, to the forces that fought in Korea and Vietnam, to our most recent urban battles in the Middle East and Southwest Asia, *the basic principles of combat in built-up areas have essentially remained unchanged in this century*. While the principles remain the same, the introduction of helicopters, fixed-wing aircraft, armor, and precision-guided munitions (PGMs) has altered some of the techniques associated with urban combat.

Twenty two modern urban battles are discussed to illustrate the trends, dominant factors, and principles of combat in urbanized areas. (The majority of this material is extracted from the 1987 study, *Modern Experience in City Combat,* produced by Abbott Associates.)

• Aachen	• Manila
• Arnhem	• Ortona
• Ashrafiyeh	• Quang Tri City I
• Ban Me Thuot	• Quang Tri City II
• Beirut Port/Hotel (I)	• Seoul
• Beirut 1982 (II)	• Sidon
• Berlin	• Stalingrad
• Cherbourg	• Suez City
• Hue	• Tel Zaatar
• Jerusalem	• Tyre
• Khorramshahr	• Zahle

A brief description of each battle is provided to illustrate that battle's significance.

a. Stalingrad (1942 - 1943). The tenacious Soviet defense of Stalingrad cost the attacking Germans dearly in every way and set up conditions for a decisive counteroffensive. This classic urban battle involved large forces and resulted in innovative urban combat techniques and the creation of the highly successful storm groups (task-organized assault units). (Length of battle: greater than 30 days) (Casualties: 1,630,000+)

b. Ortona (1943). In this Italian town, determined resistance by a battalion of the elite German 3rd Parachute Regiment against Canadian Army attackers demonstrated the difficulty of overcoming a well-prepared defense. The Canadians were unfamiliar with urban combat and had to develop urban fighting techniques during the battle. After the town was largely destroyed and the defender had extracted a high cost in time and casualties from Canadian forces, the German parachute battalion withdrew. (Length of battle: 6 - 13 days) (Casualties: estimated in the hundreds)

c. Aachen (1944). The battle for Aachen, Germany, in the fall of 1944 developed during the U.S. First Army's offensive to breach the Westwall fortifications and the vaunted Siegfried Line. Aachen, the ancient capital of Charlemagne, had symbolic political and psychological significance to both the Germans and Americans. Furthermore, it was the first city on German soil to face an assault by the Allies. This first major battle on German soil foreshadowed bitter resistance against the American attackers in subsequent battles. The German defenders surrendered only after the city was destroyed. Although the U.S. Army had achieved a clear tactical victory, the German defense of Aachen cost the U.S. First Army valuable time and delayed the planned attack to the Rhine river. (Length of battle: 14 - 30 days) (Casualties: 8,000+)

d. Arnhem (1944). On September 17, 1944, Operation Market-Garden, the largest airborne operation in history, was launched in the Netherlands. The plan was to land three airborne divisions to seize key bridges along a 100-kilometer-long corridor through which allied mechanized forces would pass as the first step in the final offensive into Germany. The British 1st Airborne Division made a surprise landing near the Dutch city of Arnhem in order to seize a bridge over the Rhine river for advancing British forces. An unexpected German armor force counterattacked and eliminated all footholds, virtually destroying the British division before a linkup could occur. (Length of battle: 6 - 13 days) (Casualties: estimated in the thousands)

e. Cherbourg (1944). By June 17, 1944, U.S. forces advancing toward Cherbourg from the Normandy beachhead succeeded in cutting off defending German forces in the Cotentin Peninsula. Four German divisions withdrew to a perimeter surrounding Cherbourg. After much fighting, particularly in strongpoints outside the city, the German garrison surrendered to the Americans. Unfortunately for the Americans, the port facilities were destroyed which denied their early use by Allied forces. (Length of battle: 6 - 13 days) (Casualties: estimated in the thousands)

f. Berlin (1945). The long, bloody Soviet offensive to seize the German capital city effectively concluded the last battle of World War II in Europe. Bitter fighting occurred, but the defense was never well coordinated due in part to poor preparation by the Germans. (Length of battle: 14 - 30 days) (Casualties: estimated in the thousands)

g. Manila (1945). Japanese Army troops evacuated Manila under pressure from advancing American forces, but the local Japanese naval commander independently decided to hold the city at all costs. Despite defending Manila with poorly trained and equipped personnel, the determined resistance resulted in a high number of casualties to attacking U.S. forces as well as the destruction of the city and much of its population. (Length of battle: 14 - 30 days) (Casualties: 22,000+)

h. Seoul (1950). Following the Inchon landing, U.S. and Republic of Korea (ROK) forces recaptured the South Korean capital from the North Koreans. The fighting was unusual in that combat was largely centered on seizure of street barricades rather than buildings. (Length of battle: 6 - 13 days) (Casualties: Marines, 2,383; others, estimated in the thousands)

i. Jerusalem (1967). Israeli forces seized Jerusalem in a well prepared and well executed operation. Despite an uncoordinated Jordanian defense, Israeli casualties in this battle were the highest of those encountered during the Six Day War. Regular Jordanian forces withdrew during the latter stages of the battle, effectively ending organized resistance. (Length of battle: 48 hours to 5 days) (Casualties: Israeli forces, 400+; Jordanian forces, estimated in the hundreds)

j. Hue (1968). On January 31, 1968, the North Vietnamese Army (NVA) and Viet Cong (VC) forces launched their Tet Offensive at targets throughout South Vietnam. As part of this operation, two NVA/VC regiments and two sapper battalions conducted a surprise attack and seized part of the walled city (Citadel) of Hue. The NVA/VC held this part of the walled city for about three weeks against determined U.S. and South Vietnamese forces before finally succumbing. The battle for Hue is considered one of the most intense and savage battles of the Vietnam War. (Length of battle: 14 - 30 days) (Casualties: Marines, 433; others, 5000+)

k. Quang Tri City I and II (1972). An objective of the North Vietnamese 1972 winter-spring offensive was the capture of Quang Tri, the northernmost major city in South Vietnam. The NVA overwhelmed the Army, Republic of Vietnam (ARVN) defenders (I). Later, the city was recaptured (II) by a smaller ARVN force using extensive artillery and air support. The large conventional forces involved on both sides made Quang Tri I and II the major urban battles of the Vietnam War. (Length of battle: Quang Tri I, 6 - 13 days; Quang Tri II, 30 days or greater) (Casualties: battles combined, 30,000+)

l. Suez City (1973). Israeli Defense Forces (IDF) attempted to seize this Egyptian city before the anticipated United Nations (U.N.) cease-fire to end the Yom Kippur War. IDF armored shock tactics led to disaster against a well-prepared Egyptian defense. High casualties forced the IDF to withdraw. (Length of battle: less than 24 hours) (Casualties: Israeli forces, estimated 100 - 500; others, unknown)

m. Ban Me Thuot (1975). This South Vietnamese highlands town was the first strategic city to fall in the final North Vietnamese general offensive in 1975 South Vietnamese forces were surprised and overwhelmed. The fall of Ban Me Thuot resulted in a rout that the North Vietnamese exploited to achieve total victory in Vietnam. (Length of battle: 24 - 48 hours) (Casualties: estimated in the hundreds)

n. Beirut I (1976). When the Lebanese civil war broke out in the spring of 1975, combat in the capital city of Beirut assumed a central role. The battle for Beirut I was a series of small, local operations between largely irregular Christian and Muslim forces fighting over control of the hotel and port districts. Combat was not decisive, but led to changes in the boundary, called the "Green Line." This separated the antagonists and lead to the stagnation of the Lebanese conflict. (Length of battle: greater than 30 days) (Casualties: estimated in the hundreds)

o. Tel Zaatar (1976). Lebanese Christian attackers encircled and methodically besieged this Palestinian camp before overcoming its defenders with a final assault. (Length of battle: greater than 30 days) (Casualties: estimated in the hundreds)

p. Ashrafiyeh (1978). The Syrian forces occupying portions of Lebanon faced a complex political situation in which the power of the Christian militia was seen as a clear threat to stability. In an attempt to weaken the Christian militia by an attack on their center of power, the Syrians laid siege to the Christian militia stronghold of East Beirut (Ashrafiyeh). This urban battle was essentially an artillery bombardment without air attacks. Syria failed to break the will of the defenders and final positions remained unchanged. (Length of battle: greater than 30 days) (Casualties: estimated in the hundreds)

q. Khorramshahr (1980). Iranian regular forces initially evacuated this port city in the face of an Iraqi offensive. Irregular Iranian forces, however, continued to fight. They offered prolonged resistance and inflicted heavy casualties. Iraq eventually won this battle, but at a high cost in time and resources that ultimately served to halt the entire offensive against Iran. The intensity of fighting during the battle for Khorramshahr earned the city the nickname, "City of Blood." (Length of battle: 14 - 30 days) (Casualties: Iraqi, 3,000 - 9,000; Iranian, estimated in the thousands)

r. Zahle (1981). Syria laid siege to the Lebanese regular forces and militia in this key crossroads town. Fighting was inconclusive and ended in a negotiated settlement whereby the Lebanese defenders evacuated the town. (Length of battle: greater than 30 days) (Casualties: estimated in the hundreds)

s. Beirut II (1982). The siege of Beirut culminated the Israeli campaign to evict the Palestine Liberation Organization (PLO) from Lebanon. Fighting under domestic and world political pressures, the IDF besieged the PLO, selectively applying heavy ground and air firepower in conjunction with psychological warfare and limited-objective ground operations. The fighting resulted in a negotiated PLO evacuation from the city. (Length of battle: greater than 30 days) (Casualties: 2,300+)

t. Sidon (1982). Israeli forces easily seized this PLO southern headquarters during the invasion of Lebanon. The IDF was fully prepared for major urban combat using lessons learned from earlier battles, but resistance was unexpectedly light as PLO forces had largely withdrawn from the city. (Length of battle: 48 hours to 5 days) (Casualties: unknown)

u. Tyre (1982). The Israeli attack on the PLO in this Lebanese coastal city was well planned, with excellent intelligence on the target. All branches of the IDF participated in an operation that included naval fire support and amphibious landings. PLO resistance was uncoordinated and easily overcome. (Length of battle: 48 hours to 5 days) (Casualties: Israeli forces, 120+; others unknown)

1005. Modern Urban Battle Analysis and Observations. The 20th-century urban warfare experience has caused us to reevaluate old factors and consider new developments that affect the way we fight in this environment. The following discussion, based on the historical analysis of these select urban battles, illustrates the importance of maneuver warfare and combined arms philosophies in the urban environment. Factors that have had an impact on the manner in which urban warfare has been conducted are:

Intelligence is imperative to success in urban warfare. "Maneuver warfare requires a firm focus on the enemy" (MCDP 2, *Intelligence*). Few subsequent tactical changes can overcome the far-reaching impact of a major intelligence error.

a. Intelligence. The historical review of modern urban battles discloses that the attacker will usually win. Failures to win generally reflect classic military errors not characteristically unique to cities. *However, of the battles studied, battles lost were attributed to errors in* *initial intelligence*. The battles of **Arnhem** and **Suez City** probably would never have occurred had the attacker known the strength and locations of the defender's forces. At **Stalingrad**, the attacking Germans were aware of the defending forces facing them in the Sixth Army's zone, but the buildup of Soviet forces in other areas of the front was not anticipated and resulted in tactical surprise at those points, diluting the German offensive push to seize the city.

Surprise is a combat multiplier and can substantially reduce the cost of urban warfare.

b. Surprise. Surprise is a combat multiplier and can substantially reduce the cost of urban warfare. It can be achieved through deception, stealth, and ambiguity (MCDP 1-3, *Tactics*). Surprise was achieved by the attacker at **Aachen** and **Ban Me Thuot** and by the defender at **Suez City**. Surprise can be an important asset to increase leverage, but, as the failure at **Arnhem** shows, not necessarily a decisive one. When surprise is employed as a means to overcome other disadvantages, it is important to maintain accurate intelligence. In urban areas, tactical surprise by the attacker can be used to preempt effective defensive preparation of a city.

c. Combined Arms. The MAGTF must capitalize on one of the key means for gaining advantage in maneuver warfare — the use of combined arms (MCDP 1-3, *Tactics*). The use of combined arms places the enemy in a dilemma. Any action the enemy takes to avoid one combat arm makes him more vulnerable to another. An analysis of categories of weapons systems found in a MAGTF helps to illustrate it is a warfighting organization well suited for MOUT:

Combat in urban areas is primarily a small unit, infantry intensive operation

(1) Infantry. Combat in urban areas is primarily a small-unit, infantry intensive operation. Restrictions on maneuver, particularly for mechanized units, increases oportunities for infiltration. Urban combat requires small-unit leadership, initiative, and skill. Decentralized actions and difficulties in command , control, and communications are typically encountered. Built-up areas, like close terrain found in other operational environments, are generally considered to be most suited for operations conducted by infantry. Infantry units can be organized, trained and equipped to negotiate urbanized terrain that restricts observation, fields of fire, and mechanized movement.

(2) Armor. The role of armor in urban warfare can be significant. Of the 22 battles studied, armor participated in 21. In three-fourths of these battles, organic tank support was a central element when special assault teams were employed. *Overall, special assault units supported by tanks were more successful than any other task organization.*

The use of tanks to the attacker inside a city has been effective only when they were protected by infantry. Tanks in support of infantry act as an "assault gun" that delivers concentrated, sustained fires to reduce held strongpoints.

The U.S. experience in **Hue** demonstrated the key role that armor can play on a combined-arms team fighting inside the city. The Marines' most effective weapons during the battle were the M48A1 Patton tank and the M-50 Ontos. Both were protected by infantry. The M48A1, with its 90-mm main gun, was used extensively to reduce fortified positions. The Ontos, an armor-protected tracked vehicle mounting six 106-mm recoilless rifles, was highly effective against concrete and steel structures. The munitions of these armored systems provided breaches that the infantry could exploit.

Armor providing direct-fire support inside a built-up area requires protection by infantry.

In contrast, during the **Suez City** battle, Israeli armor forces attacked on "armor thrust avenues" into the city, outpacing their armored personnel carrier (APC) mounted paratroop/infantry support. The Egyptian defenders lacked organic artillery (except limited antiaircraft artillery (AAA) and mortars) and had no air support and virtually no armor support. The Egyptians prepared "kill zones" on the principal avenues down which the IDF armored forces attacked. As the lead IDF armor battalion entered the second of the three road intersection objectives, the Egyptians engaged with Sagger missiles, RPGs, ZSU-23 antiaircraft guns, antitank grenades thrown from balconies, and small arms. All of the tank commanders in the lead battalion were killed or wounded. Disabled vehicles blocked the road. Vehicles veering into the narrow side streets became trapped and were destroyed. *The lack of infantry to protect the armor proved disastrous to the attacking armor battalion.*

(3) Artillery. Artillery has played an important role in most major urban conflicts. At **Aachen,** U.S. forces combined infantry with antitank teams and artillery (in a direct-fire role) down to the squad and fire team level. Also, artillery firing shells with "delay" fuzes in an indirect-fire role were used to penetrate one or more floors before exploding, thus driving the enemy to the ground where infantry and armor could attack. Artillery was also positioned to fire *perpendicular* to the direction of movement of assaulting forces. Thus, fratricide from artillery range errors was alleviated.

Artillery has two distinct roles: outside the built-up area to isolate or prevent isolation with indirect-fire; and within the built-up area to provide direct-fire support.

Artillery proved most useful for interdicting enemy supplies, enemy evacuation, and the movement of reinforcements in the enemy's rear; for physically and psychologically harassing the enemy; and in direct-fire roles within a city.

Artillery employed in the indirect-fire role has been effective in disrupting defenders in half of the studied battles. However, some indirect-fire roles have proved more effective than others in urban combat. Artillery was most effective in the interdiction of supplies, enemy evacuation, movement of reinforcements in the enemy's rear (outside the city), and for indiscriminate physical and psychological harassment of the enemy. Artillery used as an indirect-fire siege weapon, as was done at **Ashrafiyeh** and **Zahle**, proved ineffective. Artillery can also cause problems for the attacker. The rubble resulting from indirect artillery fires can create considerable obstacles for the attacker while providing the defender with obstacles, materials, cover, and concealment.

Artillery employed in the direct-fire role proved useful in the reduction of strongpoints. Self-propelled artillery was used effectively inside **Aachen** and **Stalingrad** and more recently by the IDF at **Beirut II**.

(4) Mortars. The mortar is the most used indirect-fire weapon in urban combat. The mortar's high angle of fire allows the round to reach the street level accurately without being masked by surrounding structures. During the battle for **Hue** city, the most effective fire support provided to Marines was the indirect fire from 60-mm, 81-mm, and 4.2-inch mortars. Approximately 20,000 rounds of high explosive (HE) 60-mm and 81-mm mortar ammunition were expended during the battle. The 4.2-inch mortar was used primarily to deliver riot control agent munitions. It was discovered that 4.2-inch riot control agent munitions could be fired with great effectiveness through the tops of buildings to drive the enemy out. When the enemy emerged from their concealed positions, HE rounds were fired to complete the attack.

(5) Antiaircraft Artillery. Antiaircraft artillery (AAA) was extremely useful in a ground fire role in some urban battles. AAA was used only rarely in World War II, and then generally against assaulting personnel rather than against structures. AAA has been used more frequently in more recent events, but against buildings rather than people. *The high rates of fire of modern AAA make it excellent in terms of shock and destructive potential.* However, ammunition supply can be a problem because the volume of fire. During **Hue**, the Duster, an AAA vehicle mounting twin 40-mm guns, was provided to the Marines by the U.S. Army. These guns were extremely effective in the suppression of enemy positions due to the lethality of the 40-mm round and the quantities in which it could be delivered. During the **Suez City** battle, the Egyptians used the ZSU-23 antiaircraft gun in the direct-fire role against armored columns. This gun proved to be frightening and effective. Its high volume of explosive power created shock among IDF armor personnel akin to the shock that the IDF armor thrust tactics were designed to create. During **Sidon,** the IDF used 20-mm antiaircraft Vulcans in the direct-fire role against ground targets with great success.

(6) Aviation. Historically, aviation assets have played an important role in helping to isolate the objective and to interdict the flow of the defender's supplies and reinforcements. However, aviation has been relatively ineffective when not used in conjunction with ground forces. In past wars, bombing operations have been used in attempts to reduce the defender's will to resist and destroy their physical capabilities. In the majority of urban battles, aerial bombing, by itself, did not erode the defender's will to resist, nor did it significantly degrade the defender's military capabilities.

Aviation plays an important role in interdicting movement of the defender's supplies and reinforcement into the built-up area.

The Marine air-ground task force is employed as a combined arms team. As such, the aviation combat element will not usually be called upon to conduct independent operations in MOUT. It will function as part of a MAGTF and, regardless of operating on or over urbanized terrain, will still execute the six functions of Marine aviation (i.e., offensive air support, assault support, antiair warfare, electronic warfare, air reconnaissance, and control of aircraft and missiles) in support of the MAGTF.

In future urban warfare, aviation will be even more effective due to advances in fixed- and rotary-wing aircraft, unmanned aerial vehicles, precision guided munitions, improved munitions, communications, sensors, and targeting systems. Our battle study already indicates a trend toward more extensive aviation participation in MOUT. As an example, the IDF in **Beirut II** employed bombing by fixed-wing aircraft using cluster bomb units, "smart" bombs, phosphorous, and other munitions. Attack helicopters operated on the outskirts of the built up area with impunity, and medical evacuation (MEDEVAC) proceeded swiftly and efficiently using helicopter support.

d. Combat Forces. Whether attacking or defending, the size of the force relative to the enemy can be critical to success. When provided with adequate forces, the attacker can isolate and encircle the defender and prevent a breakout or linkup. The defender can use them to create a mobile defense or to create strong reserves for counterattacks.

Other factors that impact on the size of forces required are the degree of surprise achieved and the firepower utilized (aviation, armor, artillery, mortars, etc.). They should be weighed against the sophistication of the prepared defense. Ample consideration should be given to the local population, degree of external support, and utilization of existing services (communications, water, etc.).

Regardless of the size or quality of defensive forces, the defender usually extracts large costs from the attacker in time, resources, and casualties.

In the historical examples reviewed, the defender was usually outnumbered by the attacker, the quality of the defender's available forces was inferior, and defeat of the forces defending the city was usually certain. However, regardless of the size or quality of defensive forces, the defender can extract enormous costs to the attacker in time, resources, and casualties. As was seen at **Khorramshahr**, the Iranian defenders, outnumbered 4 to 1, still held the city for 26 days. Although the Iranian defenders eventually lost the city, its defense allowed the remaining Iranian forces time to organize and redeploy. Furthermore, the winter rains that followed the battle turned much of the region into a sea of mud and largely halted further Iraqi efforts. The Iraqi army's offensive thrusts into Iran lost momentum as a result of the defense of Khorramshahr.

e. Special Assault Teams. In these battle studies, "shock units" or "special assault teams" have been used by attackers (and often by defenders) with great success. These organizations were characterized by the integration of combined arms at the battalion level and below. Control was decentralized to lower echelons (down to the squad level in some cases). Assault teams typically contained infantry with various combinations of armor, artillery, and engineers. Hence, all combat arms should plan, train, and develop common tactics, techniques, and procedures for use on urbanized terrain. Before the battle of **Aachen**, U.S. ground forces conducted intensive urban training with antitank teams and artillery pieces down to the squad and fire team levels. In contrast, the defender, in most cases having to fight with whatever forces were locally available, had no opportunity to conduct combined-arms training.

The use of combined arms warfare on urbanized terrain is imperative.

f. Time. In most cases, successful conclusion of an urban battle took two to three times longer than the initial estimates. Consequently, the additional tome resulted in the expenditure of more logistics and the loss of more personnel then initially anticipated. This often had adverse affects on the overall campaign. Well-planned urban defense, even if the defender is isolated or lacking in aviation, armor, or artillery weapons, can be time consuming to the attacker. Time can allow the defender to reorganize, redeploy, or marshal resources in other areas.

Generally, urban warfare is time consuming.

Three battles where time played an unanticipated critical role in the attacker's strategic timetable were **Aachen**, **Khorramshahr**, and **Stalingrad**. In these battles, the defenders delayed the attacker longer than was estimated, resulting in the modification of operational or strategic plans.

g. Isolation. *The attacker won all urban battles where the defender was totally isolated.* Even the partial isolation of the defenders resulted in attackers enjoying a success rate of 80 percent. Conversely, attackers won only 50 percent of the battles in which defenders were not significantly isolated, and those victories came at great cost.

No single factor is more important to the attacker's success than isolation of the urban area.

In most urban battles, some form of isolation occurred as a result of the attacker's actions. While it is unrealistic to envision complete isolation of a city until enemy forces to the rear of the city are pushed far beyond its outermost boundaries, *total isolation does not appear necessary.* The key to the attacker's success is in stemming the *unimpeded* flow of manpower, supplies, and weapons to replace the defender's losses.

h. Cost. The cost of conducting urban warfare is relative to the percentage of total expended resources, the time elapsed, and the results achieved. The cost to the attacker was considered high in the majority of urban battles. A high cost does not necessarily imply that the results were not worth the price. *The attacker and defender must thoroughly evaluate the overall cost prior to committing to an urban battle.*

From the offensive point of view, several factors can be associated with cost. First, and most importantly, isolating the urban area is critical to the attacker's success. Second, overwhelming superiority is needed if all costs are to be minimized. Third, the operation should be carefully planned. Fourth, intelligence is invaluable. Knowing where and how the city has been prepared for defense is also important. Fifth, attacking forces should understand the unique nature of urban combat. Clear tactics, techniques, and procedures for urban combat is required. Every aspect, from taking a building to using destructive technology and coordinating combined arms, must be thoroughly understood. Careful consideration of these five factors can minimize the cost of urban warfare to the attacker.

Attacker cost was generally high in casualties, time, and resources in the majority of urban battles studied.

From the standpoint of the defender, the critical variable is the defensive preparation of the city. Defensive preparations should include measures to prevent isolation of the city by the attackers. The capture of a prepared city can be made to be extremely costly. Preparations can include creating kill zones, clearing fields of fire, constructing canalizing obstacles, establishing reinforcing and fall-back positions, decentralizing command and control, and organizing multiple movement routes above ground, between rooftops, and below ground in subways and sewers. Artillery and aviation support can also be valuable force multipliers in defense.

i. Rules of Engagement (ROE). The nature of the military operation may restrict our use of weapons. The majority of urban battles since 1967 (such as **Beirut II**, **Hue**, **Jerusalem**) have had one or more of the following restrictions imposed on the attacking force:

(1) Minimizing civilian casualties and/or collateral destruction in order to:

- Avoiding alienation of the local population
- Reducing the risk of adverse world or domestic opinion
- Preserving facilities for future use
- Preserving cultural facilities and grounds.

(2) Limiting the use of specific ground or air weapons.

j. Logistics. Timely combat service support, particularly in the areas of ammunition resupply and casualty treatment and evacuation, is a critical element in MOUT.

(1) Logistics Support. Historically, combat in urban environments has seen a dramatic increase in the amounts of Class IV (shoring, sandbags, concertina wire), Class V (ammunition), and Class VIII (medical material). Unique items, such as rope, grappling hooks, and ladders are required for operations on urbanized terrain. Intense close-quarter combat requires a continuous flow of ammunition, particularly small-arms, tank, antitank, mortar, and artillery ammunition, as well as mines, grenades, and demolition explosives. Medical supplies must be readily available to treat the anticipated increase in casualties. Once battle the has been initiated, combat forces will require continuous supply.

Urban warfare requires a flexible, balanced logistics system capable of sustaining the close-quarter fighting evolutions within the built-up area.

One method of providing continuous logistics support is to establish a "push system." Essentially, this system pushes supplies to fighting units without their having to request them. A "push system" should negate the inherent delays of a "pull system," which requires units to request supplies then await their arrival. Optimally however, a balanced approach should be taken to sustain engaged forces (MCDP 4.). Combat service support, in terms of its timeliness and anticipation of the true needs of the combat units, should sustain the tempo of operations. It should avoid the delivery of unwanted or unnecessary supplies. To achieve this balanced approach, logisticians must develop flexible support plans that readily adapt to the ebb and flow of urban combat.

Mobile aid stations and surgical teams should be positioned as far forward as the tactical situation allows and be provided with dedicated evacuation vehicles.

(2) Health Service Support (HSS). Responsive treatment and evacuation plans should be established to handle the expected increase in casualties. Both immediate treatment and evacuation are critical in maintaining the morale and confidence of forces engaged in urban warfare. To meet casualty treatment and evacuation needs, mobile aid stations with surgical teams should be placed as far forward as the tactical situation allows. Dedicated vehicles and aircraft for the evacuation of casualties should be provided.

In addition to casualties resulting from physical injury on the battlefield, there are often considerable psychological casualties in modern battles, particularly urban battles. This can be attributed to the intense, sustained, and close-quarter combat associated with urban warfare. In many battles, attacking troops were not properly trained for urban warfare and felt inadequate in their ability to fight on equal terms with the defender. Many felt moral or ethical dilemmas associated with collateral damage and civilian casualties. Finally, the constant threat of being killed from any quarter coupled with the extreme fatigue of fighting in an urban environment can create psychological casualties. Leaders at all levels should be attuned to the symptoms associated with psychological casualties in order to get effected individuals prompt treatment so that they can be returned to their units.

1006. Implications of Urban Warfare. The commander charged with making decisions needs to understand the operational and strategic implications of a tactical struggle in an urban area (MCDP 1-1, *Strategy*, and MCDP 1-2, *Campaigning*.). Three urban battles (**Stalingrad**, **Hue**, and **Beirut II**) illustrate the importance of seeing beyond the tactical nature of the battle.

Commanders must see beyond the immediate tactical implications of fighting an urban battle.

The battle for **Stalingrad** had major operational and strategic implications. The entanglement of German forces at Stalingrad bought time for the Soviets to mobilize and prevent a major combined effort of two German Army Groups in the south. While Stalingrad resulted operationally in the destruction of the German Sixth Army, the strategic results were even greater. Stalingrad caused a complete change of German strategy in the east. Hitler made major changes in his General Staff, and from this point on in the war, he was a man estranged from his military leadership. The loss of enough men and equipment to field one-fourth of the German Army shook the foundations of the Third Reich.

The Battle for **Hue**, although only one of over one hundred different attacks of the Tet Offensive of 1968, had a negative impact on the will of both the American people and their political leadership. Hue marked a revolution in the coverage of war by modern mass media. It was the first time Americans could and watch an ongoing battle from their living room on the evening news. Hue was a television bonanza for almost a month. When North Vietnamese leadership directed that Hue be held for at least seven days, it was clearly *not* their intent to win a tactical battle, but to strike at the *strategic center of gravity*—in this case, the will of the American people. Although the battle for Hue was a tactical victory for the U.S., the North Vietnamese had achieved their strategic goal of making the American public question the costs associated with the war.

During the battle of **Beirut II**, the IDF's objective was to drive the PLO from Lebanon. The ability of the PLO to leverage the media to gain an advantage was one of the most significant weapons in their arsenal. Despite an Israeli tactical victory, the costs in image, prestige, allies, and, most importantly, its own national will were enormous. The results of this battle eventually caused a change of political leadership at the highest levels of government.

1007. Key Insights. Analysis of modern urban battles provides insight into the major factors affecting the course and outcome of combat. Consideration of these key insights are required in order to effectively and efficiently plan and execute MOUT:

a. MOUT is infantry intensive.

b. A tactical battle may have far-reaching implications. No longer are the strategic, operational, and tactical levels overlap, creating the situation where tactical actions can have operational and possibly strategic repercussions (MCDP 1-1, *Strategy*, and MCDP 1-2, *Campaigning*).

c. Commanders at all levels must understand the impact that media representation will have on the accomplishment of operational and strategic objectives.

d. Maneuver warfare doctrine must be applied to the environment.

e. Intelligence is imperative to success in urban warfare.

f. Surprise is a combat multiplier.

g. Armor, artillery, and aviation are effective at the outer perimeter of built-up areas for causing isolation or preventing reinforcement.

h. Armor operating inside a built-up area must be protected by infantry.

i. Artillery providing direct fire inside a built-up area can be effective in the reduction of strongpoints.

j. As force ratio increases in favor of the attacker, combat duration decreases.

k. Urban warfare is time consuming.

l. Isolation of an urban defender ultimately ensures his defeat.

m. Attack of an urban area is costly to the attacker in terms of resources and casualties.

1008. Necessity for Preparation. Since MAGTFs are usually forward-deployed forces, it is inevitable that MAGTFs will be tasked to deploy to urban areas and conduct military operations. Fighting in this environment will be violent, close, and personal. Fighting in villages, towns, and cities are likely to become more frequent for military forces as the United States responds to global crises. Forces will be expected to exercise adaptability in responding to a wide variety of missions in the world's built-up areas. Marines may soon find themselves operating in the conceptual "three block battlefield" within a city grid: feeding refugees in one block, separating combatants in the adjoining block, and returning hostile fire in a third block.

MOUT is infantry intensive. The remaining chapters and appendices of this publication focus on how the GCE fights and integrates supporting elements in an urban environment.

Chapter 2

Offensive Operations

Urban combat isolates and separates units. Operations are reduced to a series of small-unit actions, placing a premium on small-unit leadership, initiative, and skill.

2001. Introduction. The Marine Corps' manuever warfare philosophy guides the conduct of offensive operations in MOUT. This warfighting philosophy serves to guide commanders through the development, planning, and execution of missions. A detailed discussion of maneuver warfare can be found in MCDP 1, *Warfighting*, MCDP 1-3, *Tactics*, MCWP 3-1, *Ground Combat Operations* (under development), and other Marine Corps doctrinal publications.

This chapter focuses only on offensive operations in an urban environment where collateral damage is of minimal concern. It describes tactics, techniques and procedures to be employed for seizing and clearing buildings and other urbanized areas. Chapter 6 addresses some considerations associated with fighting under constraints and restraints on urbanized terrain.

Section I

Planning

2101. Considerations. Identification of the adversary's centers of gravity and critical vulnerabilities will allow the commander to focus his efforts on those portions of the built-up area essential for mission accomplishment.

a. Reasons for Attacking a Built-Up Area. A commander considers the following before deciding to attack a built-up area:

(1) Tactical Advantage. Cities control key routes of commerce and provide a tactical advantage to the commander who controls them. Control of features such as bridges, railways, and road networks can have a significant impact on future operations. Urbanized areas may be used by the enemy as a base of operations from which they launch their own offensive operations. It may be advantageous to attack those bases and separate the enemy from their support infrastructure.

(2) Political Advantage. The political importance of a built-up area may justify the use of time and resources to liberate it. Capturing a city could destroy the seat of local and national government. At the very least, it could deal the enemy a decisive psychological blow.

(3) Economical Advantage. The destruction or capture of key industrial and commercial cities with the resulting denial of production and distribution of equipment and supplies strikes at the enemy's future ability to wage war. The requirement for a logistics base, especially a port or airfield, may play a pivotal role in the enemy's ability to continue the conflict. Capture of such cities may prove extremely beneficial to the attackers, who can use these resources to their advantage.

(4) Potential Threats to Operations

(a) Enemy Threat Too Great To Bypass. Though the terrain around a built-up area may facilitate its bypass, the enemy within that urbanized area may remain a threat capable of interdicting lines of communications. This may require the enemy force to be contained or destroyed.

(b) Terrain Does Not Allow Bypass. The urbanized area may sit between two natural slopes on the avenue of approach and thus require capture in order to secure the main supply route. Additionally, the urbanized area, itself, may sit on dominating terrain that threatens combat support and CSS elements.

b. Reasons for Not Attacking a Built-Up Area. The commander considers the following reasons for not attacking a built-up area:

(1) Built-Up Area Not Required To Support Future Operations. The attacker may have adequate supply support and resources established at another site from which combat forces can be supported. Therefore, during the estimate process, commanders may assess that the urbanized area is not necessary to support future operations.

(2) Enemy Not a Threat. The commander may decide to bypass upon determination that no substantial threat exists in the built-up area that could affect the unit's ability to accomplish its mission.

(3) Time or Risk Unacceptable. The commander's intent may dictate that speed of movement is essential to the mission. Because MOUT can be time-consuming, the commander may choose to bypass the urbanized area to maintain tempo. Furthermore, the potential for numerous casualties, the expenditure of critical resources, or the restrictions placed on attacking forces may result in unacceptable risks to the commander's primary mission.

(4) Declared an Open City. The area may have been declared an "open city" because it is undefended or of religious or historical significance. By international agreements, open cities are demilitarized and must be neither defended nor attacked. (See HR, 25, Annex to Hague Convention No. IV, *Embodying Regulations Respecting the Laws and Customs of War on Land*, 18 October 1907.) The attacking force must assume civil administrative control and treat the civilians as noncombatants in an occupied country. The defender must immediately evacuate and cannot arm the civilian population. A city can be declared open only before it is attacked. Other reasons for not defending could be the presence of large numbers of noncombatants, hospitals, or wounded personnel or the city's cultural, religious, or historical significance.

2102. Commander's Estimate. Once assigned the mission to attack a built-up area, a thorough mission analysis is conducted. Marines follow the same planning process in MOUT as for other operational environments by evaluating the mission, enemy, terrain and weather, troops and support available, and time available (METT-T). Based on METT-T and application of maneuver warfare concepts, the commander can decide on a plan that defeats the enemy by exploiting identified critical vulnerabilities. The commander may not have sufficient forces to fight everywhere and so must focus on the enemy and concentrate resources to decisive ends. This may involve making conscious decisions to accept risk at some place or time and to seize opportunities at others. The challenges of an urbanized environment dictate an even greater need to fully understand maneuver warfare.

This section addresses METT-T considerations as they pertain to the offense in urban warfare. METT-T is covered in detail in MCRP 3-11.5A, *Marine Troop Leader's Guide.*

a. Mission. Mission analysis is the first step of the estimate process. One of the important mission analysis considerations in urban conflict is the scope of clearance required to accomplish the mission. Commanders and planners should ask themselves the following questions:

- Do I need to clear every building?
- Should I clear only certain blocks?
- Should I only control certain areas?
- What level of protection is required for my lines of communication?

b. Enemy. Maneuver warfare focuses on the enemy. The intelligence preparation of the battlefield (IPB) process should be used to analyze the enemy and terrain in detail. (See Field Manual (FM) 34-130, *Intelligence Preparation of the Battlefield.*) The commander decides whether threat forces are conventional or unconventional.

(1) Conventional Forces. Most potential adversaries have adopted techniques of urban combat from either the United States or the former Soviet Union. Therefore, potential adversaries will build their urban defense to counter the attack of a combined arms force. Potential enemy forces will normally organize their defense in depth. Prepared strong points will usually form the perimeter of a larger defense while reserves locate in a separate position within the perimeter. You can expect ambushes to fill gaps in the perimeter while dummy strongpoints are established to deceive the attacker. Positions for securing the entrances to and exits from underground structures and routes are normally established. Security positions will normally be located forward of first-echelon defensive positions.

(2) Unconventional Forces. Urban areas have become a haven for unconventional forces. The large noncombatant population provides cover and concealment for unconventional force operations. Conventional forces operating in MOUT will normally be placed under restrictive ROE to minimize collateral damage. Unconventional forces will often use our restrictive ROE and the noncombatant population to their advantage when devising an urban defense. (See FM 34-130, FM 7-98, *Operations in a Low-Intensity Conflict*, and Chapter 6 of this manual.)

c. Terrain and Weather

(1) Terrain. Offensive operations should be tailored to the urban environment based on a detailed analysis of urbanized terrain (see Appendix I). Commanders and subordinate leaders use **KOCOA** (**k**ey terrain, **o**bservation and fields of fire, **c**over and concealment, **o**bstacles, and **a**venues of approach) to identify important terrain factors:

(a) K: Key Terrain. Key terrain surrounding an urban area can facilitate entry or deny escape. Within the city, airports or airfields, stadiums, parks, sports fields, school playgrounds, public buildings, road junctions, bridges, or industrial facilities may be key terrain.

Critical public buildings are identified during the terrain-analysis phase of an IPB. Hospitals, clinics, and surgical facilities are important because the laws of war prohibit their attack when not being used for military purposes other than medical support. The locations of civil defense, air raid shelters, and food supplies are critical in dealing with civilian affairs. Additionally, population size, locations, and density; density of the built-up area; firefighting capabilities; the location of hazardous materials; police and security capabilities; civil evacuation plans; and key public buildings should be identified.

(b) O: Observation and Fields of Fire. Urbanized terrain is characterized by restrictive observation and fields of fire. Weapon ranges can be greatly reduced because of buildings and other manmade structures. On the other hand, high ground or tall buildings can provide perches which enhance line of sight (LOS) for observation and communications as well as for individual and crew-served weapons. This includes man-portable surface-to-air missiles.

(c) C: Cover and Concealment. Buildings, sewers, and subways can provide excellent cover and concealment for enemy and friendly forces. They also provide covered and/or concealed maneuver routes within the built-up area. The civilian population can also offer cover and concealment to enemy forces.

(d) O: Obstacles. Natural or manmade obstacles restrict or deny maneuver within the urban area. Bridges, walls/fences, canals, streams, rivers, as well as rubble created by the effects of weapons should be thoroughly analyzed. Construction sites and commercial operations such as lumberyards, brickyards, steelyards, and railroad maintenance yards are primary sources of obstacle and barrier construction materials.

These sites can also supply engineers with materials to strengthen existing obstacles or to set up antitank hedgehogs or crib-type roadblocks.

(e) A: Avenues of Approach. Avenues to the city should support maneuver and be concealed either by terrain, darkness, smoke, or a combination of the three. Avenues that canalize or choke maneuver, due to the density of built-up area or natural terrain, should be identified and avoided. Roads, rivers, streams, and bridges provide high-speed avenues for movement. **Generally, military maps do not provide enough detail for urbanized terrain analysis.** They usually do not show the underground sewer system, subways, underground water system, mass transit routes, and power plants. Local maps of intracity road networks and subway systems and city hall/department of public works blueprints of all city buildings (if available), coupled with aerial photos, should be used.

(2) Weather. As in any military operation, weather affects equipment, terrain, and visibility, but its greatest impact is on the individual Marine. Snow, ice, dust, wind, rain, humidity, and temperature extremes reduce human efficiency. Weather extremes coupled with stress and the physical strain of urban combat can be minimized with effective small-unit leadership. Weather factors include:

(a) Precipitation. Rain or melting snow may flood basements and subway systems. This is especially true when automatic pumping facilities that normally handle rising water are deprived of power. Flooding makes storm and other sewer systems hazardous or impassable. In an NBC environment, chemical agents can be washed into underground systems by precipitation. As a result, these systems may contain chemical agent concentrations that are much higher than surface areas and thus become contaminated "hot spots." Hot spot effects become more pronounced as agents are absorbed by brick or unsealed concrete sewer walls.

(b) Fog. Many major cities experience fog, especially those located in low-lying areas and along canals or rivers. Industrial and transportation areas are the most likely to be affected by fog because of their proximity to waterways. Fog adversely affects vision and optical aids. It may also be used to help conceal friendly movement.

(c) Inversion Layers. Air inversion layers are common over cities, especially cities located in low-lying "bowls" or in river valleys. Inversion layers trap dust, smoke, chemical agents, and other pollutants, reducing visibility and often creating a greenhouse effect, which raises ground and air temperatures.

(d) Temperature. Built-up areas are often warmer than surrounding open areas during both summer and winter. This difference can be as great as 10 to 20 degrees hotter throughout the year.

(e) Wind Effects. Generally, wind chill is not as pronounced in built-up areas. However, the configuration of streets, especially in the city core and outlying high-rise areas, can cause wind tunneling. This increases the effects of the wind along streets that parallel the wind direction, while cross-streets remain relatively well protected. Wind tunneling can have a negative impact on aviation support within the urbanized area by creating dangerous wind effects down streets and between buildings.

(f) Day/Night Differences. Night and periods of reduced visibility favor surprise, which in turn, may facilitate infiltration, detailed reconnaissance, attacks across open areas, seizure of defended strongpoints, and reduction of defended obstacles. Because of the difficulties of night navigation in restrictive terrain (usually without reference points and in close proximity to the enemy) forces may consider using simple maneuver plans with easily recognizable objectives.

Rotary-wing aircraft are best utilized at night when operating on urbanized terrain. The most effective method of protecting helicopters from a successful engagement by hostile ground forces is to remain unseen. During night flights, the use of night vision goggles (NVGs) should be weighed carefully. Artificial lights can render NVGs partially ineffective and can increase the possibility of mid-air collisions by degrading the goggles to the point where nearby aircraft and aerial obstructions are not seen.

(g) Aviation Weather Considerations. Weather conditions can be a critical factor in determining the amount of aviation support the aviation combat element can provide in an urbanized area. Aviation forces should consider the following:

- Presence or absence of fog, industrial haze, low clouds, heavy rain, and other factors that limit visibility for pilots

- Illumination and moon phase/angle during night vision goggle operations

- Ice, sleet, and freezing rain that degrade aerodynamic efficiency

- Updrafts and downdrafts in the urban canyon that may present an unpredictable hazard

- High temperatures and/or high-density altitudes that degrade aircraft engine performance and lift capability

- High winds (large gust spreads) and crosswinds, to include the tunneling effect, that may create localized and unpredictable hazards to aviation

- Weather conditions that create hazards on pick-up zones and LZs, such as blowing dust, sand, or snow.

d. Troops and Support Available. The commander must be aware of all of his combat power (e.g., types of weapons systems, their numbers, and their capabilities) when performing analysis for operations in a built-up area. In the initial planning phase, force size must be evaluated in relation to urban size, enemy forces, and mission assignment. In the attack of a built-up area (population 100,000+), the GCE of a MEF would be a Marine division. However, in the initial stages of an amphibious or MPF operation, a reinforced infantry battalion or regiment may provide the initial forces. No matter what the size GCE, operations will be conducted by task-organized, reinforced battalions and their companies/platoons. Therefore, by using the infantry battalion as the basis for projecting the size of an overall force, the number of regiments or divisions required to secure a built-up area can be determined. Whatever size MAGTF is required for operations in an urbanized area, much of the fighting may be generally conducted by small teams of Marines. Consequently, our success in urban fighting may largely depend upon small unit and individual tactical skills. Specifically, leaders at all levels must analyze the following factors:

- Number and type of available units
- Task organization for urban combat (See Appendix A.)
- Availability of critical weapons systems
- State of training and discipline (training for urban warfare is imperative)
- Strength in terms of men and materiel in relation to enemy and size of built-up area
- Aviation support available
- State of maintenance and supplies
- Available combat support (based on availability and mission requirements)
- Available CSS (critical logistical and maintenance items)
- Host nation support available. (If the city's occupants are friendly to the attacker, then support in intelligence, deception, and diversion may be possible.)

e. Time Available. The following issues should be considered when analyzing the time available for an attack on urbanized terrain:

- A significant amount of time is required for clearing buildings, blocks, or axes of advance.
- Marines tire more quickly when clearing buildings because of stress and additional physical exertion.

- Adequate time should be allocated for thorough reconnaissance, planning, and rehearsals. Ultimately, this can result in higher tempo in the execution of the plan.

Once an estimate of the situation (using METT-T) has been conducted, planning for offensive operations can proceed. It is important to remember that the initial estimate of the situation must be continually updated throughout the planning process and the conduct of operations to account for changes in METT-T. (See MCWP 5-1 *Marine Corps Planning* [under development])

2103. Phases of the Attack. Attacks are categorized as either hasty or deliberate. Both hasty and deliberate attacks should take advantage of as much planning, reconnaissance, and coordination as time and the situation permit. Regardless of the size of the attacking force or of the objective to be secured, the phases of the attack (i.e. reconnoiter, isolate, secure foothold, and control the objective) remain constant.

a. Deliberate Attack. A deliberate attack *is a type of offensive action characterized by preplanned coordinated employment of firepower and maneuver to close with and destroy or capture the enemy* (Jt Pub 1-02). It is a fully coordinated operation that employs the MAGTF's combined arms team against the enemy's defense. It is used when enemy positions are well prepared, when the built-up area is large or severely congested, or when the element of surprise is lost. Given the nature of urbanized terrain, the deliberate attack of a built-up area is similar to the assault of a strongpoint (see MCWP 3-1, *Ground Combat Operations* [under development]). The deliberate attack of a built-up area is conducted in the following phases:

(a) Phase I: Reconnoiter the Objective. Intelligence gathering and reconnaissance/ surveillance are critical to the planning process and success of the operation. All available intelligence resources should be used to gain vital information on the enemy. Whenever possible, a leader makes a personal reconnaissance of the objective area to collect first-hand information regarding the area to be attacked. A thorough reconnaissance yields a wealth of information to use in the continued development of the plan. Avenues of approach, observation posts (OPs), supply routes, and the emplacement positions of direct- and indirect-fire weapons systems are all examples of information that may be amassed during the reconnaissance of the objective area. Composition and structure of buildings and roadbeds, cover and concealment opportunities, and other information not apparent in a map study may have a significant impact on the plan.

(b) Phase II: Isolate the Objective. The objective can be isolated by seizing natural and man-made features that dominate the area. Isolation may also be accomplished by coordinated use of supporting arms to seal off enemy lines of communication. This phase may be conducted simultaneously with Phase III (securing a foothold). Figure 2-1 shows a task-organized, reinforced infantry battalion isolating the objective.

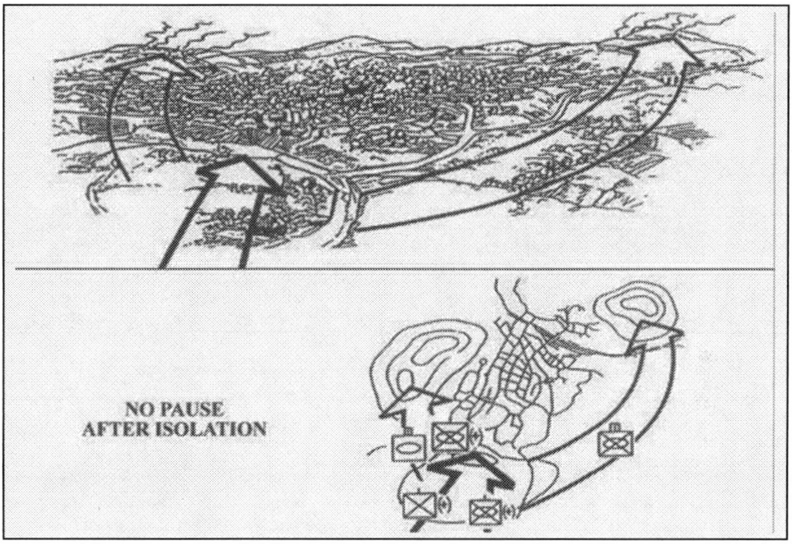

(c) **Phase III: Secure a Foothold.** Once the objective is isolated, a foothold should be secured as soon as possible in order to maintain tempo. The seizure of a foothold provides the attacking force with a position from which they can continue the assault through the objective area. The assault is supported by direct and indirect suppressive fires and smoke. Figure 2-2 (pg. 2-10) shows the battalion objective being isolated. One company is providing suppressive fires while another company seizes an initial foothold. Fire support assets are providing suppressive fires on the objective.

Figure 2-1. Isolation of the Objective Area

(d) **Phase IV: Seizing the Objective.** Once a foothold is seized and consolidated, supporting forces move to the built-up area to support the seizing of the objective area. To maintain tempo, the transition between the phases should be seamless. Once the foothold has been established, forward units continue the attack through the objective area. Supporting units assist as required. The momentum of the assault is continued until the objective area is cleared or controlled.

The assault force should establish limited objectives to ensure that the attacking forces do not get strung out along the axis of advance. Gaps may give the enemy the opportunity to infiltrate along the line of advance or make isolated friendly forces vulnerable to attack.

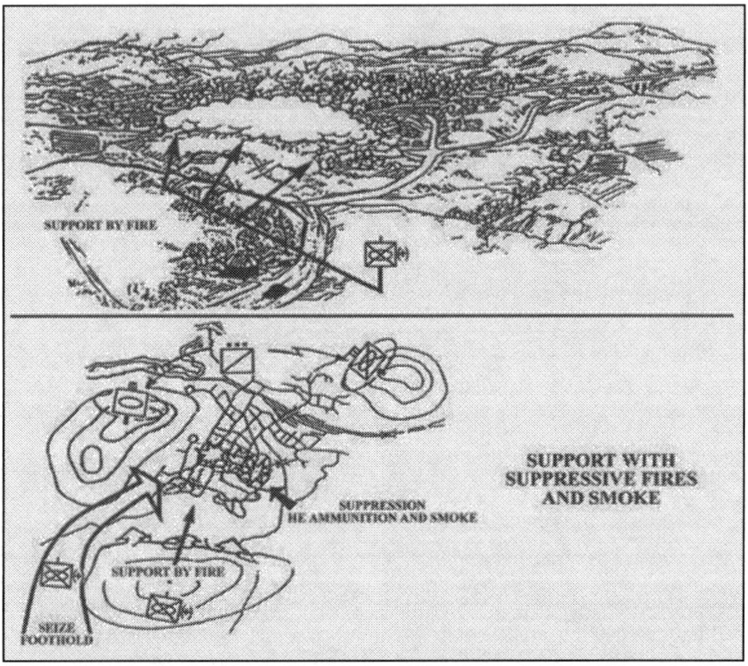

This phase also includes consolidation of the objective area and reorganization for future taskings. By setting limited objectives, the attacking forces have an opportunity to reorganize and defend against counterattacks while maintaining momentum. Once the limited objectives have been secured, the urbanized area may be sectored for detailed clearing. Clearing by sectors may help in the allocation of forces and in the prevention of fratricide. In addition to defensive preparation against counterattack, preparation should be made for follow-on missions (e.g., restoration of civilian facilities, marking and clearing minefields, etc.).

Figure 2-2. Seizing a Foothold

Seizing may consist of a systematic house-by-house, block-by-block advance through the entire zone of action, or it may be a rapid advance through a lightly defended area to seize a key objective. The difference between the two techniques is made by the commander on the basis of mission requirements and METT-T analysis. Seizing an objective in urbanized terrain requires detailed planning, coordination, decentralized command and contol, and small-unit execution. It is a continuation of the assault until all mission requirements are met.

(1) **Rapid Advance.** Rapid advance (Figure 2-3) may be used to maintain

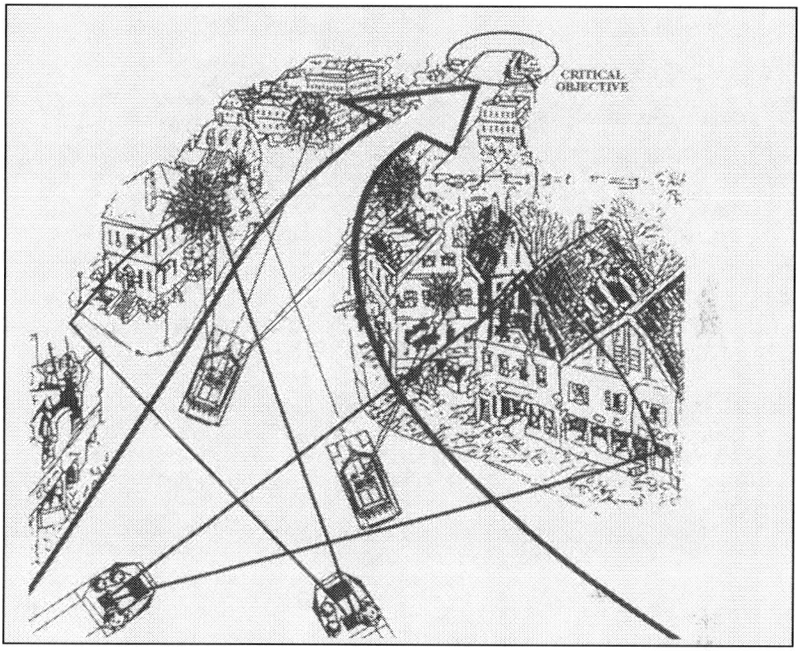

momentum through the zone of action and to seize key objectives. Some enemy positions may be isolated or bypassed in order to maintain tempo. This procedure may be used when:

- A key objective is identified (an enemy critical vulnerability or an objective that supports the friendly scheme of maneuver)
- Time is critical in reaching the objective
- Only selected buildings on the route of advance require clearing.

Figure 2-3. Rapid Advance En Route to a Key Objective

(2) Systematic Clearance. Systematic clearance (figure 2-4 on pg. 2-12) involves the deliberate reduction of all significant enemy positions throughout the objective area. It is usually used when time is not a critical factor.

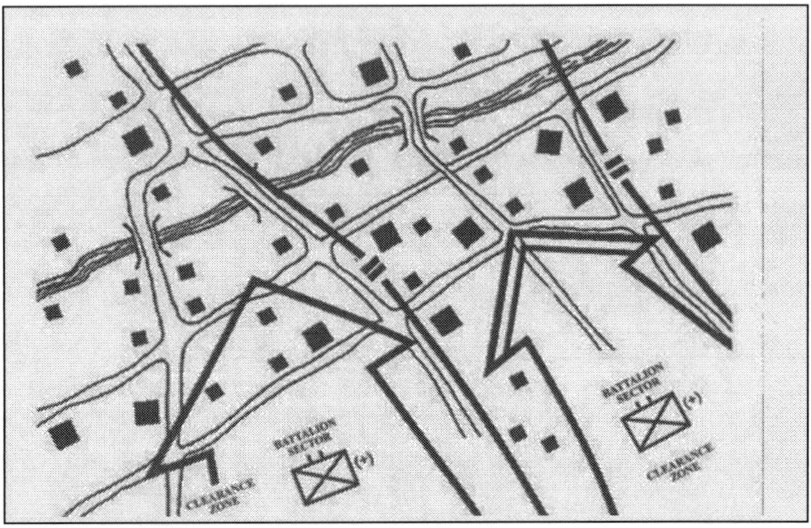

b. Hasty Attack. A hasty attack is, *in land operations, an attack in which preparation time is traded for speed in order to exploit an opportunity* (Jt Pub 1-02). Normally, hasty attacks are conducted as a result of unexpected contact during movement through the built-up area or when an opportunity for exploitation presents itself. Upon contact, the commander immediately deploys, returns fire, reports the contact and situation, develops the situation, and chooses a course of action. The hasty attack follows the same four phases as the deliberate attack; however, the time between preparation and response is compacted.

2104. Warfighting Functions. The warfighting functions (WFs) encompass all activities in the battlespace. Planners consider and integrate the warfighting functions when analyzing how best

Figure 2-4. Systematic Clearance Within Assigned Sectors

to accomplish the mission. The WFs include maneuver, intelligence, fires, logistics, command and control, and force protection.

a. Command and Control. The commander positions himself where he can best observe and influence the battle. Positioning is relative to the level of command. For example, a squad leader locates where he can observe and direct his fire teams, while the battalion commander locates where he can observe and direct his companies.

Close-quarter fighting in and around buildings makes command and control exceedingly difficult. Commanders and small-unit leaders rely on guidance and intent provided by higher

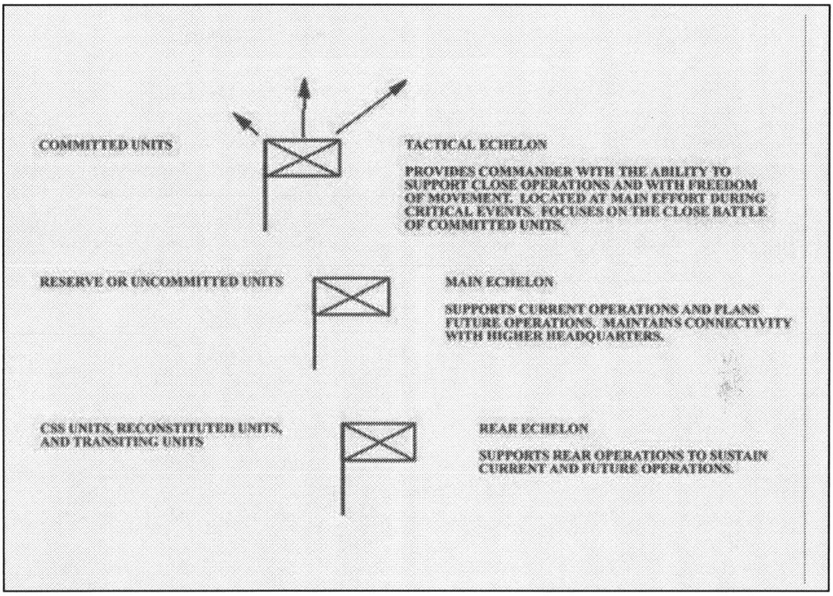

headquarters to facilitate decentralized execution while maintaining control and direction. In close-quarter fighting, commanders may have to rely upon messengers to communicate with their units and higher headquarters.

At battalion and higher levels, command and control is accomplished through three command echelons, which can be collocated or echeloned across the battlefield. The three echelons are tactical, main, and rear (see fig. 2-5). Each echelon may have a combat operations center (COC) capability. The COCs are the facilities (buildings, vehicles, or tents) used by the commander or staff at each echelon to plan, direct, control, and coordinate operations.

Figure 2-5. Tactical, Main, and Rear Echelons

(1) Tactical Echelon. The tactical echelon is normally collocated with the main echelon until the situation dictates separation. When the tactical echelon separates, designated command group members displace forward. Normally, the commander will move with the tactical echelon. The tactical echelon must have the communications and mobility necessary to allow the commander to exercise command. This echelon is usually small in size so the commander and command group can quickly displace forward to observe and influence actions. Organization of the tactical echelon should be addressed in the unit's standing operating procedures (SOPs).

(2) Main Echelon. The main echelon contains a COC, which is the nerve center for command and control of battalion through division-sized organizations. A COC consists of the command group (primary staff members with support personnel) and possesses the necessary mobility and communications for command and control of tactical operations. The main echelon COC monitors the battle, coordinates fire support, issues orders, maintains communications with subordinates and higher and adjacent commands, conducts current and future tactical plans, and assists the commander and subordinate commanders as appropriate. On urbanized terrain, the main echelon COC is normally located in buildings that provide good communications with subordinates and higher and adjacent commands. The command post (CP) is where the commander is physically located. Therefore, the CP can be located at any of the three echelons based on the commander's presence.

(3) Rear Echelon. The rear echelon's COC consists of the logistical and administrative personnel required to plan, coordinate, and execute logistics support. During urban warfare, the rear echelon COC is usually closer to combat forces than normal as it uses the city's captured buildings and facilities to provide cover and concealment for stocking supplies, conducting maintenance repair, and treating the injured.

Attack of a built-up area requires detailed planning due to restrictive terrain, close proximity to opposing forces, fire coordination requirements, and reduced communications capability. The GCE will frame a centralized plan of attack, fire support, and logistics support to conduct urban operations. However, the actual attack of a built-up area is a series of isolated, close-quarter battles carried out by small units. The execution of these attacks should be decentralized.

Communications planning procedures, as outlined in MCWP 3-40.3, *Communications and Information Systems* (under development), are applicable to the urban battle. Planners should be meticulous in their scrutiny of the environment and the effect it will have on communications. Manmade structures can create problems for single-channel radios. These structures inhibit LOS radio communications by absorbing or reflecting transmitted signals. However, the urban environment may have exploitable advantages such as the availability of electrical power, commercial telecommunications networks, and environmental control systems. Electrical power generation stations and other emergency power systems are normally found in protected structures and are probably usable. Due to their value to the enemy, these areas are more likely to be heavily defended. Enclosed areas offer excellent concealment and protection of communications and other command and control support equipment. Extensive commercial communications networks composed of miles of underground protected cable connecting central telephone exchanges are likely to be available, as well as a multitude of public service radio nets (police, fire, civil defense, taxi, etc.) complete with existing antenna systems and retransmission stations. To communicate effectively and continuously, commanders must minimize limitations imposed by the urban environment and maximize the advantage of existing civil communications.

b. Intelligence. Collection of intelligence information on urbanized terrain is difficult. Urbanized terrain provides overhead cover and concealment from observation by aircraft, reconnaissance personnel, and satellites.

(1) Information Requirements. In addition to information about the location of enemy units, nature of the terrain, trafficability of roads, etc., that all operations require, urban warfare requires detailed information on the following:

- Population density, characteristics, and culture
- Location and quality of water supply
- Width and construction of streets
- Layout of road networks
- Location, width, and load capacity of bridges
- Layout of underground networks such as subways, sewers, and utility ducts
- Types and construction of buildings
- Location of key communications and transportation facilities
- Location of canals and waterways.

Information on port facilities should include details on harbor sites, pier networks, unloading capabilities, warehouse/dockside facilities, harbor currents, and sandbars or other midstream obstructions.

(2) Collection. Accurate, detailed, and timely information is vital to the success of any urban operation. Collection operations conducted by national intelligence organizations may provide valuable information on the layout of the city, recent changes to transportation routes, political affiliations of various groups, and locations of military units. This information needs to be current and must be verified and updated as attacking forces close on the city. Reconnaissance activities such as the study of large-scale maps, aerial photographs, and background intelligence reports keep the commander's information current. The latest information on the enemy and the terrain should be continuously fed to the intelligence section using long-range reconnaissance patrols, human intelligence (HUMINT) reports, unmanned aerial vehicle (UAV) missions, and electronic intelligence. Use of reconnaissance assets should be tempered with a risk-versus-gain judgment. Reconnaissance assets should not be risked indiscriminately if the information can be collected by other means.

(3) Maps. Tactical maps should be supplemented by city street maps that contain street names and information on the location of important buildings and transportation terminals. These city street maps should be distributed to all units operating within the city. Although tactical maps are key in controlling fire support missions, simple street maps assist ground troops in maintaining their orientation within the city and in tracking buildings and areas that have been cleared.

c. Maneuver. Maneuver is the employment of forces on the urban battlespace through movement in combination with fire or fire potential to achieve a position of advantage (see MCWP 3-1). Maneuver is an essential element of combat power. Maneuver in itself cannot produce decisive results. Combined with mass, offensive action, economy of force, and surprise, maneuver provides favorable conditions for closing with the enemy. Maneuver contributes significantly to sustaining the initiative, exploiting success, preserving freedom of action, and reducing vulnerability. It is through maneuver that an inferior force can achieve decisive superiority at the necessary time and place. In many cases, maneuver is made possible only through the control of tempo and effective employment of firepower. The commander integrates supporting fires with the scheme of maneuver to create a dilemma for the enemy. Likewise, movement without fires exposes the force to effective enemy counteraction and risks losing the initiative and momentum. Maneuver that does not include violent action against the enemy will not be decisive. At all levels, successful application of this principle in the urban environment requires flexible and innovative plans and actions.

(1) Infantry/Armor Movement. Armored vehicles are vulnerable in built-up areas where streets and alleys provide ready-made fire lanes for defenders. Motorized traffic is greatly restricted, canalized, and vulnerable to ambush and close-range fire. Tanks are at a further disadvantage because their main guns cannot be depressed sufficiently to fire into basements or be elevated to fire into upper floors of buildings at close range (Figure 2-11). Whenever tanks are used in the direct-fire role, infantry must provide security against enemy ground attack. Movement of armor down narrow streets or down narrow paths through debris requires close coordination with infantry. Infantry and armor may use bounding overwatch techniques during movement (See MCWP 3-12, *Marine Tank Employment* [under development]). With this technique, infantry moves ahead of armored vehicles to clear the buildings on each side. Lead infantry units should not get so far ahead of armor that mutual support is lost. Conversely, if armor gets too far ahead of infantry, it may be quickly damaged or destroyed. Coordinated movement also contains a rear security element to protect armor from behind.

During movement down a street tanks should deploy with one tank on each side of the street focusing on the street and lower levels of buildings. *In MOUT, tanks are best deployed in pairs.* Other tanks of the tank platoon or section should move behind the infantry and fire at targets in the upper stories of buildings. When space is inadequate to deploy abreast, tanks may deploy in single file. Movement across open areas should be well coordinated to ensure mutual support. Suppression fire and smoke are used to cover force movements.

Armor can also be used in the reduction of obstacles. For example, grappling hooks and ropes attached to a tank can be used to pull down wire obstacles.

(2) Cover and Concealment. Walls and buildings can provide cover, concealment, and protection from enemy ATGM or heavy machine gun fire. Before armor enters a building, infantry should clear the building and check the ground floors to ensure that there is no basement into which an armored vehicle could fall and become trapped.

Tank and AAV generated smoke may be useful in concealing the location and movement of assaulting forces. Additionally, the thermal sights of the tank are very effective as they can use heat differentials to detect the movement or presence of concealed enemy soldiers, weapons, and vehicles.

(3) Tactical Considerations for Helicopterborne Movement. Helicopterborne operations are planned and conducted using the same combat fundamentals that apply to ground operations. However, there are considerations that apply only to helicopterborne operations. See MCWP 3-24, *Assault Support* (under development) for a detailed discussion on helicopterborne operations. These considerations include:

(a) Helicopterborne forces, once they have landed, lack vehicular mobility and heavy weapons. For this reason, it is important that the forces land on or near the objective. Such a force may be isolated for a time pending linkup with ground forces.

(b) Helicopters may be used for tactical deception. Helicopters may be used to make demonstration landings in several different zones to deceive the enemy as to the true objective of an operation.

(c) Helicopterborne assaults are normally conducted against undefended or lightly defended objectives. When attacking a well-defended objective, an LZ should be selected which provides cover and concealment for the assault force and is close to the objective.

(d) Helicopterborne forces are vulnerable to attack helicopters, fixed-wing aircraft, and surface-to-air missiles. To counter this vulnerability, friendly attack helicopters

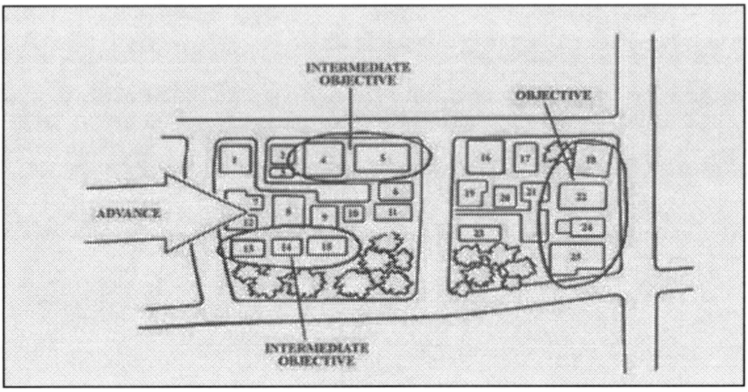

escort the helicopterborne forces during air movement, and indirect fires are used to suppress or neutralize enemy air defenses.

(4) Control Measures. The following control measures assist in the manuever of forces in MOUT:

(a) Objectives. When attacking to seize a foothold, the infantry unit may assign subordinate units the first block of buildings as their first objective. When an objective extends to a street, only the near side of the street is included. The final objective may be buildings or key terrain at the far edge of the built-up area. Key buildings, or groups

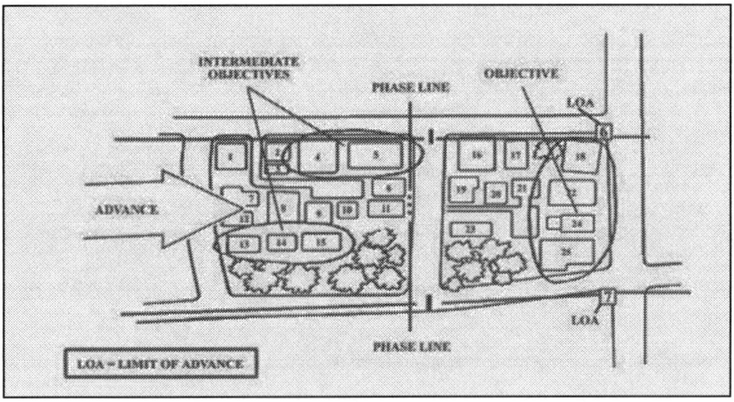

of buildings, may also be assigned as intermediate objectives. Buildings along the route of attack should be identified by numbers for clarification (Figure 2-6 on page 2-18).

During the seizing phase, bypassing buildings may increase the risk of attack from the rear or flank. It may be necessary for the unit to enter, search, and clear each building in its zone of action. A single building may be an objective for a rifle squad or, if the building is large, for a rifle platoon or company. When the commander's concept of operations is based on speed (rapid advance method) or when the force is conducting a hasty attack, a unit may be directed to bypass certain positions within its zone.

Figure 2-6. Objectives and Numbering System

(b) Phase Lines. Phase lines are control measures used to report progress or to control the advance of attacking units. Phase lines should be oriented on readily identifiable terrain features such as principal streets, rivers, and railroad lines. They should be placed on the near side of a street or open area. In an attack, a unit may have the mission to seize positions in its zone of action up to a particular phase line. (Figure 2-7).

Figure 2-7. Phase Lines

(c) Boundaries. Unit boundaries are used to define zones of action and are usually set within blocks so that a street is included in the zone. Both sides of a street should be included within the same unit's boundries.

(d) Checkpoints and Contact Points. Checkpoints aid in reporting locations and controlling movement. Contact points are used to designate specific points where units make physical contact. Checkpoints and contact points are designated by each unit as appropriate for command and control at street corners, buildings, railway crossings, bridges, or any other easily identifiable feature.

(e) Attack Position and Line of Departure (LD). A designated attack position may be occupied by forward units for last-minute preparation and coordination. The attack position is often behind or inside the last large building before crossing the LD. The LD should be located on the near side of an open area running perpendicular to the direction of attack, such as a street or rail line.

(5) Time of Attack. Ideally, the time of attack should be conducted during hours of darkness or limited visibility. Marines can exploit the poor visibility to cross open areas, to gain access to rooftops, to infiltrate enemy areas, and to gain a foothold. When attacking in unrestricted visibility, units should use smoke to conceal movement.

(6) Reserve. Reserves should be mobile and usually follow in trace of forward units. Battalion reserves normally follow one to two blocks to the rear of the lead company. If a company reserve is available, it follows within the same block so that it can immediately influence the attack. A unit with a reserve mission can be called upon to:

- Exploit an enemy weakness or friendly success
- Attack from another direction
- Attrit and/or clear bypassed enemy positions (normally requires designation of a new reserve)
- Secure the rear or a flank
- Maintain contact with adjacent units
- Counterattack

(7) Security. Each unit should ensure security of its flanks and rear (Figure 2-8 on page 2-20). Fighting in a built-up area is unique in that attacks can come from all three dimensions. Units assigned a mission to isolate can also provide security. Small units conduct security patrols and man OPs to supplement higher level reconnaissance and security operations.

Figure 2-8. Enemy Firing From Flank

(8) Engineers. Leading units should have engineers attached for immediate support. Tasks commonly assigned to engineers include:

- Preparing and using explosives to breach walls and obstacles
- Finding and exploding mines in place or helping to remove them

- Cratering roads and performing other countermobility measures
- Clearing rubble and obstacles

Requirements for engineer support in MOUT normally exceed capabilities. To compensate, all Marines should receive basic demolition training prior to conducting urban operations.

d. Fires. The attack of a built-up area may require extensive air and artillery preparation prior to the ground attack. Supporting fires suppress the defender's fire, restricts his movement, and possibly destroys his position. Consideration should be given to the rubbling effect produced by aerial and artillery bombardment. The assault should closely follow air and artillery fire to exploit its immediate effect on the defender. Maneuver units move near the final coordination line while the enemy is engaged by supporting fires. As the attacking force assaults, supporting fires lift and/or shift to block enemy withdrawal or to prevent the enemy from reinforcing their position. (See Chapter 4 for detailed discussion on air and artillery support during urban warfare.) Fire can be categorized into indirect and direct fires.

(1) Indirect Fire

(a) Artillery. Indirect artillery fire is planned to isolate objectives, to prevent reinforcement and resupply, to neutralize known and suspected CPs and OPs, and to suppress enemy defenders. Urbanized terrain requires that most indirect artillery fire use a high-angle trajectory.

(b) Mortars. Mortars are the most responsive indirect-fire weapon for the urban environment. They can be employed to hit targets of opportunity at the close ranges that are typical of combat in built-up areas.

(c) Other. Urban conflict is conducive to innovation. Many direct fire weapons can be employed in the indirect fire mode. Indirect fire weapons may also be effective at clearing the tops of buildings using variable-timed fuses.

(2) Direct Fire.

(a) Armor. Tanks are normally task organized with mechanized units for protection and security operations. The commander can employ tanks to take advantage of their long-range lethality, high-speed mobility, and survivability. They can be used outside the built-up area to cover high-speed armor avenues of approach. Tanks are normally employed in this role during the isolation phase. Tanks also support by fire the assault to seize a foothold and may support attacks to seize objectives within the built-up area. In MOUT, tanks can also be organized into special assault teams. The tank cannon's direct fire is very effective against structures. (Figure 2-9 on page 2-22 and Appendix B)

(b) Howitzers. Artillery rounds delivered by direct fire can be very effective for destroying targets in buildings. When used in this manner, the mobility, flexibility, responsiveness, and survivability of gun crews may be reduced. Artillery used in the direct fire role normally use HE shells.

(c) Infantry. Direct-fire systems, mainly machine guns, antitank guided missiles (ATGMs), shoulder- launched multipurpose assault weapons (SMAWs), and AT4s—are initially employed to support the seizure of a foothold. The positioning of antitank weapons in buildings must allow enough space for the effects of backblasts. Antitank weapons have limited as effectiveness when neutralizing targets behind walls. Designed primarily as armor penetrators, they neutralize a target only if that

Figure 2-9. Tank Direct-Fire Support Within the Built-Up Area

target is located directly behind the point of impact. Consideration must be given to weapons with minimum arming distances. Many of these factors can combine to limit the effects of weapons in the urban environment.

(d) Aviation. The vertical nature of large urban areas may block the pilot's line of sight with the target unless careful consideration is given to inbound headings and altitudes. Plan attack headings to take advantage of large open areas and street axis. Friendly ground forces must be clearly distinguishable from targets. The use of smoke, visible or infrared strobes, or other visual devices must be effectively employed by ground forces and clearly understood by friendly aircraft.

e. Logistics. Logistics support in urban operations should be timely, flexible, and responsive. Logistics planners, both in the supported units and the CSSE, should be fully integrated in the planning process. The final logistics plan must be thoroughly coordinated with the scheme of maneuver. Units may need special equipment such as ropes, grappling hooks, ladders, sandbags, and wire; medical supplies must be readied in anticipation to support increased casualties; and additional supplies must be prestocked. (See Chapter 5 for a detailed discussion on logistics and urban warfare, and MCWP 4-1, *Tactical Logistics*, for additional planning considerations.)

f. Force Protection. Force protection encompasses all those measures that preserve the fighting potential of the command. It is integrated throughout all the warfighting functions. It does not entail security operations alone. It involves the continuous implementation of active and passive measures by all units during all phases of an operation.

Some particular active and passive measures in MOUT include:

-Manmade structures in the urban environment provide cover and concealment for both the attacker and defender.

-Lines of communication can be more easily interrupted due to multiple avenues of approach.

-All units must be prepared to implement force protection in a 3-dimensional environment.

-Proximity of civilian population makes identification of non-combatants more difficult.

-ROE may be more restrictive

-Snipers and man-portable shoulder lauched air defense missiles are usually more difficult to detect.

.3

Section II

Infantry Battalion Operations

2201. Overview. This section addresses some of the operations that a task-organized and/or reinforced infantry battalion could conduct in MOUT. These operations are seldom conducted independently and are usually part of a larger joint operation. In the example given below, the battalion has seized a foothold and is operating in its sector of a town of 100,000 people. Concurrent operations are on-going through other parts of the town. The threat in the battalion sector is a company (-) with a capability to reinforce.

2202. Seizure of a Key Objective. The tactical situation may require the immediate seizure of a building, industrial complex, or transportation site that is deemed decisive terrain. Examples of such objectives may include key bridges, intersections, railheads, or facilities that constitue the center of gravity for the enemy. As an example, a battalion may be assigned the mission of seizing the only bridge that is capable of supporting armored vehicles and resupply by heavy transports along a river bisecting a town. The intent is to isolate enemy forces and secure the bridge for follow-on forces. Intelligence indicates that the bridge is lightly defended at this time. In this case, speed of attack and surprise is critical to capture the objective. Some key intelligence information is required prior to the attack. This would include the number of enemy defending the bridge, their weapons, disposition, their ability to reinforce their defense, location of LZs near the bridge, whether the bridge has been set for demolition, and the location of enemy indirect fire assets that can range the objective.

The commander decides to conduct a penetration using a combination of helicopterborne attack to rapidly seize the bridge followed by a mechanized ground attack to link-up with the helicopterborne forces (Figure 2-10). The helicopterborne force can achieve surprise and speed of movement required while the mechanized force can thrust quickly along high-speed avenues to minimize the time the helicopterborne force must hold the objective against enemy counterattack.

For the ground attack, the battalion commander makes a determination that speed of movement outweighs the risks of rapidly extending his lines of communication. He directs, on the basis of his knowledge of the enemy situation and his mission, that his maneuver forces avoid decisive engagement with the enemy along the route to minimize the time it takes to link up with the helicopterborne force. The commander provides guidance on those enemy forces that should be bypassed and those that should be isolated and reduced.

The commander task organizes his battalion for rapid movement on two axes to allow flexibility in reacting to unexpected enemy contact. Lead and rear security is conducted throughout the operation. As lead units locate enemy positions, they fix them by fire, and quickly reduce or bypass them. They also report the contact so that follow-on forces can then deal with the threat.

As the ground attack force reaches the edge of the built-up area, Marines remain mounted so as not to slow the advance. All vehicle machine guns fire to flanks and into doors, windows, and

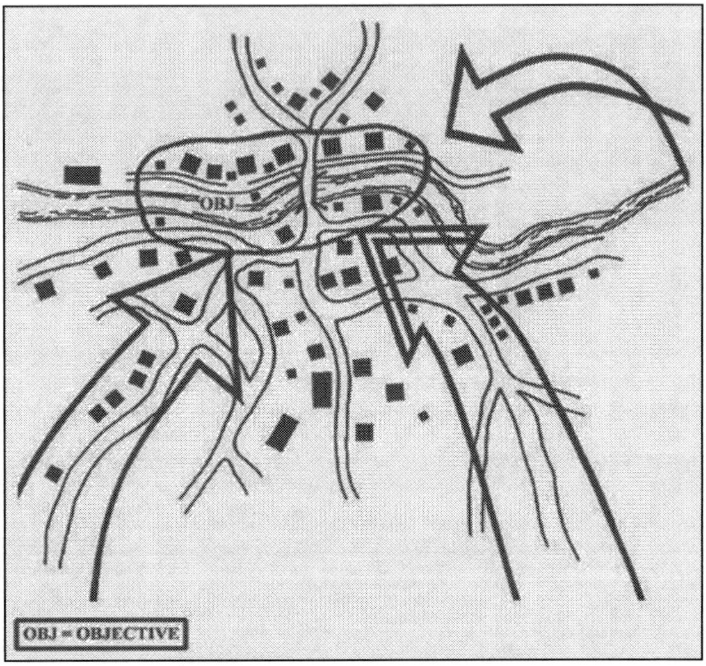

Figure 2-10. Seizure of a Key Objective

alleys to provide suppressing fires (Figure 2-11 on page 2-26). If heavy resistance is met and the column is halted, Marines dismount and the enemy and/or obstacles are cleared as quickly as possible. Once resistance and/or obstacles are cleared, the attack is immediately resumed. If necessary, units are dropped off to assume blocking positions and await follow-on clearing forces.

Once the objective is seized or linkup is accomplished, the battalion establishes a perimeter defense. The companies seize buildings around the objective and expand the size of the perimeter until it is large enough to secure the bridge. Attached engineers can then examine the bridge and clear it of any explosives.

2203. Infiltration. Infiltration is *"the movement through or into an area or territory occupied by either friendly or enemy troops or organizations. The movement is made, either by small groups or by individuals, at extended or irregular intervals. When the term is used in connection with the enemy, it infers that contact is avoided"* (Joint Pub 1-02).

Figure 2-11. Rapid Advance Movement

.3

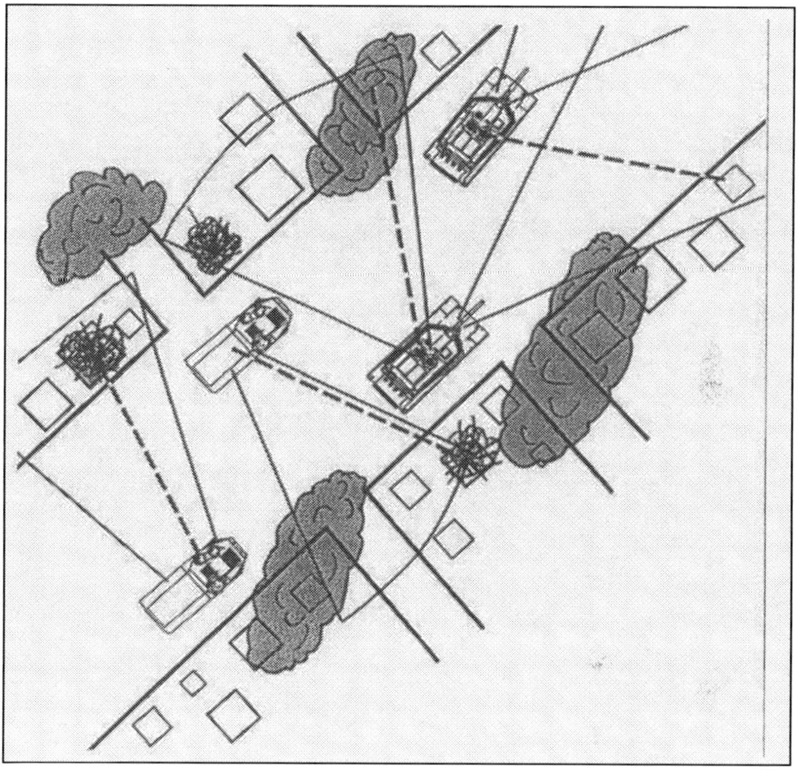

The outskirts of a town may not be strongly defended. The town's defenders may have only a series of antiarmor positions, security elements on the principal approach, or positions blocking the approaches to key features in the town. The enemy strongpoints and reserves may be positioned deeper in the city. In such a situation, an infantry battalion may be able to seize a part of the town by infiltrating companies and platoons between the enemy positions on the outskirts (Figure 2-12).

Moving by stealth on secondary streets, using the cover and concealment of alleys and buildings, the battalion may be able to seize key street junctions or terrain features, isolate enemy positions, and help following units move into the built-up area. Such an infiltration should be performed when visibility is poor.

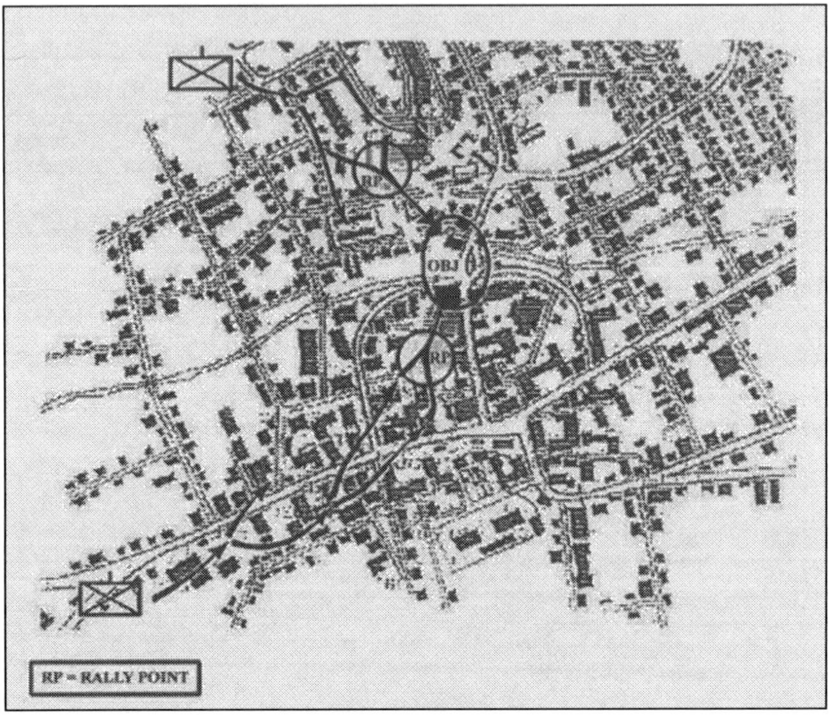

Figure 2-12. Infiltration

The infantry battalion is organized into infiltration companies with engineers attached and a designated reserve. Each company is assigned an infiltration lane that allows for infiltration by company-sized or smaller units. Depending on the construction of the built-up areas and streets, the infiltration lane may be 500 to 1,500 meters wide.

Mortar and artillery fire can be used to divert the enemy's attention and cover the sound of infiltrating forces. Armored vehicles and antitank weapons should be positioned to cover likely avenues of approach for enemy armored vehicles. Designated unit(s) provide isolation and security for infiltrating rifle companies. As the companies move into the built-up area, they secure their own flanks. Security elements may be dropped off along the route to warn of a flank attack. Engineers assist in breaching or bypassing minefields or other obstacles. Enemy positions are avoided but reported. The infiltrating companies proceed until they reach their rally points. At that time, they reorganize and make final preparations for attack of the objective.

.3

2204. Route Security. In MOUT, a battalion may be tasked to provide route security. The process of securing the route may consist of a series of small-unit actions to clear buildings along the route. Fires should be planned to isolate the route as well. It may also require the reduction of obstacles and the emplacement of flank security along the entire route. In an outlying area, the forward units may proceed by bounds from road junction to road junction. Other units could augment flank security by moving down parallel streets and by probing the flanks.

The speed of movement required and the enemy situation determine whether the unit should be mounted or dismounted. For mechanized forces, narrow streets should be avoided if possible to avoid being trapped. Designated armored vehicles overwatch the squad to its front, keeping watch on the opposite side of the street. While armored vehicle movements and positioning provide mutual support, armored vehicles should be protected by dismounted infantry.

When contact is made with the enemy, armored vehicles seek covered positions and provide direct-fire support. Supporting fires fix and isolate enemy positions. Infantry dismount then maneuver to attack. Phase lines are used to coordinate movement along the route. For example, at each phase line, the forward companies might reestablish contact, reorganize, and continue clearing (Figure 2-13).

2205. Battalion Considerations. Some of the concerns for the commander to consider include:

a. Command and Control. CP location can be critical due to line of sight restrictions in the urban environment. Consideration should be given to maintaining a highly mobile CP with links to airborne retrans stations.

b. Intelligence. Organic collection capability resides primarily from units in contact with the enemy. The Scout Sniper Platoon also provides information but they may have obstructed visibility in the urban environment. Commanders may chose to employ them more often in a sniper role. Links to higher command intelligence sources may be key to maintaining current information on the enemy. The same line of sight restrictions encountered with command and contol will also restrict intelligence dissemination. Commanders may have to rely on messengers.

c. Manuever. Maneuver of mechanized forces in an urban environment may be restricted to the size, width, and load capabilities of the streets. Consideration must be given to plan for alternate routes for any attack. Significant manmade obstacles can be encountered throughout the battlespace. Additionally, the reserve is normally kept closer to the assault force. Movement of forces by aviation is less restrictive, but may also be vulnerable to surface to air missiles and small arms threat. Availability of LZs may also be a limiting factor.

d. Fires. A key consideration on the use of fires will be the ROE and the collateral damage that can result by their use. Rubbling of buildings can complicate an already restricted maneuver capability. Weapons effects can be more pronounced and can create a hazard for friendly forces. IPB determines potential firing positions for indirect fire assets that accounts

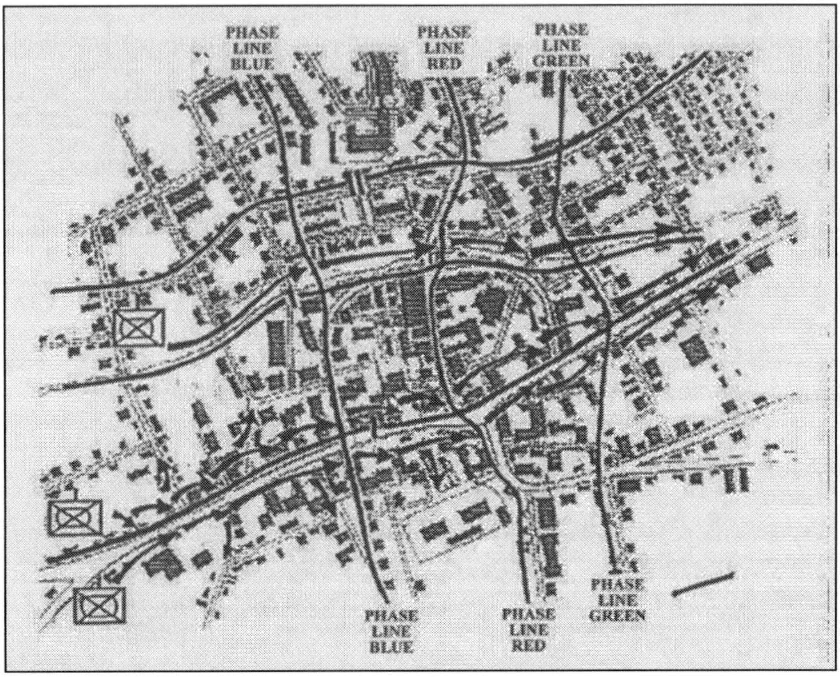

Figure 2-13. Clearing Along a Route

for the overhead masking by buildings and the ability to seat baseplates. Traditional indirect fire assets, such as artillery, may be used in the direct fire role.

e. Logistics. Historically in MOUT, the utilization of class I-V and VIII supplies is much greater. More vehicular assets should be allocated to the small unit level to transport ammunition and support the increased casualty evacuation requirements.

f. Force Protection. Lines of communication require extensive flank security in the urban environment. The commander must determine where his critical security requirements are and concentrate his force protection efforts in those areas while economizing in other areas. Manpower may further be depleted by security requirements for refugees and EPWs.

Section III

Rifle Company Operations

2301. Overview. This section addresses some of the operations the infantry battalion could assign to the rifle company in MOUT. For our focus, the rifle company is a task-organized and/or reinforced unit. Company operations will normally be conducted as part of a larger battalion operation.

2302. Attacking within a Built-up Area. Attacking within a built-up area is a task the rifle company may be assigned (Figure 2-14). This task involves the isolation, attack, and clearing of a specified block of structures. It is usually conducted in conjunction with adjacent friendly attacks or security operations. (See Appendix A.)

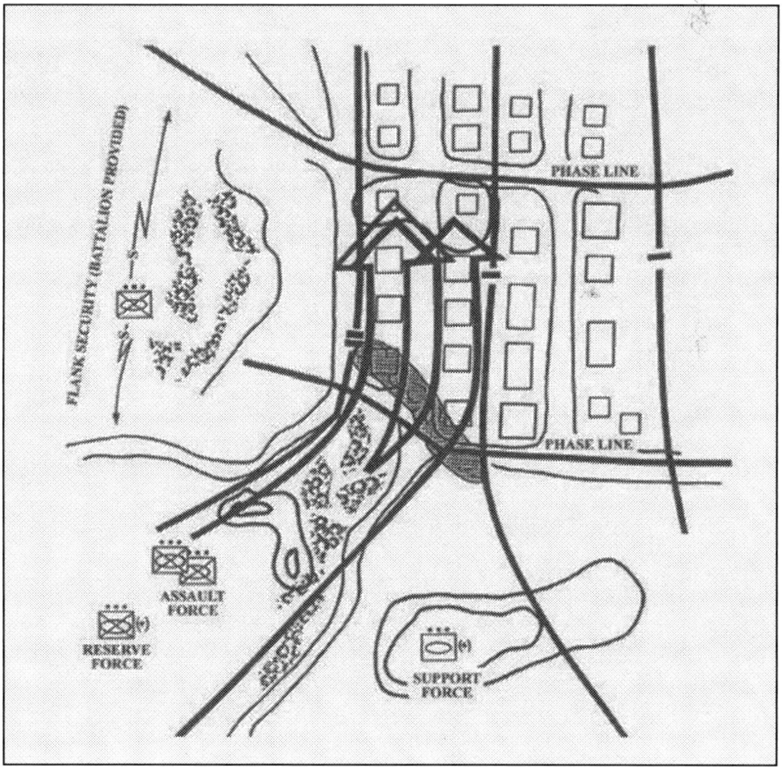

Figure 2-14. Company Attack of a Block

Before an assault, the company commander employs smoke to conceal the movements of assaulting platoons. He provides suppression fires to their flanks with direct-fire weapons. The battalion may provide additional security forces for flank protection.

(1) Concealed by smoke and supported by direct-fire weapons, an assaulting platoon attacks the first isolated building. The platoon must close on the building quickly while the enemy is still stunned by supporting fire. The company commander must closely coordinate the assault with supporting fire so that the fire is shifted at the last possible moment.

(2) The platoons continue to seize each building until the block is cleared. (See Appendix C for a detailed discussion on clearing of buildings.) After seizing the block, the company consolidates and reorganizes to repel a counterattack or to continue the attack.

The company commander directs platoon attacks on a designated front supported by direct and indirect fires. Success depends on isolating the enemy positions (which often become platoon objectives), suppressing enemy weapons, securing a foothold in the block, and systematically clearing the block's buildings room by room.

During the attack of a block, the company is usually task organized usually into four elements: *assault*, *support*, *security,* and *reserve*. The assault force is organized to conduct the actual assault of the objective. The support force provides direct- and/or indirect-fire support. The security force isolates the objective area and provides security. The reserve is organized, equipped, and positioned to exploit success. For example:

(1) The rifle company fighting in the outskirts of a city might organize as follows:

- Assault force: two platoons reinforced with the assault section
- Support force: armored vehicles, mechanized squad, and other direct- and indirect-fire weapons
- Security force: one reinforced squad.
- Reserve: mechanized platoon (-) reinforced.

(2) In a core or core periphery area (See Chapter 1), that same company might be organized as follows:

- Assault force: one rifle platoon, reinforced with engineers, SMAWs, Dragons, and a tank section (two tanks) per platoon
- Support force: One squad with direct- and indirect-fire weapons

- Security force: one rifle platoon (-)
- Reserve: one rifle platoon, mechanized.

c. Armored vehicles, machine guns, and other direct-fire weapons under the company commander's control are assigned to isolate the objective and provide suppressive fires from covered positions. These weapons should not be fired for prolonged periods from one position. The gunners should use a series of positions and displace from one to another to gain better fields of fire and to avoid being targeted by the enemy. Direct-fire support tasks are assigned as follows:

- Machine guns fire along streets and into windows, doors, alleys, etc.; however, innovative use of direct fire weapons can be key in MOUT
- Tanks, ATGMs, and AT4s fire at enemy tanks and other armored vehicles.
- SMAWs can be used to make entrances in buildings and destroy identified bunker/fighting positions.
- Tanks can fire at targets protected by walls, make entrances in buildings, and engage enemy armored vehicles with their main gun and other targets with their machine guns.
- Riflemen engage targets of opportunity.

2303. Attack of an Enemy Outpost. A motorized or mechanized force may encounter enemy outposts, checkpoints, or strongpoints. In a hasty attack on an enemy outpost, the company commander makes a quick assessment based on the mission and situation, and then reacts in accordance with the commander's intent. If not able to bypass an enemy outpost, he may have to attack and reduce it.

The company occupies positions of advantage. The tanks or other armored vehicles support the dismounted infantry in the attack by assuming support-by-fire positions from which they can fire on the outpost, prevent the enemy from escaping, and destroy any reinforcements.

The infantry platoons dismount and move to engage the enemy (Figure 2-15). They should attack from a covered route to hit the outpost at a vulnerable point. As the platoons approach the outpost, smoke may be employed to screen their movement, and supporting fires are shifted. After platoons close on the outpost, they clear the buildings quickly and consolidate. The company is then ready to continue operations.

2304. Seizure of a Traffic Circle or Major Intersection. A company may have to seize a traffic circle or major road or traffic intersection (Figure 2-16 on page 2-34). This operation consists of seizing the buildings adjacent to the traffic circle or intersection while bringing direct-fire weapons into position to fire on likely avenues of approach.

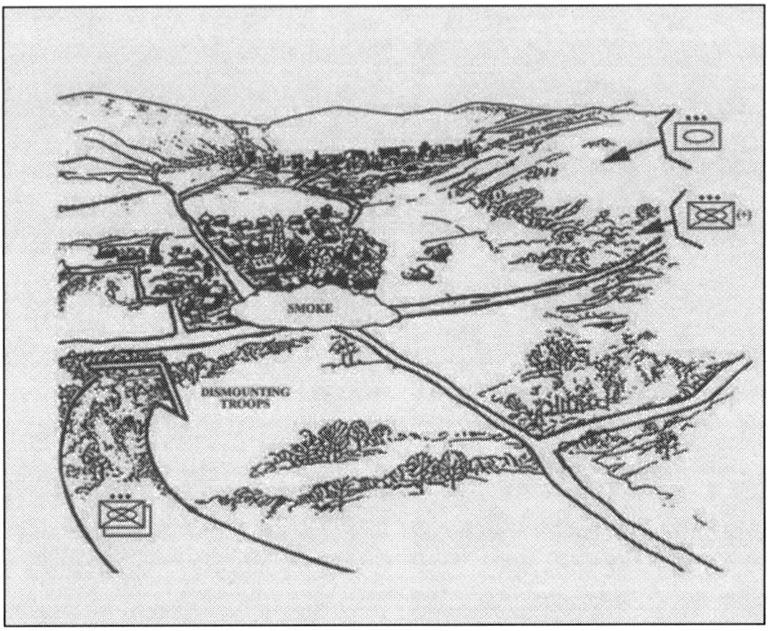

Figure 2-15. Hasty Attack of an Enemy Outpost

The company commander takes the following actions to seize a traffic circle or major intersection:

- Isolate the objective area.
- Seize and clear the buildings overlooking the traffic circle or major intersection under cover of direct-fire weapons.
- Secure the flanks against enemy attack.
- Consolidate and prepare for counterattack.

2305. Seizure of Key Terrain. The seizure of certain key terrain such as a bridge, intersection, facility, or structure may be critical to the success of an operation. Seizing key terrain for friendly use is a possible mission for a rifle company.

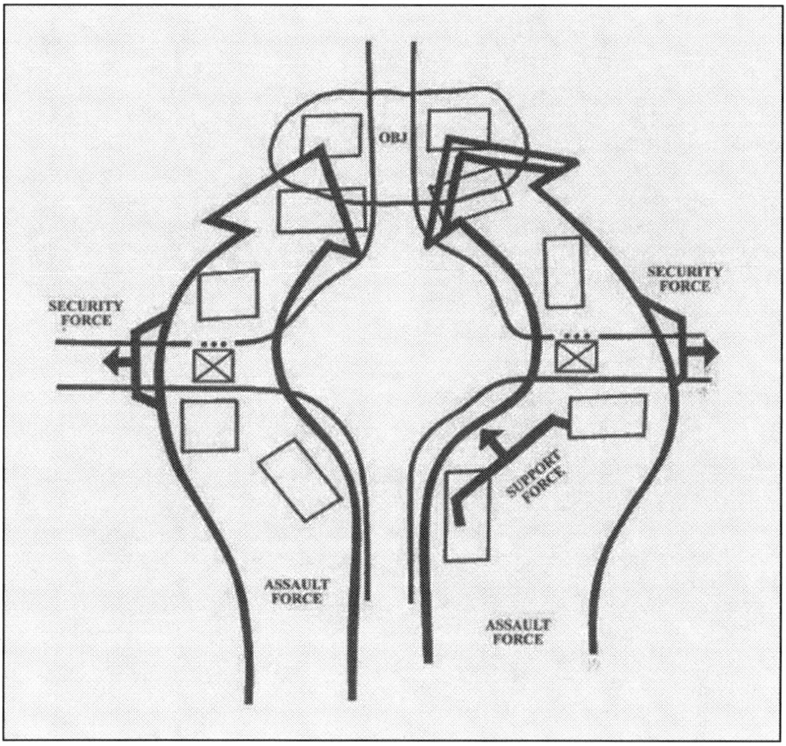

Figure 2-16. Seizure of a Traffic Circle or Major Intersection

To seize a bridge, a company should:

- Clear the buildings on the *near bank* that provide clear fields of fire for supporting weapons
- Suppress enemy weapons on the *far bank*
- Use screening smoke to obscure enemy observation
- Clear the buildings on the *far bank* that dominate the bridge
- Secure a perimeter around the bridge so that the engineers can clear obstacles or remove demolitions from the bridge.

The commander determines which buildings dominate the approaches to the bridge. Buildings that permit him to employ AT4s, ATGMs, machine guns, and riflemen are cleared. Supporting fire prevents the enemy from reinforcing units on the far bank. Enemy positions that can fire directly onto the bridge should be destroyed first. The company's forward observer (FO) should plan artillery and mortar fires to suppress the enemy's positions on the far bank.

The objectives of the assaulting platoons are buildings that dominate the far side of the bridge. One or two platoons are supported by the remainder of the company and assault across the bridge. Supporting fires are shifted and the infantry start clearing buildings. The assault continues until all the buildings in the objective area are cleared.

At this point, the engineers clear the bridge and its approaches of all mines, demolitions, and obstacles. The company commander may expand his perimeter to prepare for counterattack. Once the bridge is cleared, the armored vehicles are brought across to the far bank (Figure 2-17).

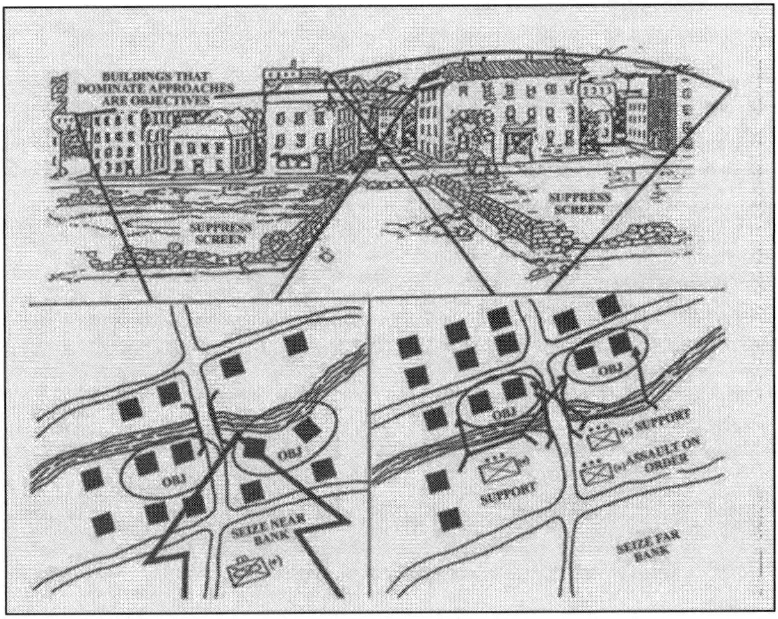

Figure 2-17. Seizure of a Bridge

2306. Movement to Contact. A rifle company may conduct a movement to contact along a road. Similarly, the company may have to reconnoiter such a route to prepare for a battalion attack. This type of mission can best be accomplished by a mechanized company with appropriate attachments. (Figure 2-18 on page 2-36).

Figure 2-18. Movement to Contact Along a Highway (Commercial Ribbon)

The speed of movement depends on the mission, terrain, and enemy situation. In open areas where rapid movement is possible, tanks should lead. In close terrain, the infantry may lead while being overwatched by the tanks. Artillery fire should be planned along the route. Engineers may accompany the lead platoon on the main route to help clear obstacles and mines as they may be encountered. Between key terrain, when contact is not likely, the company remains on their vehicles. At key terrain or when enemy contact is likely, the company moves on foot to clear enemy positions or to secure the key terrain. Armored vehicles support the infantry.

Section IV

Rifle Platoon Operations

2401. Overview. This section discusses some of the various tactical missions that might be assigned to a rifle platoon.

2402. Attack of a Building. One offensive mission assigned to a rifle platoon in MOUT is the attack of a building. The attack requires isolating the building and entering the building at the least-defended point or through a breached hole (Figure 2-19 on page 2-38). When possible, clear the building from top to bottom. Close coordination between the assault and support forces is required.

When attacking a building, a platoon should be organized into an assault force, a support force, and a security force. The assault force provides the forces that actually conduct the assault. A support force provides supporting fires and efforts that assist the assault force. A security force covers the platoon's flanks and rear. (See Appendix A.) The platoon may be augmented by armored vehicles and from the company's weapons platoon or the battalions weapons company. If a reserve is required it should be provided by the company.

The assault consists of three steps:

(1) Isolate the building and deny the enemy the ability to maneuver or reinforce.

(2) Enter the building to secure a foothold for further action.

(3) Clear the building systematically.

Clearing is performed by individual squads and teams, which leapfrog as rooms and floors are secured. Platoons tasked to clear buildings may be reinforced with combat engineers.

2403. Movement Down a Street. When moving in urban terrain, a rifle platoon normally follows the same principles of movement as in other environments. However, some movement techniques must be modified to adjust to the unique dangers encountered in an urban environment. (See Appendix A, Section VII.) This section describes the movement down the street of the lead platoon of a company.

The speed of movement depends on the enemy situation. In outlying or lightly defended areas, a mechanized platoon may proceed along the street mounted but send dismounted infantry forward to reconnoiter danger areas. In the center of a built-up area the platoon dismounts to move on foot with two squads leading—one on each side of the road, using all available cover. They move through the buildings, if feasible, to avoid exposure on the streets. The squads coordinate to provide each other with all-around observation and mutual support.

Figure 2-19. Attack of a Building

The platoon should organize and practice to move as two forces: a *maneuver* force and an overwatch force. The maneuver force (normally of one squad on narrow streets or two squads on wide streets) moves forward, scouts danger areas, and closes with the enemy. An *overwatch* force (the remainder of the platoon and its attached or supporting crew-served weapons), moves behind the maneuver force, secures the platoon's flanks and rear, and provides direct-fire support when needed. These two forces, can exchange roles from time to time to avoid overfatigue of the maneuver force (Figure 2-20).

2404. Utilization of the Reserve. In the offense, a rifle platoon may be assigned as a reserve. The reserve can be used to reinforce and attack or exploit success. Reserves are normally held at the company level.

Figure 2-20. Movement Down a Street

In outlying areas, where the terrain is relatively open, a tank platoon reinforced with mechanized assets (tank-heavy task force) may be tasked as the reserve. However, if inside the urban area, a mechanized platoon reinforced with tanks (mech-heavy task force) would most probably be directed to provide the reserve. Regardless of what unit is designated as the reserve, the following items must be considered for their employment in an urban environment:

- The reserve should plan to use a different axis of advance

- Route selection for the reserve may be more constructive since it may have to use building, side streets, sewers, etc.

- Reserves are kept closer to assault forces in MOUT than on open terrain

- Engineers should normally be attached

- Reserves should maintain a high degree of mobility

Chapter 3

Defensive Operations

The essence of defensive tactics is to place the enemy into a position that permits his destruction through the intelligent use of terrain and firepower.

3001. Introduction. In urban combat, the defender possesses key advantages over the attacker. The defender can shape the battlespace to his advantage by maximizing the natural restrictions and obstacles found in the urban environment. ROE for the attacker can add to the defender's advantage by placing restrictions on the application of force by the attacker. Knowledge of the terrain and time available for preparing defensive positions are advantages which may enable the defender to successfully resist a numerically superior force. A spirited and stubborn defense may persuade an attacker to abandon the attack. In some battles, urban defenders can be credited with repelling or decisively delaying an attacker which ultimately influenced the course of the war in their favor.

Marines may be called upon to defend a port city, an embassy located on or near a coastline, a transportation link, or the capital of a foreign government. They may also be tasked to defend from an urban area in an economy-of-force role in order to support operations elsewhere. In combat, offensive and defensive operations are inseparable. Marines need to be capable of conducting both offensive and defensive operations in an urban environment.

The fundamentals of defense do not change in an urban environment. The defenders of a city, however, usually have detailed knowledge of the terrain. This will allow them to establish an extensive defensive network that is designed to force an attacker to expend exorbitant amounts of time, supplies, equipment, and manpower. Commanders need to recognize both the advantages and disadvantages of defensive operations in an urban environment.

3002. Decision to Defend

a. Reasons for Defending Built-Up Areas. Historically, the following are some of the reasons for defending built-up areas:

(1) Denial of Important Strategy/Political Objectives. Capitals and cultural centers can be defended strictly for psychological or national morale purposes even if they do not offer a tactical advantage to the defender. Defending a city can cause an attacker to commit a significant amount of his forces which reduces his capability to attack elsewhere. The defense of a city can delay the overall offensive capability of the attacker.

(2) Retention of Key Economic Centers. In many countries, the entire nation's economic well-being may be tied to a few key cities. These key cities usually contain the country's primary industrial, transportation, and communications base. The capture of these key centers could result in the overthrow of the current government, or deny that

.3

government the ability to adequately support combat operations against enemy regular or insurgent forces.

(3) Control of Avenues of Approach. Most avenues of approach to large cities are straddled by small towns every few kilometers which must be controlled by defending forces. These areas can be used as battle positions or strongpoints to control the avenues of approach. For the attacker to utilize these avenues of approach he must sacrifice speed and expend resources. The defense of these cities or towns may constitute a large part of the overall defense of a city.

(4) Economy of Force. The tactical advantages provided by urbanized terrain allows the defender to engage a numerically superior force. The defender can conserve the bulk of his combat power so that it is available for use in other operations.

(5) Concealment of Forces. Reconnaissance and combat identification is more difficult in an urban environment. CPs, reserves, CSS complexes, and combat forces emplaced in built-up areas are much harder to detect.

3003. Reasons for Not Defending Built-Up Areas. The commander considers the following reasons for deciding not to defend built-up areas.

a. Unnecessary to the Defensive or Offensive Plan. If the built-up area is too far forward or back in a unit's defensive sector, is isolated, or is not astride an enemy's expected avenue of approach, the commander may choose not to defend it.

b. Bypassable. If the nature of nearby terrain allows the enemy to bypass the city, then it will not make a good blocking position. Some built-up areas, mainly smaller ones, are easily bypassed by existing main road and highway systems. A built-up area that can be easily bypassed normally will be, thereby effectively isolating that area's defenders from the remainder of their forces.

c. Inadequate Structures for Defense. Extensive areas of lightly built or flammable structures offer little protection to the defender. Built-up areas near flammable or hazardous industrial areas, such as refineries or chemical plants, also may not lend themselves to successful defense.

d. Adjacent Dominating Terrain. If the built-up area is small and dominated by close, prominent terrain, the commander may choose to defend on that terrain rather than in the built-up area.

e. Better Fields of Fire Elsewhere. The commander may choose to base all or part of his defense on the better fields of fire that exist outside built-up areas. An example would be an armor-heavy force defending in sectors from multiple, small, built-up areas surrounded by open or farm-type areas.

f. Open City. See chapter 2.

Section I

Defensive Planning

3101. Commander's Estimate. As in any other defensive operation, planning begins with the commander's estimate of the situation. The complxities involved with the defense of a built-up area require detailed and centralized planning. This section details the general procedures and principles for planning and organizing the defense of an urban area.

3102. METT-T

a. Mission. Mission analysis is the first step in conducting the estimate of the situation. When conducting mission analysis commanders and their staffs should analyze their assigned tasks and commander's intent to determine the scope of the defense. For example:

- Do I need to control every building or can I control the frontage from key buildings?

- Are certain parts of the urban area more significant? (i.e. main transportation centers, communication nodes government buildings)

b. Enemy. The commander must focus on the enemy and build his defense to concentrate his strength against enemy weakness. The IPB process is used to analyze the terrain and enemy capabilities. (See FM 34-130.) For example:

- If the attacker relies heavily on aviation for mobility, the defender may defend potential LZs and emplace manpad teams to counter the air threat

- If the enemy relies heavily on mechanized movement, the defender may employ additional countermobility assets.

c. Terrain and Weather

(1) Terrain. Analysis of all manmade and natural terrain features is critical for planning a defense in built-up terrain. A defending commander has the advantage of occupying the terrain on which the fight will occur, and therefore usually knows it better than the attacker.

The defender must make best use of the battlespace that he controls: ground level (streets and parks), above ground (buildings), and below ground (subways and sewers). Defenders must be aware that built-up areas are normally interdependent. A reservoir supplying water to a city may be many miles away, electricity can come from remote power stations, and the control points for these services and others, such as communications lines, may be vulnerable.

.3

(a) K: Key Terrain. Primary examples of key terrain in the urban environment are ports, airfields, power grids, communication nodes, bridges, government building complexes, or parks. We identify key terrain so that we can better select our defensive positions and also better determine the enemy's objectives.

1 **Villages.** Villages often dominate high-speed avenues of approach. If the buildings are constructed of stone, brick, or concrete, providing good protection against both direct and indirect fires, a formidable defense can be constructed

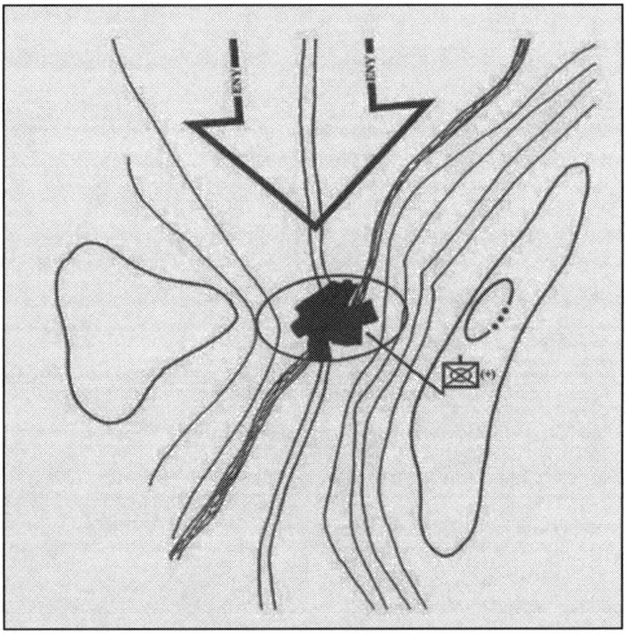

(Figure 3-1). Company-sized battle positions can be established in these small villages to block approaches into the larger urban area.

Figure 3-1. Village Defense (Chokepoint)

2 **Strip Areas.** Strip areas are built along roads or down valleys between towns and villages. They afford the defender the same advantages as villages. If effective fields of fire are available, a unit can occupy a few key positions within the strip to create an effective defense. (Figures 3-2A and 3-2B).

3 **Towns and Cities.** When a small town or city is positioned on a key enemy avenue of approach, a force can defend from the urban area while other forces

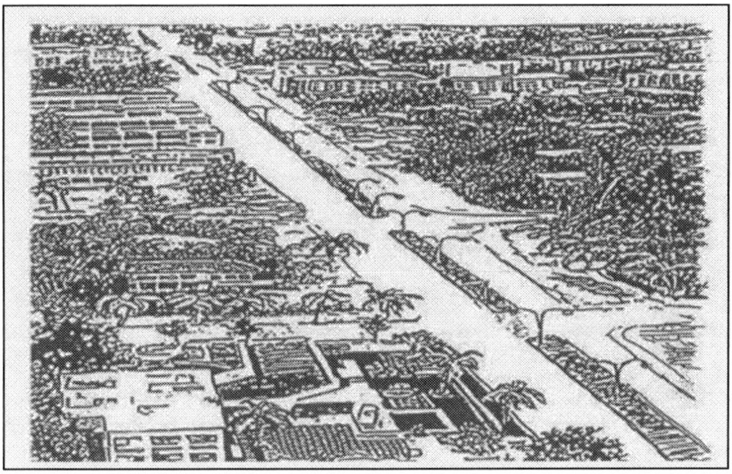

deny the enemy the ability to bypass. The town or city can be made into a strongpoint defense. (Figures 3-3A and 3-3B on page 3-6).

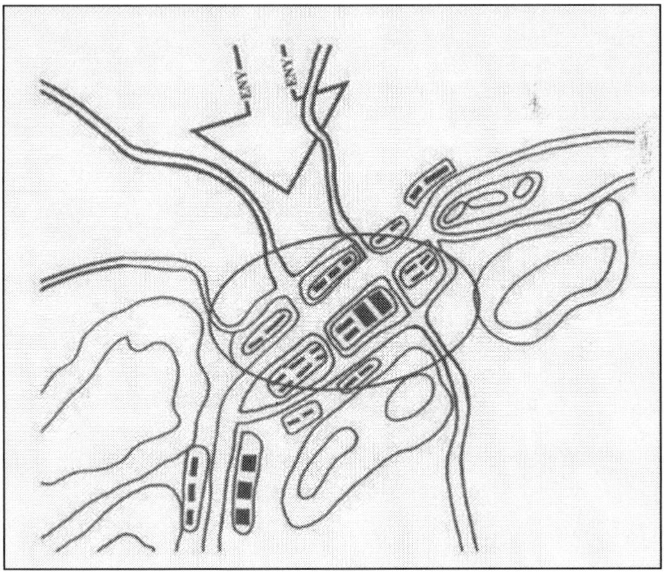

Figure 3-2A. Strip Area

.3

Figure 3-2B. Strip Area Defense

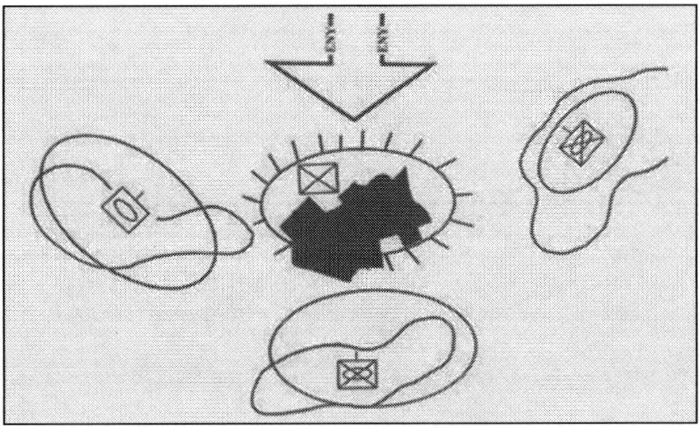

Figure 3-3A. Towns and Cities

Figure 3-3B. Defense of a Town or City

4 **Large Built-Up Areas.** Large buildings that are close together may require more forces and smaller defensive sectors than natural open terrain. The density of buildings, rubble, and street patterns will dictate the frontage of the unit.

(b) O: Observation and Fields of Fire. Attackers must generally advance by crossing streets and open areas between buildings where they are exposed to fires from concealed positions.

1 **Weapons and Range.** The defender must position weapons to obtain maximum effect and mutual support. FACs and FOs should be placed well above street level for increased observation. Fires and final protective fires should be preregistered on the most likely approaches.

2 **Limited Visibility.** The commander can expect the attacker to use periods of limited visibility to sustain or gain momentum. (See Appendix H for more detailed discussion.) The commander considers the following during periods of limited visibility:

- Unoccupied areas that can be observed and covered by fire during daylight may have to be occupied or patrolled at night.
- Remote sensors should be emplaced in dead space.
- Mines, noise-making devices, tanglefoot, and OPs should be positioned on avenues of approach to provide early warning.
- Artificial illumination to include street lamps, stadium lights, and flares should be integrated into the overall defense.
- When defenses are probed, indirect-fire weapons, grenade launchers, and hand grenades can be used to avoid disclosure of defensive position locations.

(c) C: Cover and Concealment. Although walls, floors, and ceilings can be used as protective cover, Marines should always improve these positions by using sandbags, rubble, etc. A defender can reduce his exposure by establishing routes between positions using:

- prepared breaches through buildings

- reconnoitered and marked underground systems
- trenches and sewage systems
- concealment offered by smoke and darkness to cross open areas.

(d) O: Obstacles. A city is an obstacle because it canalizes and impedes an attack. Avenues of approach should be blocked by obstacles and covered by fire. (See Paragraphs 3404.d., 3404.f., 3405.e., and Appendix F.)

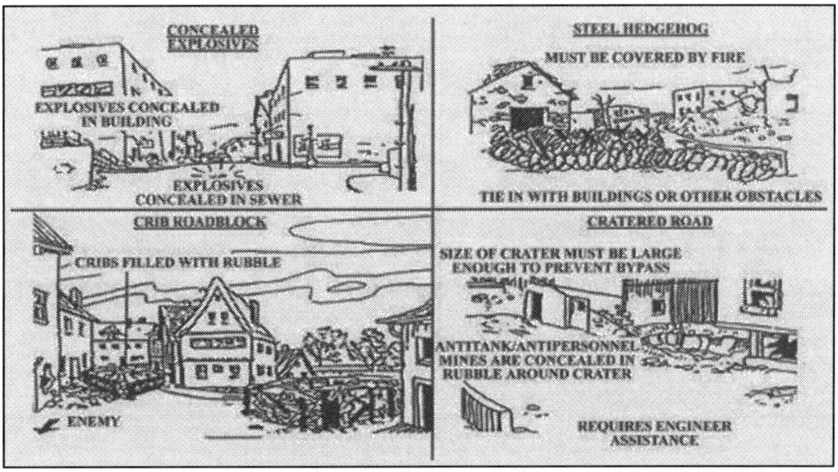

(e) A: Avenues of Approach. Avenues of approach (surface and subsurface) should be denied to the enemy in keeping with the overall defensive plan (Figure 3-4).

(2) Weather. The effects of weather as discussed in Chapter 2, Section II, pertain equally to defensive operations.

D. Troops and Support Available. The defender of a built-up area has a terrain advantage and can resist the attacker with much smaller forces. The defender stands a greater chance of success when he fights with combined arms and task organizes his forces. The discussion of troops and support available in Chapter 2, Section II, also pertains to defensive operations.

Figure 3-4. Denial of Avenues of Approach

e. Time Available. The commander establishes priorities of work, which depend on the time available. Many tasks can be accomplished at the same time, but priorities for work are accomplished according to the commander's order. An example priority-of-work sequence follows:

- Establish security and communications
- Assign sectors of responsibility, final protective fires
- Clear fields of fire
- Select and prepare initial fighting positions
- Establish and mark routes between positions (including alternate and supplementary routes)
- Emplace obstacles and mines
- Improve fighting positions.

3103. Organizing for a Defense

a. Types of Defense. The defense of a built-up area should be organized around key terrain features that preserve the integrity of the defense and provide the defender ease of movement. There are two types of defense: position and mobile. (See MCWP 3-1, *Ground Combat Operations* [under development] for details on defensive operations.) Most defenses will include a combination of position and mobile defense. The type of defense chosen is predicated on commander's intent and METT-T.

(1) Position Defense. The type of defense in which the bulk of the defending force is disposed in selected tactical localities where the decisive battle is to be fought. Principal reliance is placed on the ability of the forces in the defended localities to maintain their positions and to control the terrain between them. The reserve is used to add depth, to block, or restore the battle position by counterattack. (Joint Pub 1-02.) The position defense is usually characterized by defending key terrain. In the urban environment, this equates to the physical occupation of key public buildings, avenues of approach, transportation centers, industrial parks, etc. This defense focuses combat power to repel the attacker's advance and/or penetration while retaining the terrain. This type of defense is common for battalion-sized units and below.

The commander conducting a position defense organizes forces utilizes the same fundamentals that apply to any defense. Forces can be assigned sectors to defend and/or battle positions or strongpoints based on the commander's analysis of METT-T. Reserves are identified and located to respond quickly to exploit success.

(2) Mobile Defense. Defense of an area or position in which maneuver is used with organization of fire and utilization of terrain to seize the initiative from the enemy. (Joint Pub 1-02.) The mobile defense is characterized by its focus on the enemy, not on terrain. Terrain is still important. However, this defense is organized with a mobility capability to exploit enemy weakness. In order to conduct the mobile defense, the defender must have enough forces to defend the built-up area, and to rapidly conduct an enveloping or flanking attack.

As in the position defense, the sector, battle position, strongpoint, or any combinations thereof are also used in shaping the battlespace by canalizing the enemy into a designated location where he can be destroyed.

b. Defensive Options

(1) Defense Outside the City. The commander may defend from outside the city. This option is often selected when the terrain surrounding an urban area offers an advantage to a defender and when his defending force is of sufficient strength (See Cherbourg paragraph in Chapter 1). The presence of a large, friendly, civilian population, or buildings of historic significance or specific cultural value, would also contribute to deciding on an outside defense. A situation in which the built-up area has a preponderance of building structures that offer little protection to defenders or terrain that favors the attacker.

A commander organizes his defenses around strongpoints selected on the outer edges of the urban area when these may be suitable structures for siting weapons to gain maximum firepower range. The organization of an outer-city defense is also heavily influenced by the number and type of weapons with which the hostile force is armed. A defense against a predominantly infantry threat would differ from that planned to defeat an armored threat. The decision to defend or concentrate assets solely on the outer edge of a city are weighed against the enemy's ability to mass fires, infiltrate, or bypass strongpoints.

(2) Defense Inside the City. The commander may decide that his best defense is to defend within the city. This option is usually chosen when hostile forces possess strong, accurate, long-range fires and the defender wishes to minimize their effectiveness (See Quang Tri City example in Chapter 1). The strength of the defending force relative to the attacker may also lead to this option as an economy-of-force measure.

Structures on the outer edge of an urban area that obstruct fields of fire or that are likely to be used by an attacker should be prepared for demolition or demolished ahead of time. All defensive positions should be supplemented or covered by direct-fire weapons. Indirect fires should be planned for most likely avenues of approach. Additionally, close air support should be planned and integrated with the defense.

Security forces are placed outside the defended area to gain early enemy contact, inflict maximum casualties, and attempt to deceive the enemy as to the location of the friendly positions.

(3) Defense of a Key Sector. This option is usually employed when only a portion of an urban area be held or when strength is insufficient to defend an entire area (See the Japanese defense of Manila in Chapter 1). This option is often employed in situations where the mission requires clearing and holding a key sector, or sectors, in order to facilitate other operations.

(4) Entrapment and Ambush. This option is preferred when the defender lacks requisite strength to man organized, in-depth, defensive positions (See the highly successful Egyptian defense of Suez City example in Chapter 1). It requires the preparation of a series of blocking and/or ambush positions along major avenues of approach through an urban area. The defending force uses mines and obstacles to block alternate or possible detour routes in order to canalize the attack force into prepared ambush sites. The defender employs security forces forward of the main battle area and establishes inner or outer defensive strongpoints with extensive obstacles designed to assist in canalize the enemy into kill zones.

The entrapment and ambush option requires detailed planning and rehearsal by all echelons. Alternate plans should be prepared in the event the ambush is only partially successful or if the enemy fails to enter prepared ambush routes. Alternate ambush positions or counterattacks are planned.

Ambush planning in an urban environment does not differ significantly from planning in other environments. However, the urban environment offers an abundance of kill zones that afford the enemy little or no opportunity to escape. Ambush positions are established in selected buildings in the kill zone which offer optimum fields of fire for all weapons. Indirect fires and close air support should be planned to obtain maximum effect in the kill zone. The kill zone should also be covered with targets for indirect fire and CAS. Entrances into buildings not used as firing points should be blocked to entrap hostile forces in the kill zones.

(5) Defense in Depth. Defense in depth should be incorporated into any defensive plan. Units should be given enough terrain to allow for primary, supplementary, and alternate fighting positions as well as successive battle positions or strongpoints. This allows enough room for multiple ambush/entrapment kill zones and the positioning of a reserve.

c. Defensive Organization. The defensive battlespace is divided into three areas: the *security area*, *main battle area*, and *rear area*. Depending on the mission of the unit, forces defending a built-up area may have missions in any one or more of these areas. (Figure 3-5 on page 3-12).

.3

(1) Security Area. The defensive battle begins with a combined-arms force conducting security operations well forward of the main battle area. Security area missions can include screen , guard, and cover. Based on their assigned mission, security forces use all

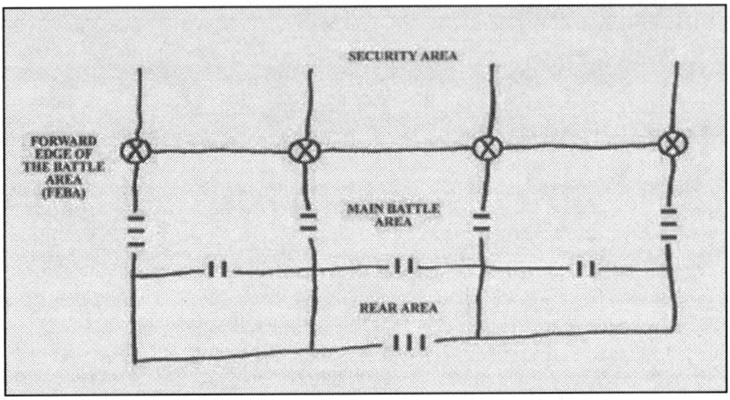

available combat power to gain time, destroy the enemy, or slow his momentum. The initial battle is fought using aviation assets and indirect fire weapons.

(a) Value of Security Area Operations. Security area operations complement to the defense by:

- Alerting the defense to the strength, location, and general direction of the enemy's main and supporting attacks
- Delaying enemy first-echelon detachments
- Initiating early engagement of enemy forces
- Deceiving the enemy as to the true location of the defender's main forces and main effort.

Figure 3-5. Organization of the Battlefield

(b) Withdrawal of Security Area Forces. The urban environment may complicate battle handover from the security force to the main battle area force. However, this transition must be accomplished smoothly to prevent the enemy from gaining momentum. The withdrawal of the security area forces must not result in an easing of pressure on the enemy.

(2) Main Battle Area. The decisive battle is fought in the main battle area. However, the

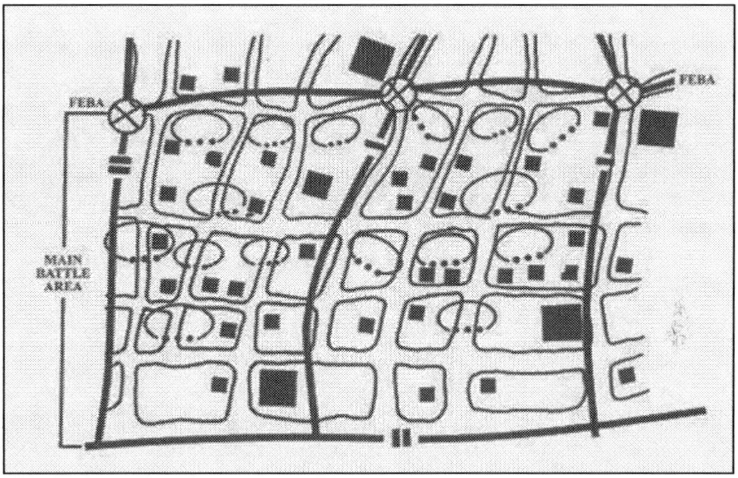

commander may deploy units on the forward edges of the city or in battle positions in depth. The defensive scheme should include forces to defend along the flanks. The commander normally employs security forces to the front to provide early warning and to deny the enemy intelligence on the unit's defensive dispositions (counterreconnaissance) (Figure 3-6).

(a) The size and location of battle positions depend on METT-T and the type of defense prepared. A sophisticated defense will often include a combination of both position and mobile defense but will always be planned in depth.

(b) A unit assigned battle positions on the forward edge of a city or town should:

- Provide early warning of the enemy's advance
- Engage the enemy at long range
- Deceive the enemy as to the true location of the defense.

Figure 3-6. Main Battle Area

(c) The defender employs all available fires to destroy and suppress the direct-fire weapons that support the ground attack. Depending on the concept of operations, weapons may engage targets at their maximum effective range or engage simultaneously for an ambush.

(d) As the attack develops, the actions of small-unit leaders become increasingly important. It is imperitive that all leaders understand their commander's intent.

(3) Rear Area. The rear area is located behind the main battle area and it contains the CSS and administrative support. Protection of these units is vital. Fortunately, the urban environment usually provides sufficient cover and concealment for CSS assets as well as covered and protected routes to the forward areas. Rear area operations are covered in detail in MCWP 3-41.1, *Rear Area Operations*.

3104. Warfighting Functions. The defensive planning sequence remains the same as in any other defensive operation. Defense planning considerations, based on METT-T, must take into account all the activities in each of the warfighting functions. The complexities involved in defense of a built-up area require detailed and centralized planning. Commanders and subordinate leaders must incorporate the following planning considerations for an urban environment when conducting a defensive operation.

a. Command and Control. The commander positions himself so that he can control the action. In an urban environment, this can be more difficult because of the close nature of fighting, prolific obstacles, poor visibility, and difficulty in communication.

Graphic control measures oriented on prominent terrain features are also used in planning and conducting combat in built-up areas. Prominent streets are ideal for use as phase lines and boundries. These and other control measures assist the commander in controlling manuever and fires throughout the battlespace.

COCs should be located in secure locations. (These can be inside or underground when possible.) (See Chapter 2, Section IV, Paragraph 2402.)

Commanders consider the effects of built-up areas on communications when they allocate time to establish communications. LOS limitations affect both visual and radio communications. Wire laid at street level is easily damaged by rubble and vehicle traffic. Also, the noise of urban combat is much louder than in other areas, making sound signals difficult to hear. Therefore, the time needed to establish an effective communications system may be greater than in more conventional terrain.

Wire is the primary means of communication for controlling the defense of a city and for enforcing security. However, wire can be compromised if interdicted by the enemy. Radio communication in built-up areas is normally degraded by structures and a high concentration of electrical power lines. Nonetheless, radio is an alternate means of communication. Messengers can be used as another means. Visual signals may also be used but are often not effective because of the screening effects of buildings, walls, and so forth. Signals must be planned, widely disseminated, and understood by all assigned and attached units. Commanders consider the following techniques when planning for communications:

(1) If possible, lay wire through buildings for maximum protection.

(2) Use existing telephone systems. (Remember, telephones are not secure.)

(3) Emplace radios and retransmission sites in buildings on top floors or atop buildings that are higher than surrounding structures. Ensure that antennas are placed on the side of buildings where the radio signal can be directed at friendly units. This will enhance LOS radio communications.

(4) Use messengers at all levels because they are the most reliable and secure means of communications.

b. Intelligence. The discussion in Chapter 2 pertains equally to a defensive operation. Intelligence gathering for defensive operations is not limited to studying only the enemy. Commanders must emphasize obtaining and using all intelligence. Commanders must emphasize the utility of the urban infrastructure to enhance his intelligence gathering and dissemination effort. The defender usually has the advantage of being familiar with the intricacies of the urban terrain that he is defending. For example, the:

- The sanitation or public works department will have knowledge of the subterranean sewer lines
- Civil engineers will have extensive knowledge of water and power distribution nodes
- Law enforcement representatives may have access to lists of people sympathetic to the attacker
- Local journalists may agree to use local sources to assist in intelligence gathering effort.

c. Manuever

(1) Positioning of Forces. Finding fighting positions in urban areas that provide both good fields of fire and cover is often difficult. The forward edges of a town usually offer the best fields of fire but can be easily targeted by enemy observation and supporting fire. These areas often contain residential buildings constructed of light material. Factories, civic buildings, and other heavy structures that provide adequate cover and are more suitable for a defense are more likely to be found deeper within the town, but they have limited fields of fire.

Because the forward edge of a town is the obvious position for the defender, it should generally be avoided. However, the defender may choose to use the edge of the town if

.3

the terrain limits the enemy's ability for engagement or if strongly constructed buildings that give defenders protection are available. The mission of such a force is to provide early warning of the enemy's advance, to engage the enemy at long range, and to deceive the enemy as to the true location of the defense. This force should withdraw in time to avoid decisive engagement. If there is limited observation from the forward edge, a force should be positioned on more favorable terrain to gain better observation and to engage the enemy at long range.

Reconnaissance by all defending elements should help select routes for use by defenders moving between positions. Movement is crucial in fighting in built-up areas. Early selection and marking of routes adds to the defender's advantages.

The commander covers probable landing and drop zones such as parks, stadiums, or large rooftops and heliports with obstacles or fire to prevent air assault. (See Appendix F.) In a large built-up area, a unit is given a sector to defend and normally establishes a series of defensive positions. Although mutual support between positions should be maintained, built-up terrain often allows the enemy to infiltrate between positions. Therefore, the defender must identify the following:

- Positions that allow suppressive fires on infiltrating routes
- Covered and concealed routes, such as subways and sewers, for friendly forces to use to move between positions
- Structures that dominate large areas
- Areas such as parks, boulevards, rivers, highways, and railroads where antiarmor weapons have longer fields of fire
- Firing positions for mortars
- Command locations that offer cover, concealment, and ease of command and control
- Protected storage areas for supplies.

Buildings that enhance the general plan of defense are selected. Mutual support between these positions is vital to prevent the attacker from maneuvering and outflanking the defensive position. Buildings chosen for occupation as defensive positions should:

- Offer good protection
- Have strong floors to keep the structure from collapsing under the weight of debris

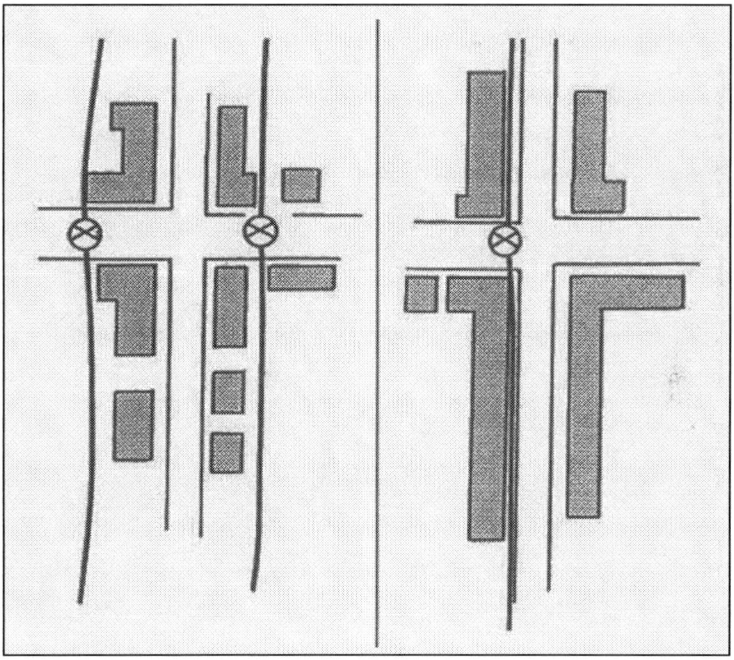

- Have thick walls
- Be constructed of nonflammable materials
- Be strategically located (corner buildings and prominent structures)
- Be adjacent to streets, alleys, vacant lots, and park sites (these buildings usually provide better fields of fire and are more easily tied in with other buildings)
- Provide for 360-degree protection and fields of fire
- Be stocked with adequate supplies as positions may become isolated and cut off for extended periods.

(2) Assign Sectors of Responsibility. Boundaries define sectors of responsibility. Sectors include areas where units may fire and maneuver without interference or coordination with other units. Responsibility for primary avenues of approach should never be split among units. In areas of semi-detached construction, where observation and movement are less restricted, boundaries should be established parallel to alleys or streets so that both sides of a street will be in a single sector. Where buildings present a solid

.3

front along streets, a boundary may have to be extended to one side of the street (Figure 3-7).

Figure 3-7. Sector Boundaries in Areas With Semi-detached and Solid-Front Construction

(3) Select and Prepare Initial Fighting Positions. Commanders select positions in depth. The unit should prepare positions as soon as they arrive and continue improving positions as long as they are occupied. Enemy infiltration or movement sometimes occurs between and behind friendly positions. Therefore, each position must be organized for all-around defense. The defender should also:

(a) Make minimum changes to the outside appearance of buildings where positions are located.

(b) Screen or block windows and other openings to keep the enemy from seeing in and tossing in hand grenades.

(c) Remove combustible material to limit the danger of fire. Fires are dangerous to defenders and create smoke that could conceal attacking forces. Defenders should store firefighting equipment (water, sand, etc.). The danger of fire also influences the type of ammunition used in the defense. Tracers or incendiary rounds should be avoided if the threat of fire exists.

(d) Turn off electricity and gas. Both propane and natural gas are explosive. Natural gas is also poisonous and is not filtered out by a protective mask. Propane gas, although not poisonous, is heavier than air. If it leaks into an enclosed area, it displaces the oxygen and can cause suffocation. Gas mains and electricity should be shut off at the main distribution nodes.

(e) Battle positions should be located so as not to establish a pattern. The unit should avoid obvious firing locations like church steeples.

(f) The unit should rehearse movements between positions.

(g) Camouflage positions.

(h) Reinforce positions with available materials, such as beds, furniture, etc.

(i) Block stairwells and doors with wire or other material to prevent or delay enemy movement. Create holes between floors and rooms to allow for covered movement within a building.

(j) Prepare range cards, fire plans, and sector sketches.

(k) Emplace machine guns in basements windows, where the guns can provide grazing fires. When basements are not used, access to them should be sealed to prevent enemy entry.

(l) Stockpile V (ammunition) and class VIII (medical supplies) items.

(4) Clear Fields of Fire. Commanders may need to rubble certain buildings and structures to provide greater protection and fields of fire. If the ceiling of a lower story room can support the weight of the rubble, collapsing the top floor of a building before the battle starts may afford better protection from indirect fires. Rubbling an entire building can increase the fields of fire and create an obstacle to enemy movement. Rubbling buildings too soon (or rubbling too many) may give away defensive locations and may destroy cover from direct fire. Rubbled buildings may also interfere with planned routes of withdrawal or counterattack.

(5) Improve Fighting Positions. When time permits, all positions, including supplementary and alternate positions, should be reinforced with sandbags and provided

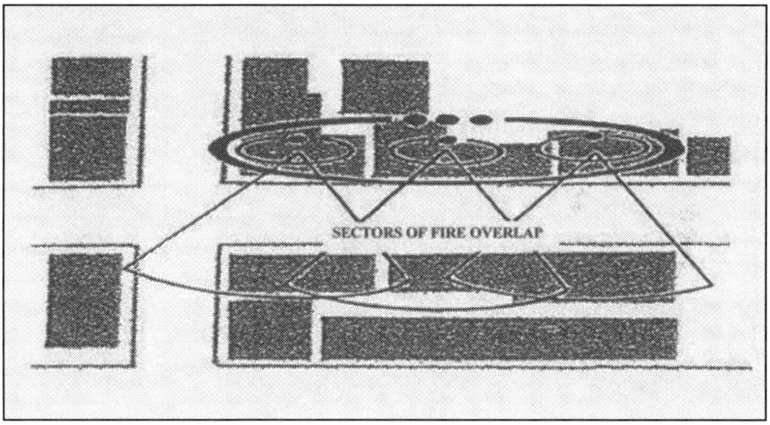

with overhead cover. Obtain support from attached engineers. (See Appendix D).

(6) Employment of Infantry

(a) The infantry battalion is normally assigned a sector to defend. However, depending on METT-T, it may be assigned to defend from a battle position or strongpoint. The battalion will usually assign its companies to either a battle position, a strongpoint, or a sector. These battle positions, strongpoints, or sectors are placed along avenues of approach to block or restrict the enemy's movement.

Depending on the type of built-up area, a company may be employed on the forward edge of the flanks of the area. This forces the enemy to deploy early without decisive engagement because it deceives the enemy as to the true location of the main defense. Other companies may then be employed in a series of battle positions and/or strongpoints in the center of the city or town. In all cases, mutual support between positions is vital. Companies and platoons should have designated alternate and supplementary positions.

(b) Once the rifle platoon commander has received his battle position or sector, he then selects the positions for his squads and crew-served weapons. Squad positions within the built-up area may be separated by rooms within buildings or be deployed in different buildings. Squad positions must be mutually supporting and allow for overlapping sectors of fire (Figure 3-8).

Figure 3-8. Sectors of Fire

(7) Employment of the Reserve. The commander's defensive plan always considers the employment of a reserve. The reserve force is organized and prepared to exploit success, to counterattack to regain key positions, to block enemy penetrations, to protect the flanks, or to assist by fire in the disengagement and withdrawal from endangered positions. For combat in a built-up area, a reserve force:

- Must be as mobile as possible
- May be a mechanized platoon, a platoon (-), or a squad at mechanized company level or a company, a company (-), or a platoon at battalion level
- May be supported by tanks or other armored vehicles
- Should be centrally located to facilitate manuever to threatened areas.

(8) Employment of Armored Vehicles. The commander plans to employ armored vehicles to take advantage of their long-range fires, heavy machine gun fires, self-generating smoke, and mobility. However, built-up areas restrict the mobility of armored vehicles and make them vulnerable to the antiarmor weapons of the enemy. Both the LAV and AAV are lightly armored and can be penetrated by heavy machine gun and antiarmor fires. Armored vehicles provide the commander with a mobile force to respond quickly to enemy threats on different avenues of approach.

(a) When armored vehicles are employed in the defense of a city, infantry should be positioned to provide them with security against close antitank fires and to detect targets for the armored vehicles. Armored vehicles should be assigned *primary*, *alternate*, and *supplementary* positions, as well as *primary* and *alternate* sectors.

(b) Armored vehicles should be located on likely avenues of approach to take advantage of their long-range fires. They may be:

- Positioned on the edge of the city in mutually supporting positions
- Positioned on key terrain on the flanks of towns and villages
- Used to cover barricades and obstacles by fire
- Placed with the reserve.

(c) Armored vehicles are normally employed as a platoon. However, sections of armored vehicles may be employed with rifle platoons or squads. This provides the armored vehicles with the close security of the infantry.

(9) Employment of Helicopterborne Forces

(a) Helicopterborne Task Force in the Defense. A helicopterborne force can defend against an infantry-heavy threat by utilizing its mobility to achieve a maneuver advantage. This force can be key in the execution of a mobile defense in both the security area and main battle area.

(b) Security Area. The helicopterborne force may be able to conduct security force operations for a larger force. Helicopters can position combat power and combat support quickly during rapidly changing situations.

(c) Main Battle Area. The mobility advantage that the helicopterborne force has over enemy infantry-heavy units may allow it to defend in greater depth in a large city. The helicopterborne force focuses on the destruction of advancing enemy forces and fights a series of battles in depth, it can attack the enemy from the front, flanks, and rear. Battle positions are selected and prepared throughout the main battle area along likely avenues of approach. Primary and alternate LZs and pick-up zones should be selected.

d. Fires. Fire support in the defense is as important as it is in the offense and has many of the same characteristics. The proximity of buildings to targets, minimum range restrictions, and repositioning requirements are all factors in the utilization of fire support in the defense. Additionally indirect fires are planned on top of and immediately around defensive positions for close support. Some of the indirect fire support considerations for the defense include:

(1) Historically, artillery fire has been used in some unique and innovative ways in an urban environment. Artillery fire support may be used in the direct- or indirect-fire mode. Artillery fire should be used to:

- Suppress and blind enemy overwatch elements

.3

- Disrupt or destroy an assault
- Provide counterfire
- Support counterattacks
- Provide direct fire when necessary
- Provide illumination in hours of darkness.

(2) Mortars at the infantry battalion and rifle company level are employed to maximize the effect of their high-angle fires. Mortars are one of the most effective indirect fire weapons in the urban environment. They should be used to engage:

- Enemy overwatch positions
- Enemy infantry before they seize a foothold
- Targets on rooftops
- Enemy reinforcements within range.

(3) Final protective fires are planned to stop dismounted assaults in front of the defensive positions. Fires within the city are planned along likely routes of advance to delay, disrupt, and destroy the enemy as he attempts to deepen a penetration.

(4) Commanders establish priorities of fire on the basis of enemy avenues of approach and threat systems that present the greatest danger to the defense. For example, during the attacker's initial advance, tanks, IFVs, and overwatching elements are the greatest threat to the defense. ATGMs should concentrate on destroying tanks first, then IFVs. Indirect fires should suppress and destroy enemy ATGMs and overwatch positions and/or elements. If enemy formations secure a foothold, priority is shifted to the destruction of enemy forces within the penetration.

(5) As the enemy attack progresses into the city, fires are increased to separate enemy infantry from its supporting armored vehicles. Friendly artillery concentrates on attacking infantry, conducting counterfire missions, and destroying reinforcements.

(6) Ample consideration must also be given to the rubbling effects of supporting arms. These effects often create massive rubbling of structures and buildings, making obstacles to both friendly and enemy movement.

(7) Mortars may be initially positioned forward in support of security forces. After their withdrawal, mortars are repositioned where they can support the in the main battle area.

(8) The commander assigns final protective fires and machine gun final protective lines (FPLs). Machine gun positions provide interlocking grazing fire and mutual support between adjacent units. FPLs are fired on planned signals. Proposed FPLs must be “walked out” to determine the extent of grazing fire available and to locate dead space. Dead space can be covered by:

- Rifle and light machine gun fire
- Grenade launchers
- Mines and boobytraps
- Indirect fires.

(9) Long-range antiarmor weapons are employed in the forward areas of the main battle area and the security area. Antiarmor weapons must be carefully positioned to prevent terrain and obstacles from reducing their range. Dismounted and manpacked antiarmor weapons should be positioned in buildings and along routes where engagement distances are reduced but where surprise and ambush are ideal.

e. Logistics. Just as in the offense, the defense utilizes a greater amount of class V and class VIII supplies in the urban environment. The defender needs to plan for multiple routes and means of supply. These can include:

- Using the city's existing transportation infrastructure to move supplies
- Prepositioning supplies in the defense in depth
- Using helicopter assets for rapid resupply
- Establishing and protecting supply routes
- Plan for utilization of existing repair and maintenance facilities
- Identify potable water sources

f. Force Protection

(1) Employment of Engineers. Engineers are employed at the battalion level or attached to companies and integrated down to squad level. Commanders must consider and assign the priority of work for engineer tasks that enhance survivability, mobility, and countermobility. Tasks that engineers can accomplish in the defense of a built-up area include:

- Constructing obstacles and rubbling for countermobility

.3

- Clearing fields of fire
- Laying mines
- Preparing routes to the rear
- Preparing fighting positions.

(a) Employment of Obstacles. Obstacles are used to shape the battlespace. (See FM 5-102, *Countermobility*, and FM 20-32, *Mine/Countermine Operations*, for details.) The four types of obstacles are:

- **Disrupting.** Disrupting obstacles are used to disrupt the enemy's timetable and force him into conducting a time-costly bypass or breach, or cause his march columns to become staggered or "telescoped."

- **Turning.** Turning obstacles are used to canalize the enemy into engagement areas or kill zones where combined-arms fires can be concentrated and flanking shots are increased.

- **Fixing.** Fixing obstacles increase the time the enemy can be fired on in an engagement or target area and to cause increased confusion.

- **Blocking.** Blocking obstacles prevent or delay the enemy from moving past a certain point.

(b) Obstacle Belts. In defensive obstacle planning, obstacles may be emplaced in a series of belts. This enhances the overall effectiveness of the obstacle plan with successive belts designed for specific purposes. The following example lays out a three-belt obstacle plan.

1 **First Obstacle Belt.** The first obstacle belt is enplaced at the nearest buildings across from and parallel to the main defensive position. This belt consists of wire and improvised barriers (including inside buildings, in subterranean avenues of approach, and outside in open areas), danger areas, and dead space. These barriers and obstacles should be heavily boobytrapped and covered by fires. This belt is designed to impede enemy movement, breaks up and disorganize attack formations, and inflicts casualties.

2 **Second Obstacle Belt.** The second obstacle belt is placed between the first belt and the main defensive position buildings, but out of hand grenade range. It is designed to impede movement, canalize the enemy into the best fields of fire, break up attack formations, and inflict casualties. This belt is not meant to stop the enemy permanently. It consists mainly of wire obstacles, improvised barriers, road

craters, and minefields. It should be heavily boobytrapped. Triple-strand concertina wire is placed along the machine gun FPL to delay the enemy in the kill zone.

3 **Third Obstacle Belt.** The third obstacle belt is the defensive position's denial belt. It consists of wire obstacles placed around and through the defensive buildings, close-in minefields, and subterranean accesses. It is designed to impede the enemy's ability to gain a foothold in the defensive area. It should be heavily boobytrapped, and integrated with mines.

(c) Emplace Obstacles and Mines. While principles for employing mines and obstacles do not change in the defense of a urban area, techniques for their employment change. For example, burying and concealing mines in streets is usually not done. To save time and resources in preparing the defense, commanders must emphasize using all available materials (automobiles, railcars, rubble) to create obstacles. Civilian construction equipment and materials should be utilized. Obstacles must be integrated into buildings and rubble areas to increase their effectiveness and to canalize the enemy. The family of scatterable mines (FASCAM) may be effective on the outskirts of a city or in parks and open areas. (See Appendix F for further discussion on obstacles, mines, and demolitions.)

(2) Air Defense. Air defense (Stinger) weapons are positioned on rooftops and parking garages in order to obtain LOS coverage for the units they are protecting. Stinger teams can be assigned the mission of protecting specific positions or placed in general support (GS) of the battalion.

(3) Fire Hazards. All cities are vulnerable to destruction by fire, especially those with many wooden structures. The defender's detailed knowledge of the terrain permits him to avoid areas that are likely to be fire hazards. The defender can deliberately set fires:

- To disrupt and disorganize the attackers
- To canalize the attackers into more favorable engagement areas
- To obscure the attackers' observation.

Section II

Infantry Battalion Defense of a Built-Up Area

This section addresses some of the missions a task-organized and/or reinforced infantry battalion could be assigned. These missions are normally conducted as part of a larger joint operation.

3201. Defense in Sector. An infantry battalion is normally given the mission of defending a sector in a city (Figure 3-9). The interior buildings in the battalion sector form battle positions for the battalion's fixed defense. The size of a sector varies, but may cover the outlying structures and can be used to provide early warning and to delay the enemy.

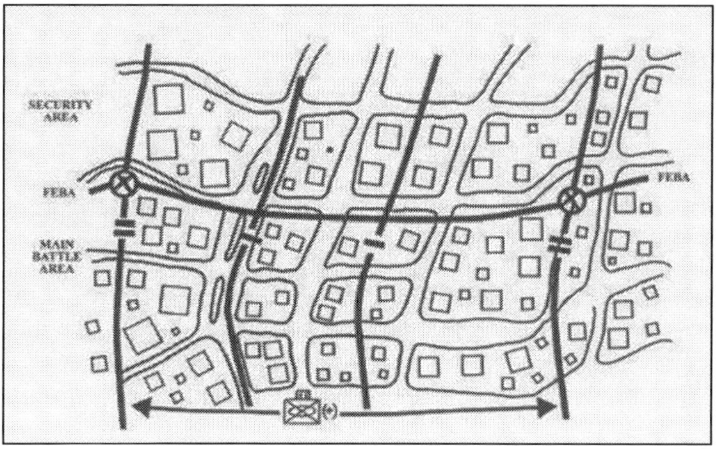

Figure 3-9. Defense of a Built-Up Sector

a. Defensive operations begin with the reconnoitering of the assigned sector to learn the terrain and to orient toward the enemy's expected advance. The main defensive positions should be formed on key terrain features, such as prominent buildings, road junctions, and positions which offer good fields of fire. Long-range antiarmor weapon systems should be deployed to engage the enemy as early as possible.

b. Depending on the situation, a reinforced rifle company may be used as a security force. The security force would operate forward of the FEBA and, depending on the mission, cause the enemy to deploy without without becoming decisively engaged. This can be done through maximum use of ambushes and obstacles while using covered and concealed routes. These routes are used for disengagement. Covered and concealed routes can be artificially made by making holes through buildings. The security force also attempts to detect the location of the enemy's main attack. Once the security force withdraws to the main battle area, it can be used as a reserve to reinforce or counterattack.

c. Based on METT-T considerations, the defense in sector (behind the FEBA in the main battle area) may consist of either company sectors, strongpoints, or battle positions. Strongpoints located on or covering decisive terrain are extremely effective in the defense.

d. Antiarmor vehicles and weapons should be used to engage enemy armored vehicles and/or cover obstacles with fire on likely avenues of approach.

e. Tanks are best used to engage enemy tanks, cover obstacles by fire, and engage in counterattacks. They should be employed in platoons where possible. However, in congested areas they may be employed in sections.

f. Artillery and mortar fire should be used to suppress and blind enemy overwatch elements, to engage enemy infantry on street approaches, provide counterbattery fire, and support counterattacks with direct and indirect fire.

g. Engineers are integrated throughout the defense. They provide support to forces in strongpoints to help prepare fighting positions. Engineers may also be attached to security forces to help in laying mines and constructing obstacles, clearing fields of fire, and preparing routes.

3202. Delay in a Built-Up Area. A delay is an operation in which a force under pressure trades space for time by slowing down the enemy's momentum and inflicting maximum damage on the enemy without becoming decisively engaged (MCWP 3-1). The delay can be oriented either on the enemy or on specified terrain such as a prominent building or industrial complex.

a. A delay in a built-up area may consist of a combination of ambushes and alternating or successive battle positions (Figures 3-10A and 3-10B on page 3-28).

(1) Ambushes are normally planned on observed obstacles. They require decentralized control and execution. Ambushes can be combined with limited-objective attacks on the enemy's flanks. They are most effective at the edge of open spaces, parks, wide streets, and so on. These attacks may be executed by dismounted infantry supported by tanks, LAVs/AAVs, or HMMWV-mounted machine guns and antitank vehicles.

(2) Battle positions should be placed where heavy weapons, antiarmor weapons, and machine guns will have the best fields of fire. Such locations are normally found at major street intersections, parks, and the edge of open residential areas. Battle positions should be carefully and deliberately prepared, reinforced by obstacles, and supported by indirect fire weapons. Battle positions are designed to inflict maximum casualties on the enemy and cause him to deploy for a deliberate attack.

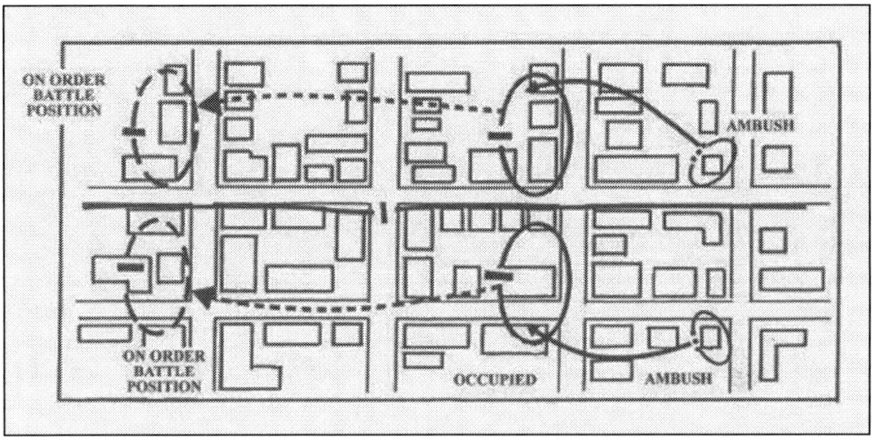

b. Tanks, LAVs/AAVs, and antiarmor weapons should prepare primary and alternate positions to improve their surivability and flexibility.

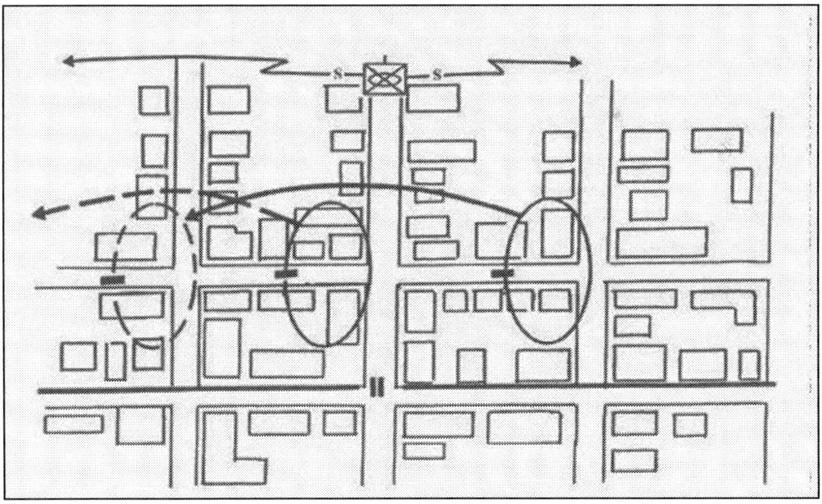

Figure 3-10A. Delay in a Built-Up Area (Successive Bounds)

Figure 3-10B. Delay in a Built-Up Area (Alternate Bounds)

c. The infantry battalion is most effective when deployed in successive delaying actions. Forces alternately engage the enemy to the maximum extent possible without becoming decisively engaged, then move on order in successive or alternate bounds. Smoke and demolitions are used to aid disengagement and cover the movement of the withdrawing force. Security units on the flank can be employed to prevent the enemy from outflanking the delaying unit. A small reserve can be used to react to unexpected enemy action and to conduct continued attacks on the enemy's flank.

d. The engineer effort should be centralized to support the preparation of battle positions, obstacle emplacement, and clearing routes.

e. The width of the infantry battalion sector depends on the nature of the buildings, obstacles along the street, and the time that the enemy must be delayed. Normally, narrow sectors supports delaying by alternate bounds while a wide sector better supports delaying by successive bounds.

3203. Defense of a Village. An infantry battalion assigned a defensive sector that includes a village may decide to incorporate the village as a strongpoint. This use is most common where the village stands astride a high-speed avenue of approach or lies between two difficult obstacles. To incorporate such an area into its defense, the infantry battalion must control the high ground on either side of the village to prevent the enemy from firing into the village. The battalion commander normally deploys one rifle company in the village to prepare a strongpoint defense

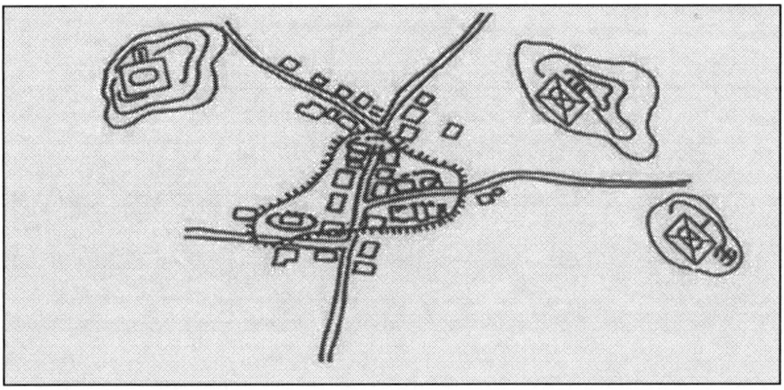

while the other rifle companies are overwatching the village, blocking bypass routes, and protecting the strongpoint's flanks. The strongpoint defense should provide a location where the enemy can be stopped and counterattacks can be launched (Figure 3-11).

Figure 3-11. Defense of a Village

a. Armored Vehicles and antiarmor weapons should be employed where the maneuver room is the greatest. This is usually on key terrain to the flanks of the village. As the security force withdraws and rifle companies become engaged, armored vehicles and weapons support by fire.

b. Although the infantry battalion's disposition should prevent large enemy forces from threatening the rear and flanks of the village, the danger of small-unit enemy infiltration requires the village be prepared for all-around defense.

c. Engineers should be assigned to the rifle company in the village strongpoint to provide continuous engineer support in preparation for the enemy's attack. The priority of engineer support should go to the strongpoint.

Section III

Rifle Company Operations

This section addresses some of the missions the infantry battalion could assign to the rifle company. For the purpose of this publication, the rifle company is a task-organized and/or reinforced unit. Assigned missions are normally part of a larger battalion operation.

3301. Defense of a Village. Once the commander has completed his reconnaissance of the village and the surrounding terrain, he develops his plan for the defense (Figure 3-12). The commander must decide whether to defend on the leading edge of the village or within the village.

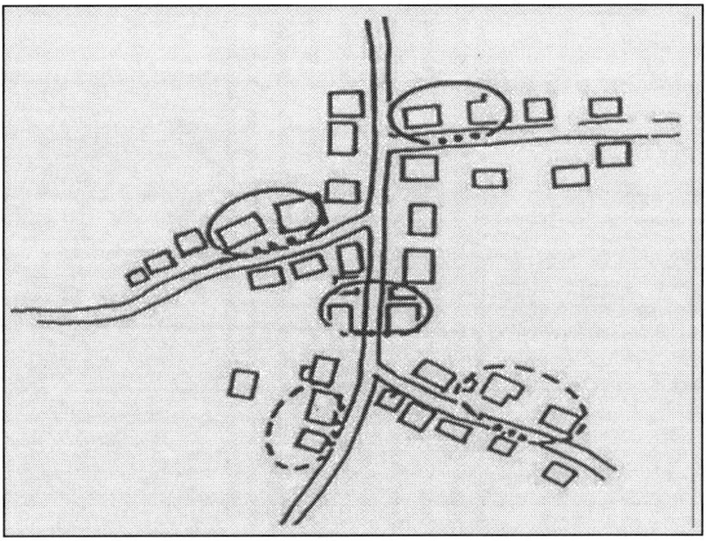

Figure 3-12. Defense of a Village

a. Several factors influence his decision. First, he must know the type of enemy that he will defend against. If the threat is mainly dismounted infantry, the greatest danger is in allowing them to gain a foothold in the town. If the threat is armor or motorized infantry, the greatest danger is that the company's defensive positions can be destroyed by direct fire. The commander must also consider the terrain forward and to the flanks of the village from which the enemy can direct fires against his positions.

b. The commander places platoons in a small group of buildings where each platoon leader establishes mutually supporting squad-sized positions. This positioning increases the area that the rifle platoon can control and hampers the enemy's ability to isolate or bypass a platoon. A rifle platoon can be responsible for the road through a village. The rest of the rifle company is positioned to provide all-around security and defense in depth.

c. Mortar positions must be protected from direct fire and allow for overhead clearance.

d. Combat vehicles should be assigned primary, alternate, and supplementary positions as well as primary and secondary sectors of fire. They should be positioned in defilade behind rubble and walls or inside buildings for cover, concealment, and movement into and out of the area.

e. If the company is reinforced with antiarmor weapons, the reinforcements should be positioned on high ground in and around the town to attain good fields of fire.

f. If the company is reinforced with tanks, the commander could place the tanks along the leading edge complement the other antiarmor weapons. The tank platoon leader selects exact firing positions and assign sectors of fire. If attacked by enemy infantry, the tanks should move to alternate positions protected by infantry. The alternate positions should allow the tanks to engage to the front and flanks with as little movement as possible. Tanks can be used as a mobile reserve after withdrawal from the leading edge of the town.

g. Obstacles are easily constructed in a built-up area. Commanders must stop enemy vehicles without interfering with his own movement in the village. Command detonating cratering charges at key street locations may be used to hamper enemy mobility. Mines should be laid on the outskirts of the town and along routes that the company will not use.

h. Engineers assist rifle platoons to prepare defensive positions. Engineers use explosives to make firing ports, mouseholes, and demolition obstacles. They also assist in the emplacement and recording of the minefields.

i. Ammunition expenditure is increased when fighting in an urban environment. Ammunition should be stockpiled at each squad position.

Other factors that the company commander may consider are:

- Resupply
- MEDEVAC
- Communications
- Firefighting

.3

- Sleep and alert plans
- Security
- Limited visibility
- Control of noncombatants.

j. The company should install a wire net and develop a plan for pyrotechnic signals to ensure adequate communications. Backup wire should be laid in case primary lines are cut by vehicles, fires, or the enemy. The commander also plans for the use of messengers .

3302. Delay in a Built-Up Area. A rifle company may be assigned the mission to delay in a built-up area.(Figure 3-13). To perform this mission, a rifle company can destroy enemy reconnaissance elements forward of the outskirts of the town, prevent their penetration into the built-up areas, and gain and maintain contact with the enemy to determine the strength and

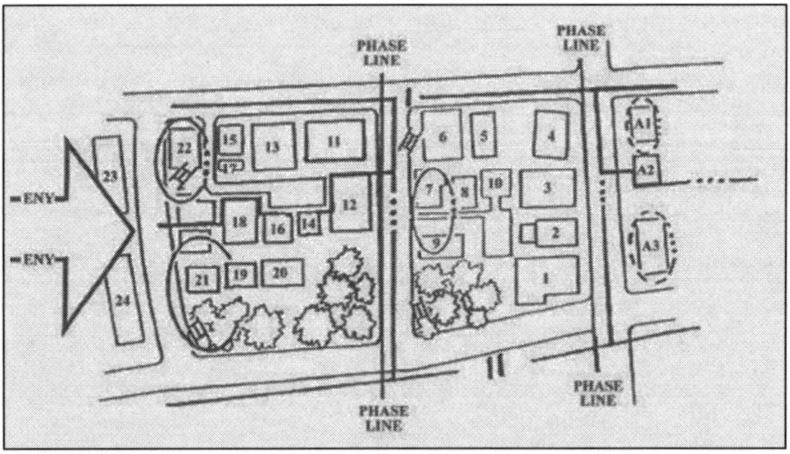

location of the main attack.

Figure 3-13. Delay in a Built-Up Area

a. Platoons delay by using patrols, OPs, and ambushes. Following each action, a platoon will disengage and withdraw.

b. The company's sector should be prepared with obstacles to enhance the delay. Engineers should prepare obstacles on anticipated enemy main routes of advance.

c. Antiarmor weapons may be positioned on the outskirts of the town to destroy the enemy at maximum range. They should be located in defilade positions or in prepared shelters. They fire at visible targets and then fall back or proceed to alternate positions.

d. Tanks in support may be used to engage enemy tanks, provide reinforcing fires, aid the disengagement of platoons, and cover obstacles by fire.

3303. Defense of a City Block. The rifle company may be assigned to defend a city block. The company actions should be coordinated with those of the security forces charged with delaying to the front of the company's position. The defense should take advantage of the protection of buildings that dominate the roads.

a. A well-organized company defense:

- Stops the attack of the enemy on the roads by using obstacles and enfilade fire
- Destroys the enemy by ambush and direct fire from prepared positions
- Ejects the enemy from footholds or fixes them for the battalion's counterattack.

b. The defensive forces should prepare ambushes on the avenues of approach, cover the obstacles by fire, and build a defense inside the buildings (Appendix D). The reserve can be tasked:

- To reinforce the fires of the defense
- To react to a danger on the flank
- To counterattack to throw the enemy from a foothold.

c. Engineers should be employed in GS to construct obstacles, prepare access routes, and assist in preparing defensive positions.

d. Tanks attached to the rifle company may be used to direct-fire support, engage enemy tanks, and support a counterattack.

3304. Defense of a Traffic Circle or Key Intersection. A rifle company may be assigned the mission to defend a key traffic circle or road junction (Figure 3-14).

a. The commander analyzes enemy avenues of approach and buildings that dominate those avenues. He plans to focus all possible firepower on the traffic circle and its approaches. He also plans for the all-around defense of the buildings that dominate the traffic circle to prevent encirclement. The commander prepares as many covered and concealed routes between these

.3

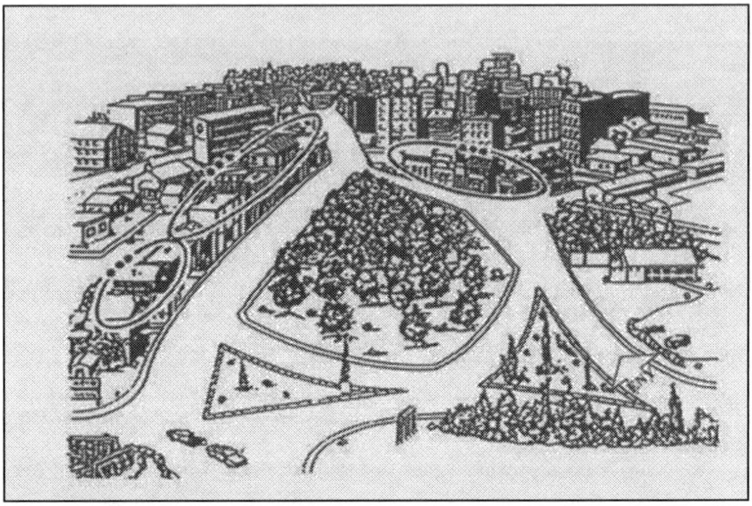

buildings as possible. These routes make it easier to mass or shift fires and to execute counterattacks.

Figure 3-14. Defense of a Traffic Circle

b. Obstacles can be used to deny the enemy the use of the traffic circle. In this case, obstacle planning should consider whether friendly forces require access to the traffic circle. Antiarmor weapons can be employed across the traffic circle if designated fields of fire exist. Tanks can be used to engage enemy tanks and provide direct-fire for counterattacks.

Section IV

Rifle Platoon Operations

This section discusses some missions the rifle company might assign to a rifle platoon.

3401. Defense of a Strongpoint. A rifle platoon may be given the mission to provide or conduct the strongpoint defense of a building, part of a building, or a group of small buildings (Figure 3-15). The rifle platoon's defense is normally integrated into the rifle company's defense. The rifle platoon must keep the enemy from gaining a foothold in its assigned buildings. It makes the best use of its weapons and supporting fires, organizes all-around defense, and counterattacks or calls for a rifle company counterattack to eject an enemy that has a foothold. The platoon leader analyzes his defensive sector to recommend to the company commander the best use of obstacles and supporting fires. (See Appendices B and D.)

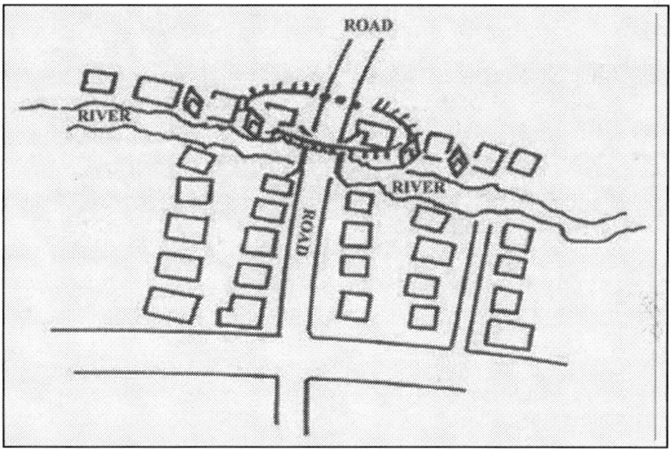

Figure 3-15. Defense of a Strongpoint

a. The rifle platoon should be organized into a series of firing positions located to cover avenues of approach, cover obstacles, and provide mutual support. Snipers may be located on the upper floors of the buildings. Unengaged units should be ready to counterattack, fight fires, or reinforce other elements of the rifle platoon.

b. Depending on the length of the mission, the rifle platoon should have available:

- Pioneer equipment (axes, shovels, hammers, picket pounders)

- Barrier material (barbed wire, sandbags)
- Munitions (especially grenades)
- Food and water
- Medical supplies
- Firefighting equipment.

3402. Defense Against Armor. Urbanized terrain is well suited to an infantry's defense against mechanized infantry and armored forces. Armored forces will usually avoid built-up areas but may be forced to pass through them.

a. Built-up areas have certain aspects that favor infantry antiarmor operations.

(1) Buildings provide cover and concealment for defending infantry.

(2) The streets restrict armor maneuver, fields of fire, and communications, thereby reducing the enemy's ability to reinforce.

(3) Rubble in the streets can be used to block enemy vehicles, conceal mines, and cover and conceal defending infantry.

(4) Rooftops, alleys, and upper floors provide good positions for flanking and overhead fire.

(5) Sewers, storm drains, and subways provide underground routes for infantry forces.

b. To prepare for antiarmor operations, a platoon should:

(1) Choose a Good Engagement Area. The best way for infantrymen to engage tanks is one at a time so that they can destroy one tank without being vulnerable to the fires of another. Good engagement areas include narrow streets, turns in the road, "T" intersections, bridges, tunnels, split-level roads, and rubbled areas. Less obvious locations can also be created by using demolitions or mines to create obstacles.

(2) Select Good Weapons Positions. The best weapons positions are places where the tank is weakest (tank deadspace) and the infantry is most protected. Forcing tanks to button up greatly reduces their ability to see and fire on close-in street and overhead targets. Figure 3-16 (on page 3-38) shows close-in tank weapon vision and engagement limitations (tank buttoned up) at street and overhead levels.

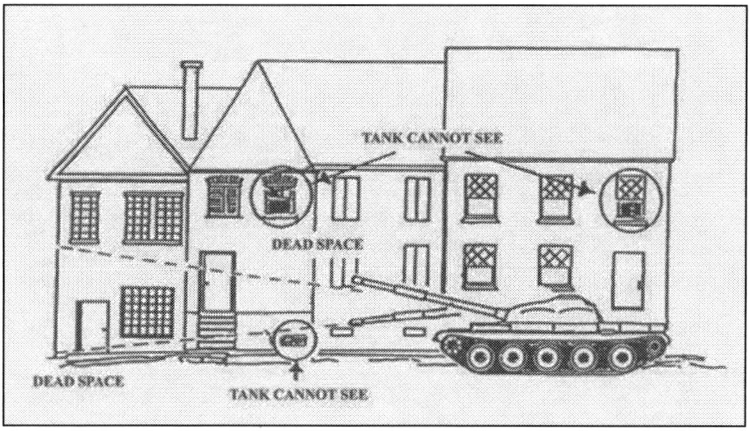

Figure 3-16. Close-in Tank Limitations at Street and Overhead Levels

(3) Assign Target Reference Points and Select Method of Engagement. After selecting the weapons positions, target reference points should be assigned to ensure adequate coverage and control fires. The target reference points should be clearly visible through the gunner's sights and should be relatively resistant to battle damage (for example, large buildings or bridge abutments, but not trees or cars). The three types of antiarmor engagement techniques are frontal, crossfire, and depth. Frontal fire is the least preferred because it exposes the gunner and it attacks the tank where the armor is the thickest. (For more information on target engagement techniques, see MCWP 3-15.5 *Antiarmor Operations*.)

Infantry forces should fire at tanks on the flanks and rear from ground level or at the top of tanks from an elevated position. A suitable antiarmor defense might be set up as shown in Figure 3-17. For a safe engagement from an elevated position, Marines should allow the tank to approach to a range three times the elevation of the weapons. However, engagement at longer ranges risks counterfire because the attacking weapon's position will not be in the tank's overhead dead space.

(4) Coordinate Target Engagement. Tanks are more vulnerable to infantry operations when they are buttoned up. The first task is to compel the tanks to button up by using direct and indirect fire. The next task is to coordinate the fires of the antitank weapons so that all targets in the engagement area are attacked simultaneously (Figure 3-18).

Figure 3-17. A Rifle Platoon's Antiarmor Defense

Figure 3-18. Coordinated Antiarmor Ambush

c. Infantry often accompany armored vehicles in built-up areas to provide them with close-in protection. Antiarmor weapons should be supported by an effective antipersonnel defense.

d. At a planned signal (for example, the detonation of a mine), all targets in the engagement area are attcked at the same time. If that is not possible, then the most dangerous target should be engaged first. Although tanks present the greatest threat, BMPs are also dangerous because their infantry can dismount and destroy friendly antiarmor positions. Rubble and mines should be used to reduce target mobility so that a target's time in the engagement area is increased.

3403. Conduct of an Armored Ambush. A rifle company can use an attached tank platoon to conduct an armored ambush (Figure 3-19). The tank platoon should be reinforced with infantry.

Figure 3-19. Armored Ambush

a. An armored ambush requires maneuver on obstacle-free roads. Obstacles outside the ambush area can be used to canalize and delay the enemy.

b. Ambushing tanks should hide approximately 1,000 meters from the expected enemy avenue of approach. A security post, located at a chokepoint, observes and reports the approach, speed, security posture, and activity of the enemy. When the enemy is reported at a trigger point or target reference point, the tank platoon leader orders the ambush.

c. The tanks move quickly from their hiding positions to firing positions, taking advantage of all available concealment. They attempt to fire flank shots on the approaching enemy at an average range of 300 to 400 meters. Firing at these ranges does not expose tanks to enemy infantry. After the engagement, the tank breaks contact and moves to a rally point with close security provided by infantry. They may then move to a new ambush site.

Chapter 4

Combat Support

Combat support units provide critical combat functions in conjunction with combat arms units to secure victory (MCRP 5-2A).

4001. Introduction. Success against a determined enemy is achieved in part by generating decisive combat power through the application of *combined arms*. This is especially true in MOUT. The MAGTF commander must effectively integrate all the capabilities of the command element (CE), GCE, ACE, and CSSE to win the battle in the shortest amount of time with the fewest casualties. The MOUT environment requires special considerations pertaining to combat support forces and capabilities.

Section I

Fire Support

4101. Indirect and Direct Fire Support

a. General. MOUT operations present unique fire support considerations. On urbanized terrain, buildings provide excellent cover and concealment to the enemy while limiting friendly observation and targeting efforts. Targets are generally exposed for brief periods of time and are often in close proximity to friendly forces. Observers will experience difficulty in finding OPs with adequate fields of observation. Terrain masking by tall buildings may restrict the delivery of indirect fires. Collateral damage and rubble effects must be considered during ammunition selection. The importance of effective communications, ROE, control measures, and procedures to prevent fratricide is magnified.

b. Offensive Support. In the offense, fire support plans should include fires to isolate the objective area, support the assault, and support the clearing action. Fires are delivered to isolate and fix the enemy and deny him the use of avenues of approach into and out of the built-up area. Most fires are normally planned and executed at the GCE or higher echelons. Fires are employed to rupture the enemy's established defenses and screen friendly maneuver, in order to maintain the momentum of the attack. Fire support is also allocated to units involved in clearing operations. Fire support plans should incorporate the employment of aerial observers and UAVs to compensate for restrictions to observation and to assist in the delivery of deep fires. Procedures for designating the forward line of own troops (FLOT), marking targets, shifting fires, and communicating in the urban environment should also be considered.

c. Defensive Support. In the defense, fire support plans address fires to disrupt and slow the enemy attack. Fires are delivered at maximum ranges along avenues of approach to separate armor and infantry forces, to canalize the enemy into killing zones, and to deny the enemy use of key facilities in the city, such as communications and transportation facilities. Most fires are planned and executed at the GCE or higher echelon. Defensive fires planned at lower echelons support fighting within the built-up area. Defensive fire support plans identify fire support coordination procedures necessary to execute the delivery of fire support.

4102. Artillery

a. Employment of Artillery. The mission of artillery does not change in MOUT. However, the method of employment is tailored to the unique requirements posed by the urban battlespace. MCWP 3-16.3, *Marine Artillery Support,* provides a detailed discussion on the employment of artillery. Considerations for MOUT include:

(1) Organization for Combat. Standard command relationships for tactical missions are used to organize for combat. During the initial phases of offensive MOUT, particularly isolation of the objective area and the initial assault to secure a foothold, control of artillery is more centralized than in other types of offensive operations. Centralized control provides better responsiveness in shifting and massing fires. As the attack progresses into the clearance phase, control of artillery becomes more decentralized with direct support (DS) artillery providing direct and indirect fires. In the defense, control of artillery should be sufficiently centralized to enable the rapid massing of fires.

(2) Movement and Positioning. Key considerations for the positioning of artillery in MOUT include:

- Mobility restrictions
- Limited availability of firing positions
- Masking of fires
- Security
- Enemy counterbattery fires

(a) Offense. Artillery is initially positioned outside or in the outskirts of the city. Artillery may displace to suitable firing positions within the city as the attack progresses. Positions should be selected that minimize masking, provide several routes for escape, and afford as much cover and concealment as possible. Potential firing positions can be found in parks, racetracks, sports fields, school yards, stadiums, parking lots, and similar areas. Special techniques for the emplacement of howitzers may be required if the ground is not suitable for normal emplacement. Some artillery positions may be maintained outside the city (see appendix B). Artillery generally makes few displacements in MOUT because it requires extensive route reconnaissance and security. If and when artillery is moved, it should be given priority of movement during displacements. Movement is conducted during night or periods of reduced visibility.

(b) Defense. Artillery is positioned on the friendly outskirts of the built-up area. Artillery displaces along predetermined routes to avoid traffic congestion.

(3) Delivery of Fires. The commander, in coordination with the fire support coordinator (FSC), considers the following when planning fire support in a built-up area:

(a) Target acquisition may be more difficult because of the increased cover and concealment afforded by the terrain. Ground observation is limited; therefore, FOs should be placed near tops of buildings.

(b) For targets in defilade, fire may be adjusted laterally so that the rounds impact on the street perpendicular to the observer's position. Airburst rounds are more suitable for this adjustment. Adjustments may have to be made by sound. When rounds impact on a perpendicular street, they are adjusted for range. When the range is correct, a lateral shift is made onto the target and the observer fires for effect.

(c) Special consideration must be given to shell and fuze combinations when effects of munitions are influenced by buildings. These include:

- Careful use of variable time is required to avoid premature arming.
- HE ammunition with impact or delay fuzes may create unwanted rubble.
- The close proximity of enemy and friendly forces requires careful coordination.
- White phosphorus (WP) may create unwanted fires and smoke.
- Fuze delay should be used to penetrate fortifications.
- Illumination rounds can be effective. However, friendly positions should remain in shadows and enemy positions should be highlighted. Tall buildings may mask the effects of illumination rounds.
- HE/variable time, HE/time fuze, and improved conventional munitions are effective for clearing enemy positions, observers, and antennas from rooftops.
- Swirling winds may degrade smoke operations.
- FASCAM may be used to impede enemy movements. FASCAM effectiveness is reduced when delivered on a hard surface.

(4) Targeting is difficult on urbanized terrain because the enemy has many covered and concealed positions and movement lanes. The enemy may be on rooftops and in buildings and may use sewer and subway systems. Targets should be planned on rooftops to clear enemy FOs, communications, and radar equipment. Targets should also be planned on major roads, at road intersections, and on known or likely enemy fortifications. Employing artillery in the direct-fire mode to destroy fortifications should be considered. Also, restrictive fire support coordination measures (such as a restrictive fire area or no fire area) may be imposed to protect civilians and critical installations.

(5) The 155-mm howitzer can be effective in neutralizing concrete targets by direct fire. A 155-mm round can penetrate 36 inches of concrete at ranges up to 2,200 meters. Restrictions may be placed on the type of ammunition used in order to reduce rubbling on friendly avenues of approach.

(6) FOs must be able to determine the location and size of dead space. Dead space is the area in which indirect fires cannot reach the street level because of buildings. This area is a safe haven for the enemy because he is protected from indirect fires. For low-angle artillery, the dead space is about five times the height of the building. For mortars and high-angle artillery, the dead space is about one-half the height of the building.

(7) Aerial observers and UAVs are effective in providing observation behind buildings immediately to the front of friendly forces. They are extremely helpful in adjusting fire. Aerial observers can also relay calls for fire when communications are degraded because of power lines or building masking.

(8) Weapons locating radars are effective in locating enemy artillery and mortar targets in an urban environment because of their high percentage of high-angle fires. However, LOS must be maintained in order to exploit this advantage. Radars should not be placed in the midst of a city due to masking. They are best located on the outskirts of the city.

(9) The use of airburst fires is an effective means of clearing snipers and manpads from rooftops. HE shells with delay fuzes may be effective against enemy troops in the upper floors of buildings, however, because of the overhead cover provided by the building, such shells have little effect on the enemy in the lower floors. (The planning and employment of artillery in offensive and defensive operations is also addressed in Chapters 2 and 3.)

b. Local Security. Artillery crews and howitzers require close security when used in the direct-fire mode. Artillery is particularly vulnerable when units are employed in isolated firing positions.

c. Communications and Information Systems (CIS). CIS must be able to satisfy the unique C2 requirements for artillery in MOUT. CIS should provide the commander with the ability to collect, process, analyze, and exchange information. Requirements such as massing fires and call for fire must be carefully planned and adapted to meet the requirements of

MOUT. For example, small units conducting independent operations must have the communications capabilities and maps to allow them to call for fire.

4103. Mortars

a. Employment of Mortars. Mortars are often the primary indirect-fire weapon for forward units in the assault or defense of a built-up area. Their high rate of fire, short minimum range, and high trajectory give mortar units the ability to fire in the tight confines of the city. Mortars can provide obscuration, neutralization, suppression, or illumination fires.

(1) Organization for Combat. Normally, 81-mm mortars are employed in GS of the battalion. The 60-mm mortar is normally used at the discretion of the rifle company commander.

(2) Movement and Positioning. 81-mm mortars employed in GS normally occupy a position within or near the infantry battalion reserve. Key considerations for the selection of positions include:

- The minimum range of the weapon
- Mask and overhead clearance
- Terrain suitable for setting base plates
- Dispersion, and accessibility.

Special techniques can be used to position and lay the 81-mm mortar. These techniques include placing buffers under base plates, using curbs and sandbags to anchor or brace mortars, and filling cans with dirt or using sandbags to emplace aiming posts. Fixed objects may be used as distant aiming points. When the depth of the defensive position is shallow, or when suitable firing positions are not available, mortars may have to be positioned behind the reserve. Positioning of mortars behind buildings greatly enhances survivability. Displacement is often executed by section. Mortars are usually not placed on top of buildings because the location lacks cover and concealment and recoil/concussion could collapse roofs.

(3) Delivery of Fires. Mortar fires are effective in the offense as well as in the defense. The indirect fires are extremely responsive. Mortars are well suited for combat in built-up areas because of their high rate of fire, steep angle of fall, and short minimum range. Mortar fires can be used to inhibit enemy fires and movement, allowing friendly forces to maneuver to a position of advantage. Effectively integrating mortar fires with dismounted maneuver is key to successful combat in a built-up area.

.3

(a) HE Ammunition. HE is used more than any other type of ammunition. Although mortar fires are often targeted against roads and other open areas, the natural dispersion of indirect fires will result in many hits on buildings. Leaders must use care when planning mortar fires to minimize collateral damage. Considerations when using HE include:

- HE ammunition is particularly effective when used on lightly built structures within cities. However, it does not perform well against the reinforced concrete found in larger urban areas.

- Point-detonating fuzes should be used. The use of proximity fuzes should be avoided because the nature of built-up areas causes proximity fuzes to function prematurely. However, proximity fuzes are useful in attacking targets such as OPs on tops of buildings.

- During World War II, the Middle East conflicts, and most recently in Bosnia, mortar HE fires have been used extensively to deny the use of streets, parks, and plazas to enemy and civilian personnel.

(b) Incendiary Fires. Employment of these munitions is planned in detail to prevent the burning of buildings and surrounding terrain that may create obstacles to friendly forces.

(c) Illumination. In the offense, illumination rounds should burst above the objective to put enemy troops in the light. Buildings reduce the effectiveness of the illumination by creating shadows. If the illumination is behind the objective, the enemy troops would be in the shadows rather than in the light. In the defense, illumination is planned to burst behind friendly forces to put them in the shadows and place the enemy forces in the light. Continuous illumination requires close coordination between the FO and the fire direction center (FDC) to produce the proper effect.

(d) Other Considerations. When planning the use of mortars, commanders must consider the following:

- FOs should be positioned on tops of buildings so target acquisition and adjustments in fire can best be accomplished.

- Commanders must understand ammunition effects to correctly estimate the number of volleys needed for the specific target coverage. Also, the effects of using white or red phosphorus may create unwanted smoke screens or limit visibility, which could interfere with the tactical plan.

- Mortar sections should plan to provide their own security.

- Commanders must give consideration to where, when, and how mortars are to displace in order to maintain immediate indirect fire support. Combat in built-up areas may adversely affects the ability of mortars to displace because of rubbling.

b. Communications. An increased use of wire, messenger, and visual signals will be required. However, wire should be the primary means of communication between the FOs and the FDC because elements are close to each other. Also, frequency-modulated (FM) radio transmissions in built-up areas are likely to be erratic. Remote siting of antennas to upper floors or roofs may improve communications and enhance operator survivability. The use of radio retransmissions can also be effective. Both in the offense and defense, commanders consider using existing civilian systems to supplement the unit's capability.

c. Magnetic Interference. Anything that uses or generates electricity generates its own magnetic field. In an urban environment, all magnetic instruments are affected by surrounding structural steel, electrical cables, and automobiles. Minimum distance restrictions for the use of the M-2 aiming circle will be difficult to overcome. However, an azimuth may be obtained from a distant aiming point.

4104. Naval Surface Fire Support (NSFS) and Naval Gunfire (NGF)

a. General. When NSFS ships are available, naval surface fires and NGF can provide some fire support.

b. Employment of Naval Surface Fire Support. The urban environment severely limits NSFS, and NGF in particular, due to its flat trajectory. NSFS can be used to supplement existing fire support but should not be planned as the primary fire support asset for MOUT. NSFS is most effective when firing on targets to isolate the battlefield and engage targets located on the seaward side of a city. Only when the ROE allows extensive collateral damage can the full effect of NSFS be brought to bear on units occupying a city.

(1) Organization for Combat. NSFS is best employed under centralized control because of the limited use of NSFS within the confines of the city.

(2) Positioning. Many of the fires delivered by NSFS ships require a careful orientation of the gun target line and may limit the size of a fire support area.

(3) Delivery of Fires. NGF can deliver a variety of ammunition in MOUT. NGF can deliver HE, incendiary, and illuminating projectiles. The NGF high-capacity projectile is very effective in producing fragmenting and penetration effects. Offensive and defensive fire support plans can incorporate NSFS fires on bridges, roads, reinforcements, artillery positions, and command and control centers. NSFS can be effectively delivered on targets along the outskirts of the city to rupture established defenses. These fires are particularly effective on buildings and on targets with a linear axis such as streets and airfields. Because of its flat trajectory, NGF employment is limited on targets within the city. Firing

.3

reduced charges at high angles does enable NGF to hit more targets but will reduce its range and accuracy.

4105. Aviation

a. Employment of Aviation. Both fixed- and rotary-wing aircraft can provide fire support in MOUT. The aviation combat element (ACE) provides the MAGTF commander with a variety of options in ordnance delivery. Aerial delivered ordnance is particularly effective against hard targets. Cluster bomb units, rockets, cannons, laser-guided bombs, and electro-optically guided missiles can be used against a variety of targets. However, the presence of civilians, the requirement to preserve key facilities, and the residual effects caused by rubble and collateral damage must be taken into consideration when planning for and using aviation in an urban environment. (See MCWP 3-24 *Assault Support Helicopter Tactical Manual* (under development), MCWP 3-23.1 *Close Air Support* (under development),and NWP 55-3-AH1, *AH1 Tactical Manual,* for greater detail on MOUT considerations for aviation.)

(1) Fixed-Wing Aircraft

(a) Offensive/Defensive Operations. During offensive operations, fixed-wing aircraft can isolate, interdict, and reduce enemy strongpoints throughout the battlespace. During defensive operations, they attrite enemy forces as they approach, attack forces attempting to bypass, or canalize the enemy to take avenues of approach that are advantageous to the defender.

(b) Other Considerations

1 **Fixed-Wing Attack of a Built-Up Area.** Air strikes can reduce or destroy vital supplies and facilities supporting the enemy. The resulting shock, concussion, and psychological effects of such attacks can reduce the efficiency and fighting spirit of enemy forces.

2 **Rubble and Debris.** The rubble and debris resulting from air attacks may increase the defender's cover and create obstacles to the movement of friendly forces.

3 **Proximity of Friendly Forces.** The proximity of opposing forces to friendly forces increases the potential for fratricide. This may require the use of precision guided munitions (PGMs) or the temporary disengagement of friendly forces in contact.

4 **Locations of Indigenous Civilians or Key Facilities.** The use of aerial delivered ordnance may be restricted by the ROE because of the presence of civilians or the requirement to preserve key facilities within a city.

5 **Limited Ground Observation.** Limited ground observation may require the use of forward air controllers (airborne) (FAC(A)).

(2) Rotary-Wing Aircraft. Attack helicopters provide a distinct advantage to the commander in MOUT. They can rapidly maneuver within the built-up area to attack a variety of targets. Urban structures offer ideal cover for attack helicopters. The ability to deliver PGMs, rockets, and 20 mm cannon fire makes attack helicopters valuable assets in reducing strongpoints, breaking up enemy concentrations, and providing suppressive fires for attacking forces. Attack helicopters also provide a highly manueverable antiarmor capability.

(a) Attack Helicopter Fire Support During Offensive/Defensive Operations

1 Provide close air support in support of the commander's scheme of maneuver

2 Provide suppression and/or destruction of enemy forces within buildings and strongpoints

3 Conduct counterattacks

4 Isolate the built-up area or objectives within the built-up area

5 Conduct armed reconnaissance and patrols.

6 Conduct rapid concentration of fires.

4106. Armored Vehicles

a. Employment of Armored Vehicles. Tanks, LAVs, and AAVs are primarily mobility assets that function best in an unrestricted, mobile environment. However, these vehicles can play an important role inside the city by using their firepower in support of offensive and defensive operations. LAVs and AAVs highly are susceptible to armor-piercing rounds and manpacked antiarmor weapons. All vehicles are susceptible to top-attack munitions. Some of the fire support roles armored vehicles may provide include:

- Suppression and/or destruction of enemy forces within buildings and strongpoints (tanks/LAVs/AAVs)
- Provide machinegun suppression fires
- Reserve or counterattack suppression fires (tanks and/or AAVs)
- Creating entry/exit points in buildings
- Isolation of the built-up area or objectives within the built-up area (tanks/LAVs/AAVs)

- Artillery-towing vehicle (LAVs/AAVs)
- Antiarmor fires
- Breaching obstacles in a direct fire mode.

b. Special Consideration. The existence of waterways, canals, and rivers in and around urban areas can provide avenues of ingress into designated objective areas. If used in conjunction with supporting fires, smoke, and/or deception operations, AAVs can capitalize on their amphibious capability and make use of waterways previously viewed as obstacles.

4107. Antiarmor Weapons. Antiarmor weapons have more utility in the defense than in the offense. Their primary task is defeating enemy armor. Secondary tasks may involve blasting holes in structures and fortified positions. As with other weapons systems, antiarmor weapons are normally employed in a decentralized fashion at the small unit level. Enemy armor should be engaged in predetermined locations that afford the most advantage to the defender. Weapon positions should be selected that provide maximum range fires. TOWs and Dragons may be fired from an enclosed area if it is a sturdy ventilated structure of appropriate size. For a more detailed discussion of antiarmor weapons, see MCWP 3-15.5, *Antiarmor Operations*.

4108. Employment of Snipers

a. General. Snipers are highly trained in field skills and marksmanship and can deliver long-range precision fire at selected targets from concealed positions. The method by which snipers are employed in urban warfare will be governed by the nature of the terrain/weather, number of snipers available, and whether the enemy employs snipers. A sniper can play an important role in MOUT. Employment of snipers influences the action by:

- Creating adverse psychological affects on the enemy by negating feelings of security
- Inflicting casualties on enemy leadership
- Enabling the infantry to seem to cover a large area, regardless of whether the terrain is physically occupied or not
- Detailing enemy positions and activity through the use of advanced optical equipment and observation techniques
- Controlling fire support
- Conducting reconnaissance in conjunction with their sniper role
- Disrupting enemy movement, observation, and infiltration, and negating the possibility of surprise.

b. Offensive Employment. Snipers should be used to gain and maintain contact with the enemy. This enhances security and prevents surprise by keeping constant and unrelenting pressure on the enemy. Some of the considerations for sniper employment include:

(1) As the "eyes" of a commander, a sniper increases the commander's flexibility by gathering and transmitting information on fire support targets.

(2) Snipers attack those targets that affect the enemy's ability to wage battle and those that will cause the maximum amount of confusion on the battlefield. Potential targets for a sniper are:

- Officers and noncommissioned officers
- Enemy snipers and marksmen
- Engineer personnel
- Personnel manning crew-served weapons
- Communications personnel and equipment
- Fire support observers and/or equipment
- Commanders of armored vehicles.

c. Combat on Urbanized Terrain

(1) Snipers can be employed to operate for extended periods of time from hidden positions to dominate and establish a "no-man's land," screen flanks, protect the rear, and deny the enemy access to certain areas or avenues of approach. Snipers can operate with the covering element to deliver accurate fires in support of search elements. Optical devices enhances their ability to detect movement and engage it instantly.

(2) Snipers can also support (by fire) infantry movement across streets. They can provide immediate precision fire on enemy machine gun nests and enemy snipers. The best countersniper weapon is the sniper. Snipers are assigned the following supporting missions:

- Delaying and inflicting casualties on the advancing enemy
- Observing and reporting potential targets
- Covering (by fire) avenues of approach and obstacles

- Harassing the enemy and causing him to deploy prematurely

- Directing supporting arms fire.

The urban environment provides an abundance of secure firing positions that are not highly accessible to countersniping.

d. Defensive Employment. In defensive operations, the sniper is best employed forward of the FEBA to provide early warning of enemy approach, disrupt it, and if possible, cause the enemy to deploy prematurely. Snipers are employed to:

- Prevent enemy infiltration
- Detect and destroy infiltrators
- Protect unit at the FEBA from surprise attacks
- Protect patrols from ambush
- Screen the flanks and rear of the defensive positions.

More detailed information regarding the employment of snipers can be found in Marine Corps Warfighting Publication (MCWP) 3-15.3, *Scout Sniping* (under development).

Section II

Other Combat Support

4201. Engineers

a. General. The nature of the urban battlespace requires extensive use of engineers. In both the offense and defense, the proper employment of engineers can be decisive. Engineers must also be prepared to repair and maintain the internal support facilities of the city in order to support further operations. MAGTF commanders plan to use all available engineer assets, including engineers in the GCE, ACE, and CSSE, as well as any naval construction forces (NCFs).

b. Engineer Organization. Engineers are almost always task organized. The compartmentalized, closed-in nature of the city, which limits observation, mobility, and communications, will dictate the best organization for the mission. Teams should be formed to perform specific tasks for small units and must be positioned for quick employment. In MOUT, engineer support is key to maintaining the mometum of the attack. Engineer teams will usually move with the units they are supporting. Two-man demolition teams may be assigned to support a rifle platoon or squad to breach obstacles or to neutralize booby traps and improvised explosive devices (IEDs).

Commanders must carefully analyze mission requirements, the alignment of enemy forces, the condition of the battlespace, and available assets, to ensure an organization that will provide optimum engineer support. Most of the actual engineer work is done at the small unit level.

c. Engineer Tasks

(1) Offense. Urban combat is by nature a very slow, deliberate affair. Even if the enemy has not prepared the area for a defense, it will take time to secure the city. A well-constructed urban defense will slow or stop any attack unless impediments to movement can be quickly removed or bypassed. Properly organized and positioned engineers can reduce delays by performing a variety of tasks which include:

(a) Engineer reconnaissance to determine the level of effort required to penetrate outer and inner defenses, clear obstacles, and identify bypass routes.

(b) Bridge repair and construction on critical avenues of approach. Bridge construction may also open additional avenues of approach that the enemy is not prepared to defend.

(c) Breaching obstacles both outside and inside the city. This includes breaching minefields and neutralizing booby traps and IEDs.

(d) Assault breaching into buildings.

(e) Reducing enemy strongpoints using demolitions and heavy equipment.

(f) Preparing helicopter landing zones (HLZs).

(g) Assist in the preparation of defensive positions in the event of a counterattack.

(2) Defense. An urban area can be made nearly impregnable with proper organization of the battlespace and skillful preparation by engineers. Obstacles in outlying areas can delay/disrupt the enemy and canalize him into engagement areas. The streets and alleys inside a city can also be use to canalize enemy forces into engagement areas and kill zones. In preparing the defense of an urban area, engineers perform several critical tasks to include:

(a) Engineer preparation of the urban battlespace. This should begin well outside the city with complex obstacles in support of security forces. Selective demolition of bridges and roads and other countermobility measures can force the enemy into taking avenues of approach that favor the defense.

(b) Constructing obstacles forward of and between defensive positions to block, disrupt, fix, and turn enemy attacks. A majority of the obstacles will be directed at interfering with the movement of personnel and vehicles, but some effort must be directed at preventing vertical and subterranean movement. Material designed especially for obstacles and barricades, such as barbed wire and tetrahedron(s), is necessary, but improvised barriers can be just as effective and are readily available. Vehicles and building rubble are the most common improvised barriers.

(c) Enhancing mobility by opening and/or marking preplanned routes into attack positions for counterattacks or alternate positions to protect a flank.

(d) Providing technical advice on which buildings are more suitable for use as strongpoints or which rooftops are strong enough for helicopter landings can be vital. Any reinforcement of building roofs should be supervised by the engineers.

(e) Preparing tank positions that offer them cover and concealment.

(f) Assisting infantry units in clearing clear fields of fire, especially for longer range weapons like the TOW.

(g) Playing a potential role in the maintenance of critical urban services such as power, water, and communications. This function may also be performed by naval construction units.

(h) Assisting in the preparation and maintenance of railroads, airfields, and ports. This function may also be performed by naval construction units.

4202. Assault Support Helicopters

a. Combat Assault Transportation. The mobility provided by helicopters can be a significant advantage in the urban environment. In many modern cities, office buildings may have helipads on their roofs. Other buildings, such as parking garages, are usually strong enough to support the weight of a helicopter. The delivery of Marines onto a building can also be accomplished by rappelling, fastroping, or jumping out of the helicopter as it hovers just above the roof. The rooftops should be inspected to ensure that no obstacles exist, such as electrical wires, telephone poles, antennas, or mines and wire. The use of helicopters in this role is usually predicated on the antiair threat. Helicopterborne forces are highly vulnerable to small arms fire, AAA and manpads.

b. Small-Scale Helicopterborne Assaults. Small units may be helo-lifted onto the rooftop of a key building. Depending on the construction of the roof, rappelling or fastroping Marines from the helicopter may be more advantageous than landing them on the rooftop. Fastroping provides a rapid and safer means of insertion than jumping from a low hover. With practice, Marines can accomplish a fastrope insertion with minimum exposure.

c. Large-Scale Helicopterborne Assaults. For large-scale assaults, rooftop landings are not practical; however, open spaces (parks, parking lots, sports arenas) within the built-up area may be used. Depending on METT-T, helicopterborne forces may assault into an LZ in one wave or scheduled waves.

d. Mobility of Forces and Supplies. In the urban environment, movement of forces by helicopters may become a major requirement. Units engaged in house-to-house fighting normally suffer more casualties than units fighting in open terrain. These casualties must be evacuated and replaced quickly with new forces. At the same time, roads are likely to be crowded with resupply and evacuation vehicles and may also be blocked with craters or rubble. Helicopters provide a means of overcoming these obstacles.

e. Helicopterborne Assault Techniques. Helicopter assaults into enemy-held territory are extremely difficult (Figure 4-1 on page 4-16). One technique is to fly nap-of-the-earth down a broad street or commercial ribbon while attack helicopters and assault helicopter door gunners suppress buildings on either side of the street. Artillery preparations can be incorporated into the helicopter assault plan to suppress rooftop and building antiair threat. Feints and demonstrations in the form of false insertions can deceive the enemy as to the intended objective. Helicopters often deliver forces to the last covered position, short of the fighting, then return without exposure to enemy direct fire. Similar flight techniques can be used for transportation of supplies and casualties.

.3

Figure 4-1. Helicopterborne Assault of a Built-Up Area

f. Combat Roles. Below are some of the roles that assault support helicopters may play in urban warfare:

- Conduct insertion and extraction of forces (See MCRP 3-11.7A, *Helicopter insertions and Extractions* [under development])
- Provide command/control and aerial retransmission
- Support CSS operations
- Provide air reconnaissance.

4203. Employment of Reconnaissance Forces. Reconnaissance units traditionally make maximum use of rural camouflage and terrain. Area, zone, and route reconnaissance are conducted from well-concealed vantage points through clandestine insertion by foot, vehicle, rubber boat, or helicopter. These methods of insertion may still apply in village and strip areas for collection of information, sensor employment, conducting OP radio relays, or a combination of missions. In major urban areas, undetected insertions are difficult to accomplish because of the high density civilian population.

The use of reconnaissance assets must be used wisely. Reconnaissance assets must not be risked indiscriminately if the information can be collected by other means. Consideration should be given to flank and rearward employment of ground reconnaissance units, in addition to close (0 to about 2,000 meters), distant (approximately 2,000 - 20,000 meters), and deep (from about 20,000 meters and beyond, plus preassault) missions. Reconnaissance of bypassed pockets of resistance, avenues of approach, canals, bridges, subways, and sewer systems (possibly in conjunction with combat engineers) is feasible. Observation, sniper, and listening posts manned by reconnaissance personnel are also missions that reconnaisance personnel perform. The limited combatant diving capability of division reconnaissance may be employed to advantage in many coastal cities.

4204. Military Police (MP). MPs can play a significant role in the urban environment. They may conduct battlefield circulation control, area security, enemy prisoner of war (EPW) and law and order operations. These operations require continuous coordination with MAGTF units as well as with host nation military and civilian police.

a. MP units support area damage control operations. With the increased possibility of rubbling, MP units report on and block off affected areas and reroute movement to alternate road networks.

b. When conducting EPW operations, MPs collect prisoners as far forward as possible and transport them to the rear. They operate temporary collecting points and holding areas to briefly retain EPWs and civilian internees. EPW operations are important in built-up areas because the rate of capture can be higher than normal.

c. Commanders must realize that MP support may not always be available and that infantry units may have to assume certain MP functions. These functions include:

(1) Providing convoy escort and security for lines of communications.

(2) Controling roads, waterways, and railroad terminals that are critical chokepoints in the main supply routes.

(3) Providing security of critical sites and facilities, including communications centers, government buildings, water and electrical supply facilities, command and control nodes, airfields, storage facilities, and other essential areas.

(4) Provide refugee control in close cooperation with host nation civil authorities.

(5) Collect and escort EPWs.

4205. Communications

a. General. Manmade structures create many challenges for CIS in an urban environment. Other CIS challenges include extensive commercial and communications networks and public service radio nets. Commercial communication networks are commposed of miles of underground protected cable which connect and control telephones and computer networks. Public service radio nets, (including police, fire, civil defense, and taxi) compete with existing communication systems. To communicate on an effective, continuous basis in the urban environment, commanders must minimize communication limitations imposed by this environment and maximize the advantage of existing civil communications capabilities.

b. Radio. Manmade structures will decrease the effectiveness of VHF radio communications. The physical locations of radios are critical for effective communications. Some specific locations to avoid include powerlines, underground areas, and steel bridges.

Very high frequency (VHF) radios provide the primary communication capability at the battalion level. It is imperative that these radios be employed as effectively as possible. Communicators can use the following techniques to enhance VHF communications equipment:

- Relocation of VHF radios to overcome phase cancellation due to numerous dead spots
- Use of the higher end (50 - 75 MHz) of the available frequency spectrum
- Use of retransmission facilities to improve the reliability of communications
- Use of radio remotes.
- Use of field expedient antennas.

In setting up retransmission stations, antennas should be camouflaged to blend in with urban surroundings and to prevent them from being targeted by enemy fires. Antennas can be concealed by such objects as water towers and steeples.

In larger cities with many tall buildings, it may be necessary to construct a terrain profile of the area of operations. A communications plateau can be deduced by calculating the average height of the structures in a zone. Operation of antennas above the median line of this plateau will result in better wave propagation and aid in retransmission.

c. Wire Communications. Wire may be used extensively in urban environments. It offers two primary advantages over radio communications:

- It is not susceptible to interference and absorption suffered by radio systems.
- Security of the overall system is greater because wire systems are essentially point-to-point.

Maximum use should be made of existing facilities for concealment. Sewers, subway tunnels, cable stairwells, elevator shafts, intact buildings, and pole lines offer ideal concealment for military wire systems. Precautions must be taken to lessen the possibility of vehicles tearing lines and disrupting communications. Existing telephone poles and tunnels can be used to keep wire above or below the street level.

Maximum use of existing commercial facilities should be made to supplement or replace military CIS. Telephone exchanges are linked together by underground cable systems and constitute a key resource in the urban communications system. These facilities are normally windowless buildings constructed of steel-reinforced concrete and provide ideal CPs and communications centers. The physical location of these exchanges, along with circuit diagrams and telephone directories, should be determined as soon as possible. Action should be initiated to secure these facilities early. Other facilities that contain extensive internal/external telephone systems may also be used as command centers. These include most municipal buildings, department stores, local police stations, and large business offices.

d. Messengers. Messengers provide a secure and flexible means of communication, particularly in an electronic warfare (EW) environment. However, careful selection of messenger routes is necessary to avoid pockets of enemy resistance. Disorientation in an urban environment occurs more often and more easily than in a rural environment. Marking facilities, transportation networks, and city streets on overlays often facilitates messenger service. Selected primary and alternate routes can often interface with designated logistics/resupply routes. Early restoration of street signs and recognizable landmarks are is essential.

e. Telephone Networks. Captured civil or military communications facilities can be utilized to augment organic communications capabilities. A civilian phone system can provide a reliable means of communication. If a telephone encryption device is not available, codes and authentication tables should be used.

Telephone technology differs from place to place, but there are universal concepts and practices in their operation. The most important concept to the military planner is network hierarchy.

Network hierarchy is a concept of organization by which a telephone company can rapidly and expeditiously allocate calls through a particular group of offices. As an ancillary function, it represents a control feature by which traffic can be rerouted to avoid congested areas or damaged facilities. This network hierarchy could play an important role in the planning and execution of the scheme of maneuver and fire support.

4206. Employment of Sensors

a. Sensor systems (seismic, infrared, acoustic, or magnetic) are normally employed in a rural context but also have traditional applications in village and strip area terrain. Use of sensors in urban areas is largely undocumented. In major urban areas, the capability for emplacement by air over select areas is reduced. Emplacement in, under, or around urban structures may also

.3

reduce microphone efficiency. Tactical employment will be degraded by such obstacles as concrete and steel structures, high-density population, and traffic.

b. Training in camouflage techniques for microphones and sensor terminals in an urban environment should be conducted. Operators should receive sufficient training to distinguish background-generated clutter from enemy foot, vehicular, and mechanized traffic. Use of and reliance on sensors may be seriously reduced during offensive operations in major urban areas. However, greater use of sensors will occur in subterranean defensive operations. Sensors should continue to be employed as far forward as possible and be centrally controlled.

c. Sensor employment should be coordinated with counterintelligence (CI) efforts for placement, readouts, and recovery. Sensors can also play a role in rear areas to augment rear security operations. Sensors, in themselves, are only component elements of an integrated intelligence system. Their employment is limited only by the imagination.

4207. Employment of Air Defense. Basic air defense doctrine does not change when units operate on urbanized terrain. The employment principles of mix, mass, mobility, and integration all apply to the employment of air defense assets. The commander must consider the following when developing his air defense plan:

a. Enemy air targets such as principal lines of communication and road and rail networks and bridges are often found in and around built-up areas.

b. Good firing positions may be difficult to find and occupy for long-range air defense missile systems in the built-up areas.

c. Movement between positions is normally restricted in built-up areas.

d. Long-range systems can provide air defense cover from positions on or outside of the edge of the city.

e. Radar masking and degraded communications reduce air defense warning time for all units. Air defense control measures must be adjusted to permit responsive air defense within this reduced warning environment.

f. Positioning of air defense weapons in built-up areas may be limited to more open areas without masking, such as parks, fields, and rail yards.

g. Stinger teams provide protection for infantry battalions the same as in any operation. When employed within the built-up area, rooftops normally offer the best firing positions.

h. Heavy machine guns emplaced on rooftops can also provide additional air defense.

Chapter 5

Combat Service Support and Legal Aspects of Combat

During combat in built-up areas, the terrain and the nature of operations create unique demands on the CSS system. Meeting these demands will require innovative techniques and in-depth planning.

Section I

Combat Service Support

5101. Introduction. CSS is "the essential capabilities, functions, activities, and tasks necessary to sustain all elements of operating forces in theater at all levels of war" (Joint Pub 1-02). The CSS organization throughout the MAGTF, down to battalion level, must provide a responsive CSS system that can perform all functions and tasks associated with meeting identified CSS requirements. This responsibility does not change during urban operations. Forward support for combat forces continues to be the basic concept governing CSS operations. No significant changes in doctrine or organization are required. However, the characteristics and nature of urban combat do affect how CSS is provided.

5102. Combat Service Support Resupply, Maintenance/Repair, and Replacement

a. Resupply of Ammunition. Combat in built-up areas is characterized by extremely high ammunition expenditure rates. Not only do individual Marines expend more ammunition, but they also use greater quantities of munitions such as smoke, concussion, and fragmentation grenades; AT4s; claymore mines; demolitions; and so on. The ammunition consumption rate for the first day of combat in a built-up area can be up to four times the normal rate. Even though this rate decreases during succeeding days, consumption remains high. Leaders should plan to meet these high consumption rates. The plan must include how ammunition and demolitions are to be moved forward to the companies. AAVs should be designated for the movement of ammunition if rubble or glass prevents wheeled-vehicle traffic. Marines may need to organize into carrying elements if streets are blocked by rubble.

b. Resupply of Fuel. The amount of bulk fuel needed by forces during combat in built-up areas is greatly reduced. Combat vehicles normally use less fuel in built-up areas because they travel shorter distances and perform less cross-country traveling. Engineer equipment and power generation equipment may use more fuel, but requirements are small. A unit may not use much fuel daily, but when it does need fuel, a problem exists in delivering bulk fuel to the vehicle. In open terrain, a vehicle that has run out of fuel can be recovered later. In built-up areas, the same vehicle is probably going to be lost quickly. Leaders should plan for and provide the means of moving limited amounts of bulk fuel forward to combat units.

c. Maintenance and Repair. Maintenance contact teams must operate well forward to support units fighting in built-up areas. Although some maintenance operations may be consolidated in civilian facilities, many vehicles will have to be repaired near fighting positions. Battle damage assessment (BDA) and repair procedures allow mechanics to analyze, repair, and return damaged vehicles to a serviceable condition. Other considerations:

(1) Combat in built-up areas generates a high demand for tires.

(2) The dust and rough handling that are characteristic of combat in built-up areas also place great strains on communications and night vision devices.

(3) The unit armorers and their small-arms repair kits provide only limited maintenance. S-4s should plan for increased weapons maintenance demands.

d. Replacements. Units conducting combat in built-up areas must expect high casualty rates. Casualty reports must be prepared scrupulously and forwarded via the S-1 to the battalion personnel officer located at the consolidated administration center (CONAD). Other considerations:

(1) The S-4s must plan to expedite the evacuation of wounded from the built-up area. Location of battalion aid stations and evacuation routes must be planned and disseminated to the lowest level. Higher casualty rates should be expected and may require the stockpiling of medical supplies and augmentation of additional HSS personnel.

(2) The personnel officer quickly processes replacements and coordinates their movement forward via the S-1 officer.

(3) Proper accountability of personnel at all levels is required. Timely and accurate personnel accountability and strength reporting support decisionmaking and initiate the replacement cycle. Leaders maintain accountability through the use of battle rosters and by establishing procedures for periodic reporting of numbers to higher commands. During combat, strength reports are provided on request or as significant changes occur.

5103. Critical Classes of Supply. Requirements for supply and resupply will vary from those encountered during operations in other tactical environments. The critical classes of supply listed below will assume greater importance during combat inside a built-up area.

a. Class I (Rations). The process of ordering and moving rations to forward units is complicated by the dispersed nature of combat in built-up areas and its increased caloric demands on Marines. Hot meals should be provided when practical.

(1) Combat in built-up areas not only causes great stress on Marines, but also requires great physical exertion. This combination of stress and exertion quickly causes dehydration. Unless potable water is continuously provided, Marines will seek local sources, which are usually contaminated by petroleum, oils, and lubricants (POL) runoff, sewage, bacteria, or unburied corpses. Marines who are not provided sufficient quantities of potable water become casualties as a result of drinking from contaminated sources or from dehydration. Waterborne contaminants can quickly render entire units combat ineffective.

(2) Water and other liquid supplements such as coffee, tea, or soup that must be forwarded to exposed positions may need to be backpacked at night.

b. Class II (General Supplies). Combat in built-up areas increases wear and tear on combat uniforms and footgear. Supply officers should increase on-hand stocks of uniforms, boots, and individual combat equipment such as protective masks and armored vests. Nuclear, biological, and chemical (NBC) protective suits become damaged quickly when worn in the urban environment. Extra stocks of these and protective mask filters should be kept on hand. (For further information on NBC considerations, see Appendix G.) Limited amounts of other Class II and IV (barrier materials) items may be available locally. These should be gathered and used if authorized and practical. Local shops may provide such items as hand tools, nails, bolts, chains, and light construction equipment, which are useful in preparing a defense or reducing enemy-held positions. The unit's organic wire communications net may be augmented with locally obtained telephone wire and electrical wire.

c. Class III (POL). Bulk fuel may have to be brought forward from fuel tankers by using 5-gallon cans. One man can carry a fuel can long distances, even over rubble, if it is lashed to a pack frame. Supplies of bulk Class III items and some prepackaged POL may be available at local gas stations and garages. These may be contaminated or of poor quality. The S-4s should coordinate to have fuel tests performed.

d. Class IV (Barrier Materials). If a unit is defending a built-up area, the requirements for Class IV materials are less than in other areas. This class of supply is probably the most available locally. After coordinating the effort with higher headquarters, S-4s and supporting engineer officers gather materials for use in strengthening the defense. Cargo trucks, wreckers, or recovery vehicles from maintenance platoons or engineer units can be used to load and move barrier materials. The defense of a built-up area may require concertina wire and/or barbed wire to restrict the enemy infantry's movements. Barriers can be built of abandoned cars and buses, which are dragged into position, turned on their sides, and chained together through the axles.

e. Class V (Munitions). Combat in built-up areas causes ammunition to be expended at extremely high rates. Commanders should plan for early resupply of explosives, grenades, and ammunition for small arms, direct fire, and indirect fire.

(1) In the defense, the S-4 should prestage as much ammunition as practical in dispersed storage areas. These storage areas should be protected, and they should be easily accessible to the forward defensive positions. In the offense, attacking forces should not be overburdened with excessive ammunition. Mobile distribution points may be set up as low as at company level.

(2) Leaders should plan to continuously deliver ammunition to the leading units as they advance. Ammunition may be carried by armored vehicles close behind the advancing forces or by designated carrying elements. Modern ammunition, particularly missiles, is characterized by extensive amounts of packing material. S-4s must remove the ammunition depot overpack before the ammunition is transported forward. Resupply by helicopter (prepackaged slingloads) may be feasible.

(3) Removing the overpack from large amounts of ammunition can be a time-consuming process. It may require being augmented by available Marines. If carrying elements are used to move ammunition forward, a Marine can carry about 75 to 90 pounds by using a pack frame. Bulky and heavier loads can be carried by lashing them to litters and using teams of two to four men. Loads of up to 400 pounds can be carried moderate distances using four-man teams.

Note: DO NOT use medical corpsmen to carry ammunition forward as described above—it is a violation of the Geneva Accords.

f. Class VIII (Medical Supplies). Because of the decentralized nature of combat in built-up areas, medical supplies should be dispersed throughout the unit, not just consolidated with the aid station and the individual corpsman. Marines should carry additional bandages, and units should have additional splints and stretchers.

5104. Health Service Support. The regimental and battalion surgeons are responsible for planning and executing HSS within their respective units. The most critical functions during combat in built-up areas include preventive medicine, trauma treatment, and evacuation. In addition, there should be a plan for the treatment and evacuation of NBC-related casualties that could occur in combat in built-up areas.

a. Combat in built-up areas exposes Marines not only to combat wounds, but also to the diseases endemic to the AO. Commanders must enforce preventive measures against the spread of infectious diseases. The unit surgeon advises the commander on how best to implement the use of prophylactics.

b. Corpsmen should be placed at strongpoints, at battle positions, and in units likely to be fighting in somewhat isolated positions.

c. Battalion aid stations must be placed farther forward than in operations in more open terrain. Protection offered by urban structures will permit this forward location, but the limited range of observation will require that personnel involved in evacuation be aware of their surroundings and the threat.

d. Corpsmen attached to rifle platoons are trained in the treatment of traumatic injuries; however, they can quickly become overwhelmed by the number of casualties needing care. If time and training are available before the conduct of urban operations, the commander should increase first aid training. Immediate first aid support by a buddy to an injured Marine may be the difference between life and death and will greatly assist the corpsmen. The aid station should plan to care for the increased casualties that are inherent to combat in built-up areas, as the incidence of crushing injuries, eye injuries, burns, shrapnel wounds, and fractures increases.

e. The difficulties encountered when evacuating casualties from urbanized terrain are many and require innovative techniques and procedures. The planning for evacuation on urbanized terrain must include special equipment requirements, use of litter teams, use of air MEDEVAC, use of an ambulance shuttle system, communications requirements, and techniques for locating casualties.

(1) Special equipment requirements include ropes, pulleys, skid litters, axes, crowbars, and other tools used to break through barriers.

(2) Although litter teams are labor intensive, they are required for evacuation from buildings, where casualties can occur on any level. Also, rubble in the streets, barricades, and demolition of roads will require a heavy reliance on litter teams.

(3) Communications presents one of the biggest obstacles to casualty evacuation. Urbanized terrain renders LOS radios ineffective. Also, individual Marines normally do not have access to radios. Therefore, a wounded Marine within a building may be difficult to find and evacuate. The unit SOP should contain alternate forms of communication such as colored panels or other forms of markers that can be displayed to hasten rescue when the battle is over. Also, a systematic search of the area after the battle may be required to recover casualties.

f. The use of local medical facilities, hospitals, professional medical help, and medical supplies may be available during combat in large, built-up areas. The commander must adhere to the guidelines established within the theater as to when and how these facilities can be used. If civilians are wounded in the unit's area, the commander is responsible for providing them with aid and protection without disrupting military operations. A commander cannot confiscate civilian medical supplies unless he makes provisions to provide adequate replacements if civilians are wounded.

g. The commander is responsible for the evacuation of deceased personnel to the nearest mortuary affairs collection point, whether they are U.S., allied, or enemy personnel, or civilians. (See Joint Pub 4-06, *JTTP for Mortuary Affairs in Joint Operations.*) Some general considerations for the handling of deceased personnel include:

(1) The theater commander is the approval authority for hasty burial.

(2) The deceased person's personal effects must remain with the body to assist in the identification of the body and to facilitate shipment of personal effects to the next of kin.

.3

Retention of personal items is considered looting and is, therefore, punishable by the Uniform Code of Military Justice (UCMJ).

(3) When operating under NBC conditions, the bodies of deceased personnel should be decontaminated before removal from contaminated areas to prevent further contamination and casualties.

(4) Care must be exercised when handling deceased personnel. Improper handling of deceased personnel can result in a significant decrease in unit and civilian morale.

5105. Personnel Services. Timely and accurate personnel services are just as important during combat in built-up areas as in any other operation.

a. The S-1 plans for all personnel services that support and sustain the morale and fighting spirit of the battalion. Among the most important of these services are:

- Replacement operations
- Strength accounting
- Casualty reporting.

These functions allow the commander to maintain accountability of his forces and to redistribute personnel or change taskings for subordinate units to maintain combat power and tempo.

Chapter 6

Noncombatant Considerations in Urban Operations

Noncombatants can have a significant impact on the conduct of military operations.

Section I

6101. Introduction. Commanders must be well educated in the legal aspects of military operations involving noncombatants. The urban environment contains numerous legal pitfalls in dealing with the civilian population. Adherence to the ROE and the standards of the Geneva Accords will keep the commander on firm legal ground.

6102. Noncombatant Impact. Large concentrations of civilians can greatly impede tactical operations. Noncombatants may have the following effects on military operations:

a. Mobility. Noncombatants civilians, attempting to escape the battlespace, can block military movement. Commanders plan routes to be used by civilians and seek the assistance of the civil police in refugee control.

b. Firepower. The presence of noncombatants can restrict the use of firepower. Areas may be designated no-fire areas to prevent noncombatant casualties. Other areas may be limited to small-arms fire and grenades. The control of fire missions may be complicated by the requirement for positive target identification. Detailed guidance on the use of fires in the presence of noncombatants is promulgated by the MAGTF commander. In the absence of guidance, the general rules of the law of land warfare apply.

c. Security. The presence of noncombatants increases security requirements in an urban environment to preclude:

- Noncombatants entering defensive areas
- Pilferage of equipment
- Sabotage
- Terrorism

d. Obstacle Employment. The presence and movement of Noncombatants will influence the MAGTFs commander's obstacle plan. Minefields may not be allowed on designated refugee routes or, if allowed, must be guarded until the passage of refugees is completed. The use of boobytraps may be curtailed until noncombatants have been evacuated. Commanders need to review and understand current national and international treaties concerning the employment of mines and boobytraps.

6103. Command Authority. The limits of command authority over government officials and noncombatants must be established. The host government's responsibility for its populace and territory can affect the commander's authority in civil-military matters. In less secure areas, where the host government may be only partly effective, the commander may have to assume greater responsibility for noncombatants.

6104. Source Utilization. Operations in highly populated areas may require the diversion of men, time, equipment, and supplies to accomplish humanitarian tasks. If host government agencies collapse, the impact on military resources could be substantial.

6105. Health and Welfare. The disruption of noncombatant health and sanitary services may sharply increases the risk of disease among both noncombatant and military personnel.

6106. Law and Order. Marines may have to augment civilian forces to protect life and property and to restore order. Marines may also have to secure vital government facilities for the host nation. (For more information on how to control civilians violating civil law, see FM 9-15, *Civil Disturbances*.)

6107. Public Affairs Officer and Media Relations. Free media access to units in the field is desirable; however, operational security, existing guidelines, and ROE considerations have priority. All members of the media should have an escort officer. Escort officers may be detailed from line units when public affairs personnel are unavailable. Enforcing established guidelines helps prevent negative publicity which could jeopardize the operation or national and strategic objectives.

6108. Civil Affairs Activities and Psychological Operations (PSYOP).

a. General. CA and PSYOP are mutually supporting and may have essential roles in MOUT. They can be critical force multipliers. CA and PSYOP may offer the possibility of victory without the destruction, suffering, and horror of protracted urban conflict.

(1) CA activities are activities performed by commanders, staffs, department of defense elements and units, and foreign military forces that (1) embrace the relationship between military forces and civil authorities and populations in areas where military forces are present; and (2) involve application of civil affairs functional specialty skills, in areas normally the responsibility of civilian government, which enhance conduct of civil-military operations (Joint Pub 1-02). CA activities are inherently civil-military in nature. They assist in coordinating military efforts to strengthen host government legitimacy, and help prevent or reduce violence by bridging critical gaps between civil and military sectors. MAGTF responsibilities for CA activities are determined during planning. U.S. Army CA units can be requested to support the MAGTF. The Marine Corps Civil Affairs Group (CAG), located in the Marine Corps Forces Reserve, may be activated and assigned to the MAGTF.

(a) CA activities are characterized by applications of functional specialties in areas normally the responsibility of indigenous government or civil authority. CA activities may extend to assumption of government functions required in an occupied territory during or immediately subsequent to hostilities. Refer to Joint Pub 3-57, *Doctrine for Joint Civil Affairs*.

(b) CA support to a MAGTF would normally include procuring local goods and services for military use, reducing civil interference in military operations, and assisting a MAGTF commander in fulfilling moral and legal obligations to the local government and noncombatant populace. See MCWP 3-33.1, *MAGTF Civil Affairs*, for more detailed information.

(2) Psychological operations are planned operations to convey selected information and indicators to foreign audiences to influence their emotions, motives, objective reasoning, and ultimately, their behavior. The purpose of psychological operations is to induce or reinforce foreign attitudes and behavior favorable to the originator's objectives (Joint Pub 1-02).

PSYOPs are planned and conducted by the MAGTF. PSYOPs are integrated and coordinated down to the lowest tactical level. PSYOPs can weaken enemy resistance and reduce its combat effectiveness. PSYOPs can also help prevent civilian interference with military operations.

PSYOPs may include the use of television, radio, posters, leaflets, and loud speakers, to disseminate propaganda and information. Television, including video tapes, is one of the most effective mediums for persuasion. In areas where television is not common, televisions may be distributed to public facilities and selected individuals. See MCWP 3-36.2/FM 33-1, *Psychological Operations*, for more detail.

6109. Refugee Control. The control of refugee movements in combat operations is primarily a host nation function.

The most important principle in refugee control is reliance on host government forces (military/civilian) whenever possible. U.S. forces will have limited refugee control capabilities. If a host government is incapable of performing the function, MAGTFs and MPs may be required to conduct any or all of the following refugee control measures:

- Enforce standfast orders or prohibit other unauthorized movements
- Screen and take custody of enemy agents or other hostile personnel discovered among the refugees
- Screen refugees for suspicious activities (including detection of infiltration by guerrillas, enemy agents, and escaping members of hostile forces)

.3

- Perform traffic control on designated refugee routes
- Escort refugee columns
- Direct or divert refugee columns to secondary roads and through areas not used by military elements
- Perform police duties such as traffic control or security at refugee assembly areas, collecting points, and centers or camps
- Enforce prescribed restrictive measures such as curfews, identification and pass systems, restricted area restraining lines, no-passage lines, checkpoints and roadblocks, travel permits or manifests, prohibition on the use of vehicles, priorities of movement, and limits on personal belongings.

Section II

6201. Commander's Legal Authority and Responsibilities

a. General. Leaders at all levels of command are responsible for protecting noncombatants and minimizing collateral damage to their property. Looting, vandalism, and brutal treatment of civilians are strictly prohibited by the Geneva Accords. Individuals who commit such acts may be severely punished. Civilians, and their religions and customs, must be treated with respect. Women are especially susceptible to many forms of abuse, and special care must be taken to protect them. Urban warfare offers many opportunities for abuses of the civilian population. This section focuses on those civilian-military confrontations that are most common in the urban environment.

b. Control Measures. Commanders may enforce control measures to conduct operations, maintain security, or ensure the safety and well-being of the civilians.

(1) Curfew. A commander with the mission of defending a town could establish a curfew to maintain security or to aid in control of military traffic. However, a curfew would not be legal if imposed strictly as punishment.

(2) Evacuation. A commander can require civilians to evacuate towns or buildings if the purpose of the evacuation is to use the town or building for essential military purposes, to enhance security, or to safeguard those civilians. The commander must specify and provide security for those being evacuated. Food, clothing, and sanitary facilities should be provided at the destination.

(3) Forced Labor. The Geneva Accords prohibit the use of civilians in combat operations. However, they may be required to perform some forms of forced labor prior to and after the conduct of combat operations. Guidelines for use of civilian labor should be published. The commander may force civilians over 18 years of age to work if the work does not oblige them to take part in military operations. An example of forced labor that is permitted would be maintenance of public utilities as long as those utilities are not used in the general conduct of the war. Jobs can also include services to the local population such as care of the wounded and burial. Civilians can also be forced to help evacuate and care for military wounded, as long as doing so does not involve any physical danger. Prohibited forced labor include digging entrenchments, constructing fortifications, transporting supplies or ammunition, or acting as guards. Only civilians who are volunteers can be employed in such work.

c. Civilian Resistance Groups. Another situation that commanders might encounter is combat with a civilian resistance group. The combatant status of these groups varies with their organization, methods, and adherance to the Geneva Accords. For example:

(1) Civilians accompanying their armed forces (with an identity card authorizing them to do so) are treated as EPWs when captured. Examples would include civilian members of military aircraft crews, war correspondents, supply contractors, and members of labor units or of service organizations responsible for the welfare of the armed forces.

(2) Captured civilians of a nonoccupying territory who have taken up arms against an invading enemy without time to form regular armed forces are considered EPWs if they wear a distinctive insignia that can be seen at a distance, carry their weapons openly, and operate according to the rules and customs of warfare. Other civilians who provide assistance to such groups may not be entitled to status as combatants, depending on whether they are actually members of the resistance group. They are normally best treated as combatants until a higher authority determines their status.

(3) Armed civilian groups that do not meet the criteria of a legal resistance group or individuals caught in the act of sabotage, terrorism, or espionage are not considered legal combatants. If captured, they will be considered criminals under the provisions of the law of land warfare. They should be detained in a facility separate from EPWs and should be quickly transferred to the MPs. Reprisals, mass punishments, hostage taking, corporal punishment, pillaging, or indiscriminate destruction of property are prohibited.

(4) The law of land warfare lets the commander control the civil population under the conditions already described using his own organic resources. However, language and cultural differences between U.S. and foreign personnel make it good practice to use local authorities such as the police for such purposes. Use of the police does not relieve a commander of his responsibility to safeguard civilians in his area.

d. Protection of Property. Like civilian personnel, civilian buildings and towns normally have a protected status. As such, they are not legal targets. They lose their protected status if authorities determine that the enemy is using them for military purposes. All doubt as to whether a town or building is a military target must be resolved before it can be targeted. For example:

(1) If the enemy is using a building or a portion of the town for military purposes—such as a supply point or a strongpoint—that building or that portion of the town is a legal target. However, engagement of the target should be limited to the buildings or portion of the town being occupied by the enemy.

(2) Normally, religious, historical, and cultural objects and buildings are not legal targets. They are sometimes marked with symbols to signify their status. Medical facilities are identified under the internationally recognized Red Cross, Red Crescent, Red Lion, or Red Star of David symbols. The fact that such symbols are absent does not relieve a commander of his responsibility to protect objects he recognizes as having religious, cultural, medical, or historical value.

(3) The misuse of such objects by the enemy is grounds to disregard their protected status. A demand should be made for the enemy to stop his misuse of the protected object before any action is taken. However, if an enemy FO or sniper uses a church for an OP, a commander would be justified in destroying it immediately because of the immediate danger posed to his troops. Another example is a religious shrine that is used as a telephone switchboard; in this case a warning would be appropriate because it would take some time to dismantle the system. Attacks on these types of targets should be limited to the least amount necessary to neutralize the enemy installations.

(4) The destruction, demolition, or military use of other buildings is permitted under the law of land warfare if required by clear military necessity. Destroying a house to obtain a better field of fire would be a legal act. But destroying it as a reprisal would not. However, firing on any houses that are occupied or defended by an enemy force is usually legal.

Chapter 7

The Urban Environment and Restrictions to Operations

"The responsibility of officers . . . and the training necessary are of a very different order from their responsibilities and training in ordinary military duties. Instead of striving to generate the maximum power with the forces available, the goal is to gain decisive results with the least application of force and the consequent minimum loss of life."
USMC, 1940 (FMFRP 12-15, *Small Wars Manual*)

Section I

Military Operations Other Than War

7101. Introduction. One of the most likely missions that U.S. Marines will undertake abroad will be military operations other than war (MOOTW). These missions typically will take place in the Third World and are normally associated with urban areas. MOOTW focuses on deterring war, resolving conflict, promoting peace, and supporting civil authorities in response to domestic crises. It is important to remember that political considerations permeate at all levels. Although some types of MOOTW, such as peace enforcement, may have the same characteristics of war, they normally have more restrictive rules of engagement (ROE).

Urban environments are inherently complex, and sometimes unpredictable. Local governments may have limited functions. Services may be overburdened, leading to allocation on the basis of personal connections and corruption. Services may be provided by a mixture of formal, high-technology and informal, labor-intensive systems. Large portions of the population may be poor, young, and underemployed. Nevertheless, there may be a high degree of social organization and interaction. Word of mouth and informal organizations may be as important as mass media and formal organizations to the communication structure of the area. Indigenous populations may have experienced centuries of foreign interventions and become quite adept at innovative ways of communicating. These factors and others often combine to make MOOTW a challenging and frustrating endeavor.

7102. Types of MOOTW. MOOTW encompasses a wide variety of military operations, that may take place in an urban environment. Forward deployed MAGTFs are ideal forces to respond to contingency operations involving MOOTW. These operations include:

- Combatting Terrorism
- DOD support to counterdrug operations
- Enforcement of Sanctions/MIO
- Enforcing Exclusion Zones

- Humanitarian Assistance
- Military Support to Civil Authorities
- Nation Assistance/Support to Counterinsurgency
- NEO
- Peace Operations
- Recovery Operations
- Show of Force Operations
- Strikes and Raids.

7103. Recent Experience and Lessons Learned. Recent Marine Corps urban experiences—Lebanon (1982 - 1984); Grenada (1983); Panama (1989); and Somalia (1992 - 1994)— are examples of MOOTW. Although each operation was unique, the lessons learned are similar. For detailed information, commanders may obtain after action reports through the Marine Corps Lessons Learned System (MCLLS). Some of the lessons learned include:

(1) Demonstrate a powerful military presence. Where and when possible, use tanks, armored vehicles, heavy automatic weapons, fixed-wing aircraft, and attack helicopters.

(2) PSYOP, CI, HUMINT teams, and linguists, are essential to communications and intelligence gathering.

(3) Topographic detachment is useful to provide necessary sketches and maps.

(4) UAVs can be used for security, reconnaissance, and surveillance.

(5) Whenever entering an AO, isolate the area, patrol, and use intelligence to evaluate specific buildings or groups of buildings for search-and-clear operations. PSYOP, CI, HUMINT teams, and linguists can be used, as appropriate, to support patrols.

(6) Take advantage of the night by using night vision devices.

(7) The most effective means of dealing with snipers is with counter snipering operations.

(8) Snipers are also a tremendous asset to overwatch patrol movements, provide intelligence, and direct fire support. (See MCWP 3-15.3.)

(9) Random patrolling, traffic control points, random checkpoints/roadblocks, and ambushes disrupt enemy movements, reduce looting, seize curfew violators, and convince the population that order has been restored.

(10) Use armored vehicles to move quickly to new checkpoints and set up roadblocks or ambushes. They also increase protection for infantry. Their night vision sights also provide increased security and recognition during periods of limited visibility.

(11) Extensive patrolling provides current combat information used to maintain situational awareness. It helps to familiarize units with the AO and presents a strong military presence. (See Appendix A, Section VII)

(12) Tanks, armored vehicles, or attack helicopters increase military presence and can provide immediate, accurate gunfire.

(13) Before search and clear of a building, use loudspeakers to explain your intentions to noncombatants and enlist their cooperation.

(14) Use a minimum amount of force when dealing with noncombatants and their property. This will assist you in obtaining them to cooperate and may prevent turning them into active combatants. Assign engineers to search-and-clear forces, as necessary.

(15) MPs can instruct small unit leaders on the proper method of conducting vehicle, individual, and room search techniques. In some cases, it may be appropriate to attach MPs to search-and-clear forces. (See MCWP 3-34.1, *Military Police in Combat.*)

(16) Maintain a strong, mobile reserve.

(17) Avoid becoming predictable.

7104. Urban Terrorism. Terrorism is "the calculated use of violence or threat of violence to inculcate fear; intended to coerce or to intimidate governments or societies in the pursuit of goals that are generally political, religious, or ideological" (Joint Pub 1-02). Terrorism is a criminal act that is symbolic in nature. Its intent is to influence an audience other than the victim.

The likelihood of Marine involvement with terrorism increases in the urban environment. Terrorism in an urban environment can occur across the spectrum of conflict. Terrorists are an integral element in an insurgency and also play a major role in unconventional warfare. Terrorists can disrupt economic functions, expose a government's incompetence, eliminate opposition leaders, and elevate social anxiety. (See MCRP 3-02D, *Combating Terrorism*; and MCRP 3-02E, *The Individual's Guide for Understanding and Surviving Terrorism.*)

In countering this threat, Marines should determine whether it is internally or externally directed terrorism. Terrorism rooted externally must be severed from its roots. Against internal terrorism,

.3

Marines should attempt to penetrate the infrastructure and destroy the leadership of the terrorist groups.

The keys to defeating terrorists are awareness, education, and intelligence in order to deny, deter, delay, and detect terrorist acts. Rapid coordination among military units, military services, local police, and other government agencies is essential in denying the terrorists targets and refuge.

a. Nature of Urban Terrorism. Terrorism is generally an urban phenomenon. Incidents of terrorism are increasing in the international community, and in the United States, and pose a significant threat for U.S. military forces. Military personnel in noncombatant roles are considered terroist targets. For example, the attack on the Marine compound in Lebanon can be considered a terrorist attack instead of a military action because the Marines were officially in a noncombat activity (a multinational peacekeeping activity).

b. Terrorists in the Urban Area. Cities are target rich environments for terrorists. They offer ample cover and concealment. They contain large amounts of resources, transportation, communications, and potential recruits or sympathizers to their cause. Cities also provide the media attention essential for terrorists to accomplish their mission.

c. Combating Terrorism

(1) Counterterrorism. Counterterrorism includes the full range of offensive measures to prevent, deter, and respond to terrorism (Joint Pub 1-02). Population protection and control to combat the terrorist threat may be a major effort within the urban environment. The ability to protect the general population from attack and intimidation is essential.

(2) Antiterrorism. Antiterrorism includes defensive measures used to reduce the vulnerability of individuals and property to terrorist acts, to include limited response and containment by local military forces (Joint Pub 1-02). The extent of these measures varies based on assessments of the local threat. These include personnel awareness and knowledge of personal protection techniques. They also include crime prevention and physical security programs to "harden" the target, making installations and personnel less appealing as terrorist targets.

Section II

Restrictive Conditions

7201. Introduction. Commanders can expect to encounter restrictions on the use of force in MOOTW, especially in the urban environment. While we tend to focus on a traditional image of urban combat, MOUT extends beyond the house-to-house, close-combat effort to seize a city. Recent history provides many examples of the need for precise application of force to minimize collateral damage. MOOTW generally requires the application of force in varying degrees. Future missions in urban areas will likely follow this same pattern. While Marine forces must be prepared for more intensive urban combat, they must also train to apply only the level of force necessary to accomplish the mission. *Our tactics may have to rely more on innovation than firepower.*

Basic tenets of manuever warfare remain the same in MOOTW, but the tactics, techniques, and procedures may have to be modified to stay within established ROE and to avoid unnecessary collateral damage.

7202. Rules of Engagement. ROE are "directives issued by competent military authority which delineate the circumstances and limitations under which United States forces will initiate and/or continue combat engagement with other forces encountered" (Joint Pub 1-02). ROE for Marine forces are usually set by the JTF commander. They are based on National Command Authorities (NCA) guidance, the mission, the threat, the laws of war, and host nation constraints on force deployment.

MOOTW focuses on deterring war, resolving conflict, promoting peace, and supporting civil authorities in response to domestic crises. It is important to remember that political considerations permeate at all levels. Although some types of MOOTW, such as peace enforcement, may have the same characteristics of war, they normally have more restrictive rules of engagement (ROE). Restraint requires the careful balancing of the need for security, the conduct of operations, and the political objective. Excessive force antagonizes those parties involved, thereby damaging the legitimacy of the organization that uses it while possibly enhancing the legitimacy of the opposing party. Ensure all personnel understand the ROE and incorporate the practice into training exercises at all levels.

ROE must be fully integrated in the MAGTFs plans. As an example, table 7-1, presents the ROE used in Operation United Shield:

ROE Used for Operation United Shield
Nothing in these Rules of Engagement limits your right to take appropriate action to defend yourself and your unit.
a. You have the right to use deadly force in response to a hostile act or when there is a clear indication of hostile intent.
b. Hostile fire may be returned effectively and promptly to stop a hostile act.
c. When US forces are attacked by unarmed hostile elements, mobs and/or rioters, US forces should use the minimum force necessary under the circumstances and proportional to the threat.
d. Inside designated security zones, once a hostile act or hostile intent is demonstrated, you have the right to use minimum force to prevent armed individuals/crew-served weapons from endangering US/UNOSOM II forces. This includes deadly force.
e. Detention of civilians is authorized for security reasons or in self-defense.
Remember: 1. The United States is not at war. 2. Treat all persons with dignity and respect. 3. Use minimum force to carry out mission. 4. Always be prepared to act in self-defense.

Table 7-1. ROE Used in Operation United Shield

7203. Planning. Commanders and planners thoroughly study each situation to identify the restrictive measures necessary to prevent needless civilian casualties and damage while also protecting the lives of the Marines who will execute the mission.

a. Mission/Command Guidance. The mission statement itself may determine much about the required level of force for a given operation. A humanitarian relief operation may require only individual weapons for security and personal protection. A raid on an insurgent communication center may need more. The requirements for an NEO conducted in a permissive environment would be different from those in a nonpermissive one. The key is to conduct a thorough analysis of the mission and determine the necessary combat power and how it must be applied to accomplish the mission. Many of these missions will be accompanied by a great deal of additional guidance concerning operational restrictions. These restrictions may come from the NCA, State Department, Department of Defense (DoD), or the joint task force (JTF) commander, and other sources. Such guidance must be carefully studied and incorporated into the plan. The commander can impose restrictions on his own forces that are more stringent then those dictated by higher authority. A careful analysis of the ROE will provide key insights into other restraints that the commander may wish to impose on his own forces.

b. Intelligence/Threat. Obtaining detailed and specific intelligence on the target and the area is essential to minimize collateral damage. Information on the physical layout of an area is often as important as information about the enemy. This is especially true for an NEO where avoiding conflict is a primary concern.

c. Location. The geographic area into which operations are to be conducted will normally dictate some restrictions on operations. In an urban environment, where the city is located in a desert, a common restriction is placed on the destruction of water producing facilities. If the operation is conducted where there is a large Hindu population, special protection may be given to cattle. If the operation is on an island nation, restrictions may be placed on destroying port or airfield facilities. In an urban environment with a large amount of shanty towns, restrictions may be placed on the use of incendiary devices.

d. Restrictions on Force Employment. The *right of self-defense is inherent in all military operations*. Marines should not be placed in a situation where they are not provided the means and authority to protect themselves or others. Beyond that, the restrictions placed on the forces conducting a particular operation depend on a variety of factors. These include:

- NCA guidance
- Type of mission
- Nature of the threat
- Political environment
- Host nation attitude.

e. Lethal Fires. Most restrictions in MOOTW are normally concerned with the application of fires. For instance, a NEO in a permissive environment may not require the landing of fire support assets However, a raid may require supporting arms to be landed with detailed instructions on their use. Aviation in MOOTW is commonly restricted in the amount and type of ordnance it can deliver because of the potential for collateral damage. However, in both a permissive and nonpermissive environment, supporting arms still should be planned. The most common types of fire support assets that will be restricted in the urban environment include artillery mortars, tanks, armored vehicles, and fixed-wing attack aircraft.

f. Search and Clear Buildings. Marine forces organize to search and clear a building under restrictive conditions in the same ways as discussed in Appendix C. However, because noncombatants may be occupying rooms within the building, clearing procedures must be modified as follows:

(1) Squad organization remains the same and is reinforced as appropriate (see Appendix A), except assault elements become search elements. Search elements enter and search rooms in the same manner as assault elements enter and clear rooms, as discussed in Appendix A.

(2) ROE must be identified and known by all personnel before entering or clearing a room or building.

(3) PSYOP or CA teams can help remove noncombatants before a search or battle starts.

(4) The building or area of search should be isolated with security forces and the intent to search the building should be communicated via loudspeakers.

(5) Assault forces move to the building and begin systematically to search room by room and floor by floor. Marines should carry ammunition and other equipment as in normal urban operations. However, Marines will be restricted in the use of their weapons and other equipment as dictated by the ROE. If permitted, a riot control agent (such as CS gas) can be used to assist in the clearing of a building. Additionally, shotguns and stun grenades are more appropriate than rifles and machine guns or fragmentation grenades for this type of restricted, close-combat operation.

g. Communications and Information Systems (CIS).

(1) CIS must be able to satisfy the C2 requirements of MOOTW. CIS systems must provide the commander with the ability to rapidly collect, process, analyze, and exchange information in order to keep the potential for violence to a minimum. These systems must make available the information needed, when it is needed, whenever it is needed. When supporting arms are on standby, fire support coordination and air traffic control nets are particularly critical.

(2) CIS officers (G/S-6) must carefully analyze the mission, situation, threat, and specific characteristics of the urban environment to design reliable, flexible, and responsive CIS. These systems should include single and multichannel radios, tactical telephones, and tactical satellite communications. A detailed backup CIS plan should also be prepared. The success of CIS in MOOTW depends on designing, planning, and employing CIS that satisfies the commander's C2 requirements.

Appendix A

Organization for Urban Combat and Fundamental Combat Skills

Successful combat operations in built-up areas depend on training in fundamental skills and proper employment techniques and procedures.

Before conducting urban operations, Marines must be trained in fundamental combat skills to conduct urban operations. Training for this unique environment will reduce casualties because properly prepared Marines will be able to fight more effectively. This appendix supersedes all other fundamental combat skills taught for fighting in urbanized terrain. Urban combat will require Marines to use their best initiative and judgment in applying the techniques and procedures described in this appendix.

Section I

Organization

1. Structure. Basic infantry organization does not change when the battlefield is moved into an urban area. However, infantry units may be task organized or reinforced down to squad level in order for small units to fight and win in this violent, three-dimensional battlespace. Several new terms that address organization and employment have been added to clearly define mission assignments. Figures A-1 through A-3 show platoon through fire team organization breakdown.

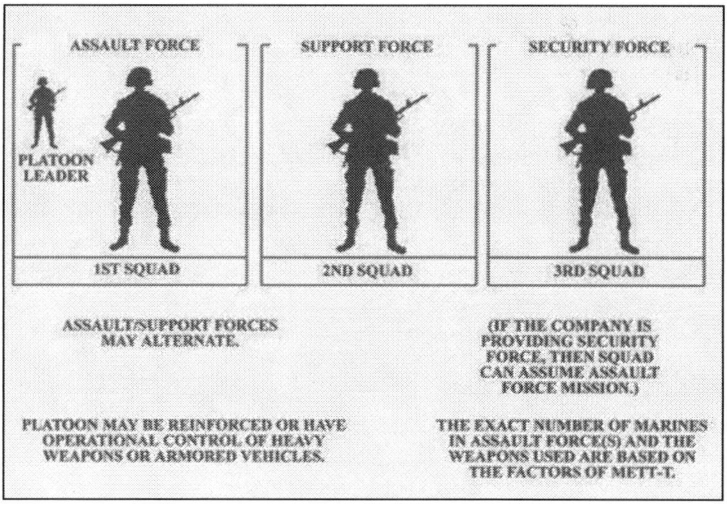

Figure A-1. Task-Organized Marine Platoon

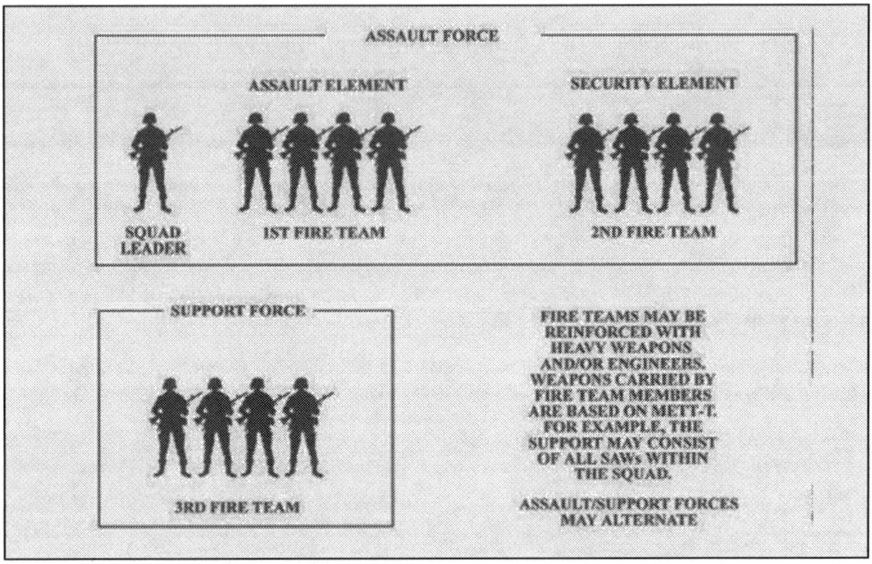

Figure A-2. Task-Organized Marine Squad

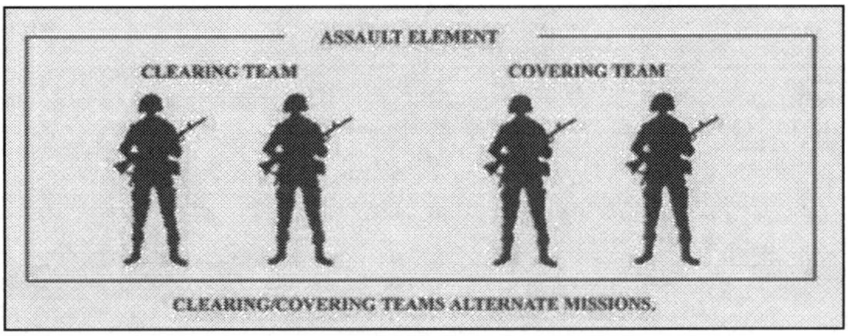

Figure A-3. Task-Organized Marine Fire Team

Section II

Weapons Handling and Firing Techniques

2. Weapons Carries. During MOUT, a Marine often finds himself very unexpectedly engaging targets at close ranges. Weapons carries provide a safe and effective method for handling the service rifle. For additional in-depth discussion of weapons handling and firing techniques, refer to MCWP 3-01X, *Field Firing of the M16A2 Rifle*.

a. Tactical Carry. The tactical carry is used when no immediate threat is present. It permits control of the rifle while moving and still allows quick engagement of the enemy. The buttstock of the rifle is placed alongside the body at approximately hip level, and the barrel is angled upward approximately 45 degrees in the general direction of the enemy (Figure A-4).

Figure A-4. Tactical Carry

b. Alert Carry. The alert carry is used when enemy contact is likely. Engagement of the enemy is faster from the alert carry than from the tactical carry. The buttstock of the rifle is placed in the shoulder with the muzzle angled down approximately 45 degrees and pointed in the likely direction of the enemy (Figure A-5).

In the alert carry, the Marine keeps both eyes open and scans for any threat with the weapon always in line with the Marine's LOS (this is called "guns 'n' eyeballs"). When a threat is spotted, the Marine quickly snaps his weapon into a firing position, looks over the rear sight aperture, and aligns the tip of the front sight post center mass on the target. Once the threat has been eliminated, the Marine returns to the alert carry and continues to scan for targets. Accuracy is the primary goal.

Figure A-5. Alert Carry

c. Ready Carry. The ready carry is employed when contact with the enemy is imminent. It allows for immediate target engagement. The buttstock of the rifle is in the shoulder with the muzzle of the rifle pointed in the direction of the enemy (Figure A-6). As in the alert carry, sight alignment and sight picture are achieved as the shot is fired. Marines should always strive for a clear tip of the front sight post and center mass hold.

Figure A-6. Ready Carry

d. Short Stocking. Rifles may be "short stocked" to increase their maneuverability in an enclosed area and to reduce possible "target indicators" for the enemy (muzzles sticking around corners, for example). The weapon is held in such a manner as to reduce the length of the weapon without sacrificing too much accuracy (Figure A-7). The stock is positioned so that the pistol grip is behind the Marine's head. The Marine may use his index finger or thumb to manipulate the trigger. The handguard of the weapon is placed against the Marine's cheek, and a firm stockweld is attained. The Marine should use the tip of the front sight post. Short stocking may be conducted from either side. However, care should be taken not to mask the ejection port cover. The short stocking technique may be used in the tactical, alert, or ready carry.

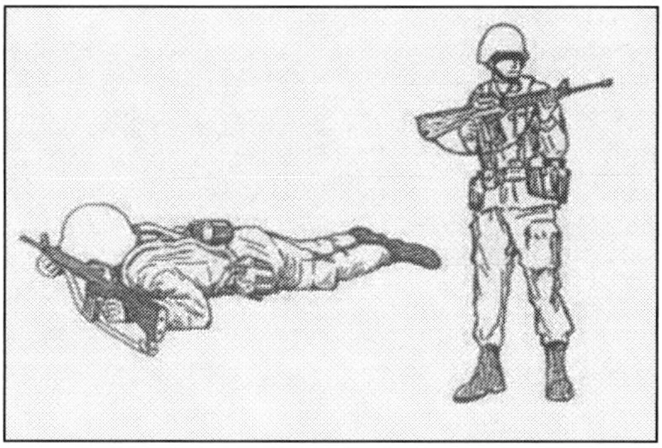

Figure A-7. Short Stocking

e. Weak-Hand Carries. Marines may carry their weapon on the weak-hand side to obtain maximum cover and fire from the weak-hand position to avoid exposing themselves. However, firing from the weak-hand position may reduce a Marine's ability to fire accurately. This may be acceptable when Marines are providing supporting fires, but in close kill-or-be-killed engagements, rapid, accurate firing from the strong-hand position is preferred.

3. Firing Techniques. The speed and uncertainty of combat require Marines to act without hesitation while achieving accurate target engagement. Several firing techniques are listed below. For an in-depth discussion of firing techniques that can be applied to MOUT operations, see MCWP 3-01X.

a. Pieing. Pieing is an effective technique for clearing dead space inside rooms and buildings to gain security of hallways, stairwells, mouseholes, and so on. It is conducted by using the ready carry position or short stocking technique. The weapon is aimed at a sector of a window, doorway, corner, or hallway and slowly moved at different angles, sectoring off the window, doorway, corner, or hallway until each sector is cleared of any threat. As soon as a hostile threat is seen, immediate, accurate fire can be placed on the threat. Figures A-11, A-34, A-35, A-38, A-47, and A-49 show examples of the pieing technique.

b. Aimed Quick Fire. The Marine's initial focus is on the target. As the rifle is brought up, the firing eye looks through or just over the rear sight aperture, and the front sight post is used to aim at the target. The Marine fires two quick shots. Focus remains on the front sight post throughout the aiming process.

Immediately after target engagement, the Marine scans for additional targets. To scan, perform the following steps:

(1) Lower the weapon to look over the sights.

(2) Place the trigger finger straight along the receiver.

(3) Scan the area for targets and assess the situation. Wherever the head moves, the muzzle moves (eyes, muzzle, target). Keep both eyes open to increase the field of view.

c. Pointing Quick Fire. The pointing system is based on the phenomenon that when a person looks at an object and simultaneously points a finger at it, the finger aligns itself on the point of focus of the eyes with no conscious effort on the part of the individual. When a Marine looks at an object and simultaneously brings his rifle to his shoulder, the rifle in effect becomes an extension of the pointed finger. Consequently, it aligns itself naturally with the object on which the shooter is focusing.

When a target appears, the Marine will keep both eyes open, concentrating intensely on a small, specific focal point near the base of the target mass. The rifle is brought simultaneously to the hollow of the shoulder. The head is held high, stock welded to the jaw. The eyes are 2 - 3 inches over the top of the sights, staring intently at the target. As soon as the rifle is brought to the shoulder, two quick shots are fired. Focus on the target is not broken during the interval between initially seeing the target and discharging the rifle.

d. Instinctive Shooting. There may be situations in which a Marine is surprised and may need to react immediately. If possible, the Marine should engage the threat by using the tip of the front sight post. However, *speed* may be more important. The Marine's weapon and body are quickly "pointed," and the target is engaged. It is important that the body be turned with the weapon in order to achieve a natural point of aim. Simply pointing the weapon will usually result in a miss. Once the first two shots have been fired and the Marine regains the initiative, the weapon should be quickly moved to the Marine's shoulder and the tip of the front post used for sighting subsequent shots.

Section III

Movement

Movement techniques within built-up areas are the first fundamental skills that Marines should master. These movement techniques should be practiced until they become habitual. To reduce exposure to enemy fire, Marines should avoid silhouetting themselves, avoid open areas, and select their next covered position before moving.

4. Crossing a Wall. Each Marine should learn the correct method of crossing a wall (Figure A-8). After reconnoitering the other side, the Marine quickly rolls over the wall, keeping a low silhouette. The speed of his movement and a low silhouette deny the enemy a good target.

Figure A-8. Crossing a Wall

5. Observation Around Corners. The area around a corner should be observed before the Marine moves beyond it. The most common mistake the Marine makes at a corner is allowing his weapon to extend beyond the corner before observing, thereby exposing his position. Using the short stocking technique (Paragraph 2.d.) reduces exposure. Additionally, Marines can use a technique called "popping the corner" to reduce exposure time (Figure A-9). The Marine will get into a prone position near the corner of a building or obstacle around which he needs to observe. The weapon is short stocked, and the muzzle is pointed in the direction the Marine is looking. This allows the Marine to engage a target, if necessary, when he observes around a corner. The Marine will crawl to the corner but not expose himself. He will raise his upper body onto his elbows. The Marine will then push his body forward with his feet and legs without moving his elbows. His upper body, with the weapon ready, will move forward. The final position will expose the weapon, the Marine's helmet, and a minimal amount of the Marine's face. The Marine's forearms will come to rest on the deck giving him a low profile, the ability to observe around the corner, and the immediate capability to engage targets with his weapon.

Figure A-9. Correct Technique for Observing Around a Corner

6. Movement Past Windows. Windows present another hazard to Marines. The most common mistake made when passing a window is exposing the head to enemy observation from inside the room. If a Marine shows his head, an enemy gunner inside the building could engage him through the window without exposing himself to return fires.

a. The correct technique for passing a window is for the Marine to stay below the window level. He makes sure he does not silhouette himself in the window; he "hugs" the side of the building. An enemy gunner inside the building would have to expose himself to covering fires to engage the Marine (Figure A-10A).

Figure A-10A. Moving Past Windows

b. The most common mistake Marines make when passing a basement window is not being aware of it. A Marine should not walk or run past a basement window because he presents a good target to an enemy gunner inside the building. When using the correct procedure for negotiating a basement window, the Marine stays close to the wall of the building and steps or jumps past the window without exposing his legs (Figure A10-B).

c. An alternate method is for the pointman to pie off the window (or other opening) as he approaches it. As the pointman moves toward the window, the second Marine (behind the pointman) steps out to provide security to the direct front. The pointman continues to cover the opening by pieing, all the while using the near edge for cover. The pointman covers and sectors off that part of the interior that can be viewed from the outside. The second Marine moves with the pointman and maintains security to the front, eventually moving past the pointman as the pointman clears the inside corners of the opening. Once the opening is cleared, the second man may assume the duties of point, or when subsequent Marines move up to secure the opening, the original pointman may reassume those duties. Figure A-11 illustrates this alternate method.

Figure A-10B. Marine Passing Basement Windows

7. Use of Doorways. Doorways should not be used as entrances or exits because they are normally covered by enemy fire. If a Marine must use a doorway as an exit, he should move quickly through it to his next position, staying as low as possible to avoid silhouetting himself (Figure A-12). Preselection of positions, speed, a low silhouette, and the use of covering fires should be emphasized when exiting doorways.

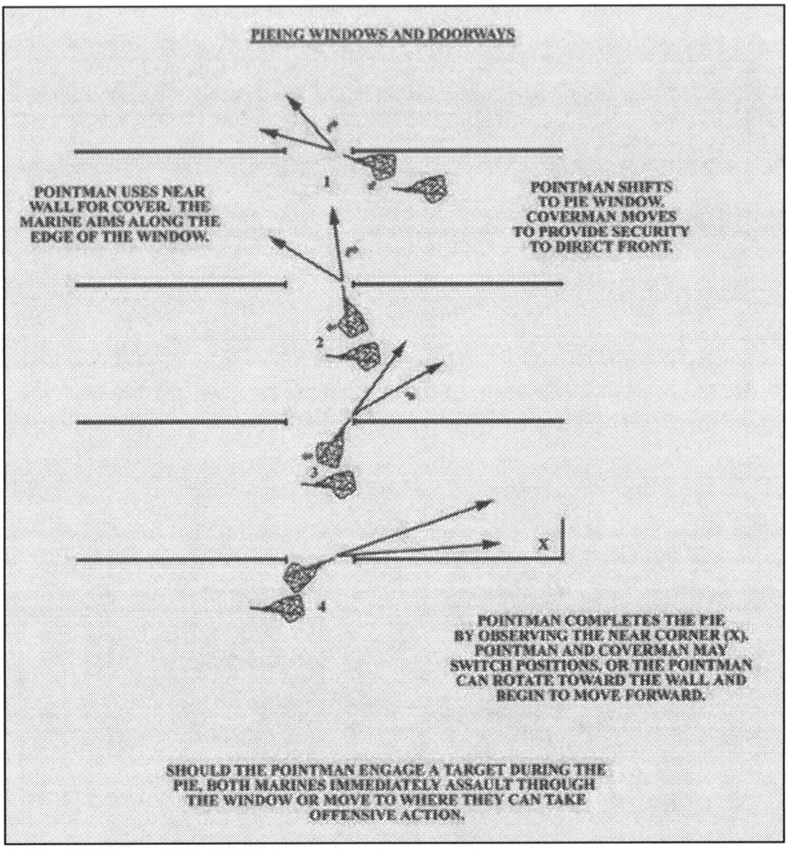

Figure A-11. Alternate Technique for Moving Past Windows

8. Movement Parallel to Buildings. Marines and small units may not always be able to use the inside of buildings as a route of advance. In such cases, they should move on the outside of the buildings (Figure A-13). Smoke, covering fires, and cover and concealment should be used to hide movement. When correctly moving along the outside of a building, the Marine hugs the side of the building, stays in the shadow, presents a low silhouette, and moves rapidly to his next position (Figure A-14). If an enemy gunner inside the building fires on a Marine, he exposes himself to fire from other squad/fire team members who are covering him. Furthermore, an enemy gunner farther down the street may have difficulty detecting and engaging the Marine.

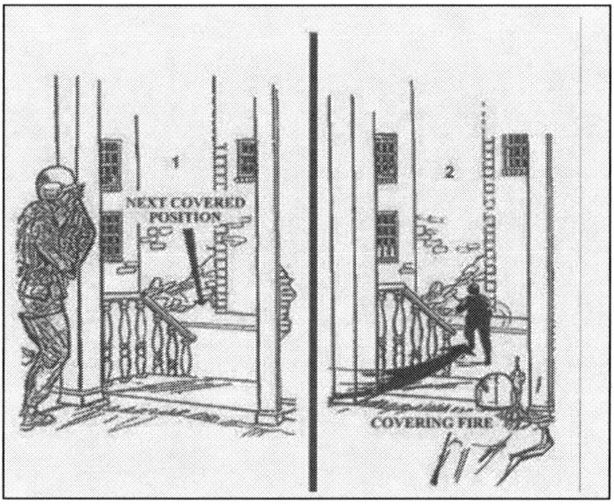

Figure A-12. Exiting a Doorway

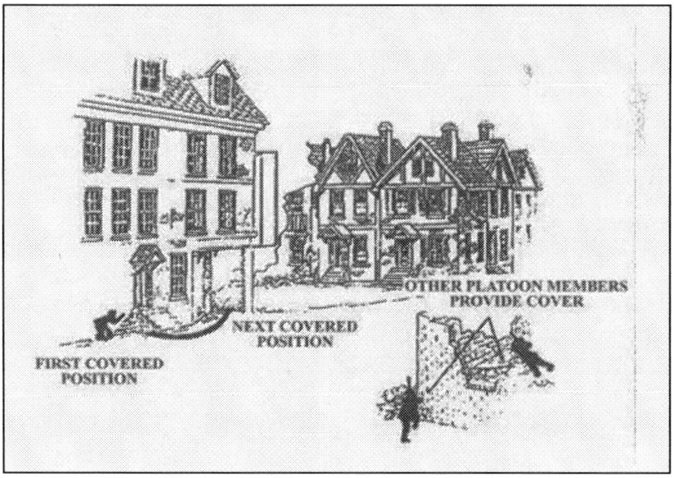

Figure A-13. Moving Outside a Building

Figure A-14. Selection of the Next Position

9. Crossing Open Areas. Open areas such as streets, alleys, and parks should be avoided. They are natural kill zones for enemy crew-served weapons. They can be crossed safely if certain fundamentals are applied by Marines and small-unit leaders.

a. When using the correct procedure for crossing an open area, the Marine develops a plan for his own movement. (Smoke from hand grenades or smoke pots should be used to conceal the movement of all Marines.) The Marine runs the shortest distance between buildings and moves along the far building to the next position. By doing so, he reduces the amount of time during which he is exposed to enemy fire.

b. Before moving to another position, the Marine should make a visual reconnaissance and select the position that offers the best cover and concealment. At the same time, he should select the route that he will take to get to that position.

c. When moving from position to position, each Marine should be careful not to mask his supporting fires. When he reaches his next position, he should be prepared to cover the movement of other members of his assault force or element.

10. Assault Element Employment. Moving as an assault element from building to building or between buildings presents a problem because an assault element presents a large target to enemy fire. When moving from the corner of one building to another, the assault element should move across the open area in groups of two, maintaining their integrity within their respective clearing or covering team (Figure A-15). The assault element leader (fire team leader) is responsible for the secure movement of his unit. Smoke should be used to screen the assault element's movement. The covering team will move first to establish security on the far side. Their movement is protected by the fire of the clearing team. Once the covering team is in position, they provide cover as the clearing team moves across and prepares, then enters, the structure. Moving from the side of one building to the side of another presents a similar problem, and the technique of movement employed is the same (Figure A-16).

There may be situations in which it is best to move in larger groups. An example is that Marines crossing in pairs may key enemy snipers in to the movement, setting them up for good shots at the next pair to cross. Leaders should remain aware of the situation and threat and plan movements accordingly.

Figure A-15. Assault Element Movement

11. Movement Inside a Building. When moving within a building that is under attack (Figure A-17), Marines should avoid silhouetting themselves in doors and windows. If forced to use a hallway (Figure A-18), Marines should move by hugging the wall to avoid presenting targets to the enemy.

Figure A-16. Movement to an Adjacent Building

Figure A-17. Movement Within a Building Under Attack

Figure A-18. Moving Down a Hallway

Section IV

Entry Techniques

Marines should enter a building with minimum exposure. They should select an entry point before moving toward the building; avoid windows and doors; use smoke to conceal their advance to the building; use demolitions, tank rounds, and so on to make new entrances; precede the entry of the clearing team with a grenade; enter immediately after the grenade explodes; and be covered by the covering team.

12. Upper Building Levels. Clearing a building from the top down is the preferred method. Clearing or defending a building is easier from an upper story. Gravity and the building's floor plan become assets when throwing hand grenades and moving from floor to floor.

a. An enemy who is forced to the top of a building may be cornered and fight desperately or escape over the roof, but an enemy who is forced down to the ground level may withdraw from the building, thus exposing himself to friendly fires from the outside.

b. Various means such as ladders, drainpipes, vines, helicopters, or the roofs and windows of adjoining buildings may be used to reach the top floor or roof of a building. In some cases, one Marine can climb onto the shoulders of another and reach high enough to pull himself up. Another method is to attach a grappling hook to the end of a scaling rope so that a Marine can scale a wall, spring from one building to another, or gain entrance through an upstairs window.

13. Use of Ladders. Ladders offer the quickest method to gain access to the upper levels of a building (Figure A-19). Units can get ladders from local civilians or stores or obtain material to build ladders through supply channels. If required, ladders can be built with resources that are available throughout the urban area; for example, lumber can be taken from inside the walls of buildings (Figure A-20). Although ladders may not permit access to the top of some buildings, they may offer security and safety through speed.

14. Use of Grappling Hook. A suitable grappling hook and rope are selected to meet the need. The grappling hook should be sturdy, portable, easily thrown, and equipped with hooks that can hold inside a window. The scaling rope should be 5/8 of an inch to 1 inch in diameter and long enough to reach the objective window. Knots are tied in the rope at 1-foot intervals to make climbing and holding easier. Marines should follow the procedures outlined below.

a. When throwing the grappling hook, the Marine should stand as close to the building as possible (Figure A-21). If the Marine stands close to the building, he is less exposed to enemy fires and he does not have to throw the hook as far.

Figure A-19. Using Ladders To Get to Upper Levels

Figure A-20. Getting Lumber From Inside the Walls

Figure A-21. Grappling Hook Thrown at Close Range

b. Making sure there is enough rope to reach the target, the Marine should hold the hook and a few coils of rope in the throwing hand. The remainder of the rope, in loose coils, should be in the other hand. The rope should be allowed to play out freely. The throw should be a gentle, even, upward lob of the hook, with the other hand releasing the rope as it plays out.

c. Once the grappling hook is inside the window (or on the roof), the Marine should pull on the rope to obtain a good hold before beginning to climb. When using a window, the hook should be pulled to one corner to ensure a good "bite" and to reduce the Marine's exposure to lower windows during the climb.

d. The use of grappling hooks is the least preferred method for gaining entry to upper levels of buildings. Grappling hooks should be used only as a last resort and away from potential enemy positions. This method may potentially be used on adjacent buildings that offer concealed locations and a roof that connects to enemy positions.

15. Scaling Walls. When forced to scale a wall during exposure to enemy fire, all available means of concealment should be used. Smoke and diversionary measures improve the chances of a successful exposed movement. When using smoke for concealment, Marines should plan for wind direction. They should use fire, shouts, and fake movement to distract the enemy.

a. A Marine scaling an outside wall is vulnerable to enemy sniper fire. Marines moving from building to building and climbing buildings should be covered with friendly fire. Areas between buildings offer good fields of fire to the enemy. Properly positioned friendly weapons can suppress and eliminate enemy fire. The M203 grenade launcher is effective in clearing the enemy from rooms inside buildings (Figure A-22).

Figure A-22. Employment of the M203 Grenade Launcher

b. A Marine scaling a wall with a rope should avoid silhouetting himself in windows of uncleared rooms or exposing himself to enemy fires from lower windows. He should climb with his weapon slung over the firing shoulder so that he may quickly bring it to a firing position. He should clear the lower room with a hand grenade before going outside the window. Before doing this, the Marine first loosens the safety pin so that he needs only one hand to throw the grenade. The objective upper story window should not be entered before a hand grenade has been thrown in.

c. The Marine enters the objective window with a low silhouette (Figure A-23). Entry can be head first; however, a preferred method is to hook a leg over the windowsill and enter sideways, straddling the ledge.

16. Rappelling. Rappelling (Figure A-24) is an entry technique that Marines can use to descend from the rooftop of a tall building into a window. (See Army manual TC 21-24 for more information on rappelling.)

17. Entry at Lower Levels. Buildings should be cleared from the top down. However, it may be impossible to enter a building at the top; therefore, entry at the bottom or lower level may be the only course of action. When entering a building at the lower level, Marines avoid entry through windows and doors because both can be easily boobytrapped and are usually covered by enemy fire.

Figure A-23. Entering the Objective Window

Figure A-24. Rappelling

a. Ideally, when entering at lower levels, demolitions, artillery, tank fire, antiarmor weapons fire, or similar means are used to create a new entrance to avoid boobytraps. Quick entry is required to follow through on the effects of the blast and concussion.

b. When the only entry to a building is through a window or door, supporting fire should be directed at that location. (Specific lower level entry techniques are shown in Figure A-25.) If no supporting fire is available, AT4s can be employed instead.

c. Before entering, Marines throw a hand grenade into the new entrance to reinforce the effects of the original blast. When making a new entrance in a building, they consider the effects of the blast on the building and adjacent buildings. If there is the possibility of a fire in an adjacent building, Marines coordinate with adjacent units and obtain permission before starting the operation. In wooden-framed buildings, the blast may cause the building to collapse. In stone, brick, or cement buildings, supporting fires are aimed at the corner of the building or at weak points in the building construction.

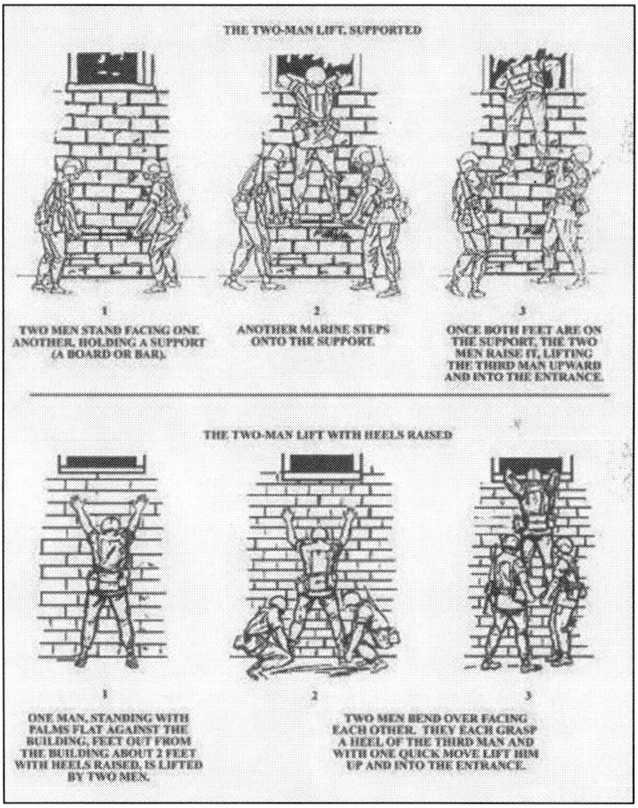

Figure A-25. Lower Level Entry Techniques

Figure A-25. Lower Level Entry Techniques (Continued)

Figure A-25. Lower Level Entry Techniques (Continued)

18. Hand Grenades. Combat in built-up areas (mainly during the attack) requires extensive use of hand grenades. The Marine should throw a grenade before entering rooms or negotiating staircases, mouseholes, and so on. This usually requires the use of both hands and employs both the overhand and underhand methods of throwing.

a. Two techniques may be used in the throwing of a grenade. The *preferred technique* involves throwing the grenade hard enough that it bounces or skips around, making it difficult to pick up. The hard-throw, skip/bounce technique may be used by Marines in training and combat. The *least preferred technique*, which involves cooking-off the grenade for two seconds then throwing it to prevent the enemy from grabbing the grenade and tossing it back, may be used by Marines, as appropriate, during actual combat but not in training.

b. Nonverbal or verbal alerts are used, as appropriate, before throwing a grenade. A visual showing (preferred) of the grenade to be thrown is made to assault element members, and a visual acknowledgment from them is received. A nonverbal alert may ensure that the enemy is surprised when the grenade is thrown. If the situation demands, a voice alert can be used, but the element of surprise may be lost. When or if a voice alert is used, the voice alert is "**FRAG OUT**;" when an enemy grenade has been identified, friendly forces shout, "**GRENADE**."

c. The construction material used in the building being cleared influences the use of grenades. In some situations, concussion grenades may be preferred over fragmentary grenades during offensive operations or when defending from hasty defensive positions. If the walls of a building are made of thin material, such as sheetrock or thin plyboard, the Marine should either lie flat on the floor with his helmet pointing toward the area of detonation or move away from any wall that might be penetrated by grenade fragments.

d. Marines should throw grenades into an opening before entering a building to eliminate any enemy that might be near the entrance (Figure A-26). Using the M203 grenade launcher is the best method for putting a grenade in an upper story window.

e. When a hand grenade must be used, the Marine throwing the grenade should stand close to the building, using it for cover. At the same time, the individual and the rest of the element should have a planned area to which they can move for safety if the grenade does not go through the window, but falls back to the ground.

f. The Marine throwing the grenade should step out far enough to lob the grenade into the upper story opening. The Marine's weapon should be kept in the nonthrowing hand so that it can be used if needed. The weapon should never be laid down outside or inside the building. Once the grenade has been thrown into the opening, assaulting forces should move swiftly to enter the building. This technique should be employed only when the window has already been broken. Otherwise, the chances are high that the thrown grenade will bounce off of the window and fall back onto the ground without going into the room.

Figure A-26. Hand Grenade Thrown Through Window

WARNING

After throwing the grenade, the Marine should immediately announce, "FRAG OUT," to indicate that a grenade has been thrown. He then takes cover because the grenade may bounce back or be thrown back, or the enemy may fire at him.

g. If Marines must enter the building by using the stairs, they first look for boobytraps. Then they throw a grenade through the stairwell door, let it detonate, and move quickly inside. They can use the staircase for cover.

h. The best way to enter a building is to breach the exterior wall. Again, a grenade should be thrown through the hole while using all available cover, such as the lower corner of the building (Figure A-27).

i. Another way to enter a room is to blast mouseholes with demolitions. In moving from room to room through mouseholes, Marines should use grenades to clear the room as they do in moving through open doorways.

j. Although buildings are best cleared from the top down, this is not always possible. While clearing the bottom floor of a building, Marines may encounter stairs that must also be cleared. Once again, grenades play an important role. Before climbing the stairs from the bottom floor, Marines should first inspect for boobytraps then toss a grenade to the head of the stairs, if appropriate (Figure A-28). The use of a grenade in these situations is based on stair/building construction and building layout. *Throwing grenades up staircases is extremely* *dangerous, and extreme caution and common sense should be used.* In most situations, throwing a grenade up a staircase is not recommended. (Again, the grenade is being used in the initial building entry or clearing of the bottom floor of a building having a staircase.) If a grenade is thrown too hard or the staircase is too steep, the grenade can roll back down the staircase. For this reason, if a grenade is thrown, it should be thrown underhand to reduce the risk of it bouncing back and rolling down the stairs. A nonverbal or, as appropriate, voice alert is used when throwing the grenade. Once the first grenade has detonated, another grenade should be thrown over and behind the staircase banister and into the hallway, destroying any enemy hiding to the rear.

Figure A-27. Entering Through a Breached Hole

Figure A-28. Tossing a Grenade Up a Stairway

k. After the stairs have been cleared, assaulting forces move to the top floor and clear it. Upon clearing the top floor, forces move downstairs to clear the center and bottom floors and to continue with the mission.

Note: Because large quantities of hand grenades are used when clearing buildings, a continuous supply must be available to forces having this mission within a built-up area.

Section V

Clearing Techniques

This section will discuss basic clearing techniques; however, the fundamentals may have to be modified depending on the various situations and building/room layouts that Marines may encounter. Caution should be taken when using a stacked type of position as personnel may be concentrating in a danger area where they could be exposed to fires penetrating walls or explosive effects from boobytraps or grenades thrown by the enemy. *Marines should always remain aware of the protective, or lack of protective, capabilities of the materials comprising the walls, floors, ceilings, and doors within the areas they are clearing.*

19. Clearing a Room, Door Closed or Open, From Stacked Positions. The following discussion and series of figures will focus on the two-man clearing team in the conduct of clearing a room. The covering team is positioned by the assault element leader to provide security as the clearing team begins its procedures to enter and clear a room.

a. Shooter Number Two is behind Shooter Number One (stacked position). This positioning is the same if the door is already open or nonexistent (Figure A-29).

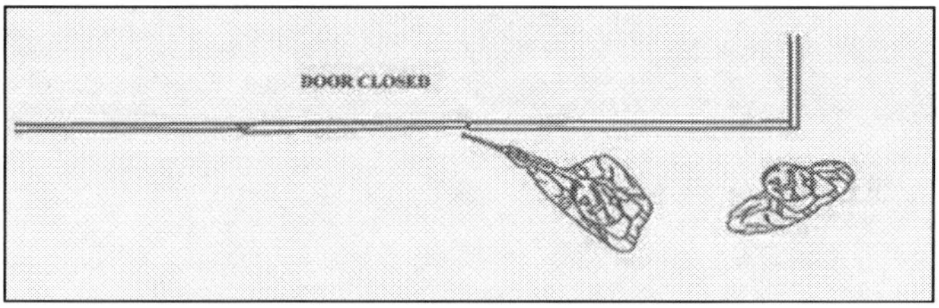

Figure A-29. Clearing a Room, Door Closed

b. Shooter Number One determines the direction in which the door opens then quickly informs Shooter Number Two and forces the door open.

When a door is the only means of entering a room, Marines should beware of fire from enemy soldiers within the room and of boobytraps. When opening a door, Marines should not expose themselves to fires through the door. The two-man clearing team forces the door open by directing a short burst of automatic fire through the door around the latch (Figure A-30) and then kicking it open. Another method of opening the door is to use a shotgun with 12-gauge slugs to shoot the hinges and/or the latch. When the doorknob is locked and/or deadbolt locks are present, shooting the hinges of the door is recommended. The door may then be opened by kicking or using a battering ram.

Figure A-30. Shooting the Door Open

Marines can also use an ax or demolitions if they are available. As a last resort, Marines can kick the door open with a front kick (Figure A-31) or mule kick. This is the least preferred technique because it is difficult and tiring to the Marine. It also rarely works the first time, thereby giving any enemy units within the room ample warning (and giving the enemy time to shoot through the door). In situations where restrictions on firing and limiting structural damage have been imposed and/or boobytraps are not expected, using the door handle to open the door may be appropriate (apply METT-T).

Figure A-31. Opening a Door With a Front Kick

c. Once the door is open, Shooter Number Two immediately throws a hand grenade into the room. Shooter Number One maintains cover for Shooter Number Two as he prepares to throw the grenade into the room. Shooter Number Two then returns to the original stacked position behind Shooter Number One (Figure A-32). If Marines have no grenades, the procedures remain the same except for throwing the grenade. *Note: When the door is already open or nonexistent, the stacked position is always used.*

WARNING

Because fragments from M67 fragmentation grenades may injure Marines outside the room, they should not be used. Marines should use MK3A2 offensive hand grenades instead. Also, the technique of cooking-off hand grenades can be deadly unless properly performed.

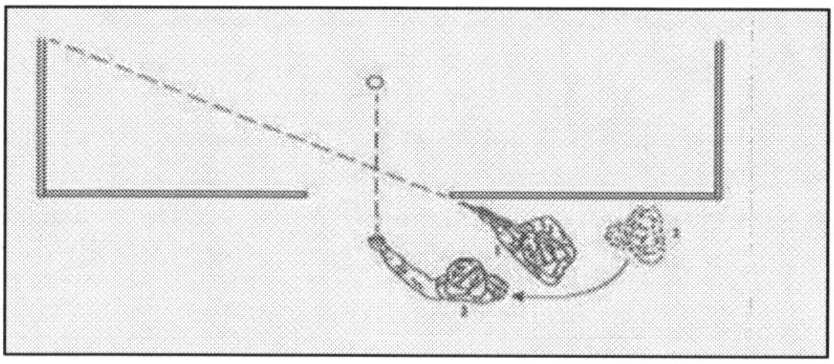

Figure A-32. Throwing a Grenade Into a Room

(1) Throwing Grenades. Two techniques are available for using grenades in room clearing. The *preferred technique* for Marines is to throw a concussion hand grenade into the room so hard that it skips and bounces, making it difficult for the enemy to pick up and throw back. The skip/bounce technique should be used by Marines during training and combat. The *least preferred technique* is to cook-off a concussion hand grenade by removing the grenade's safety pin, releasing the safety lever, counting off two seconds by thousands (one thousand and one, one thousand and two), and then throwing the grenade into the room. Cooking-off the grenade will be used only as appropriate during combat.

(2) Nonverbal and Verbal Alerts. To alert all that a grenade will be thrown, a visual showing of the grenade is made to assault element members, and a visual acknowledgment from them is received. A nonverbal alert may ensure that the enemy is surprised when the grenade is thrown. If the situation demands, a voice alert can be used, but the element of surprise may be lost. When or if a voice alert is used, the voice alert is "**FRAG OUT**;" when an enemy grenade has been identified, friendly forces shout, "**GRENADE**." This allows Marines to distinguish between warnings for outgoing and incoming grenades.

d. After the grenade explodes, Shooter Number One steps across the threshold and clears his immediate area (Figure A-33). He engages targets from the ready carry position. Shooter Number Two follows immediately behind Shooter Number One, buttonhooks, and clears his area.

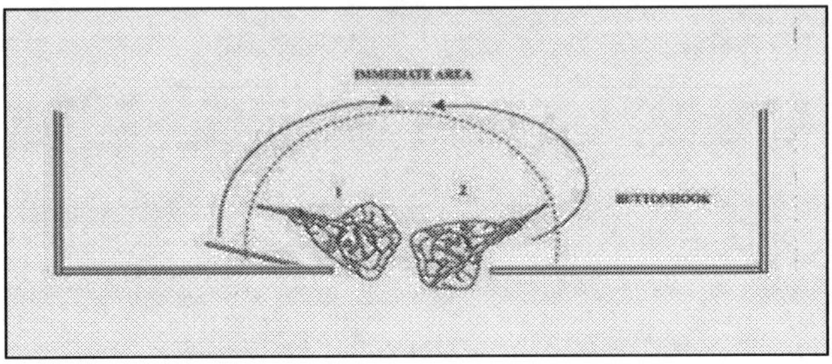

Figure A-33. Shooters Enter the Room

e. Both shooters clear the immediate area and along their respective walls, starting from the nearest respective corner and continuing to the farthest respective corner (Figure A-34). Shooters use the pieing technique to systematically clear the room by sector. (See Paragraph 3.a. of this appendix.)

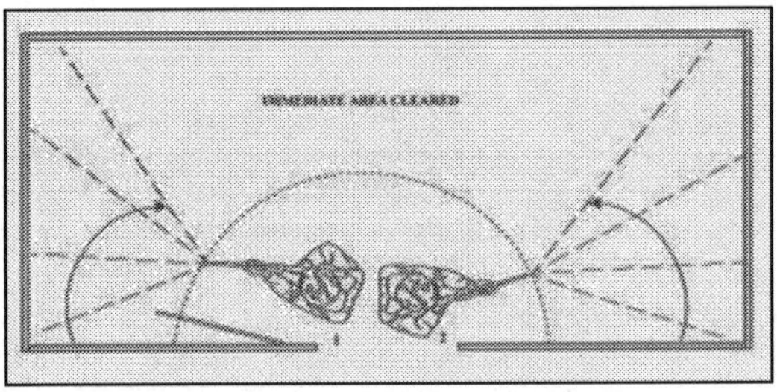

Figure A-34. Clearing a Room

f. Both shooters then establish a dominant position in the room one step away from the wall and two steps into the room and clear the room by sector, pieing to the opposite side of the room (Figure A-35). Meanwhile, the covering team (two-man team) in position outside the room being cleared provides security.

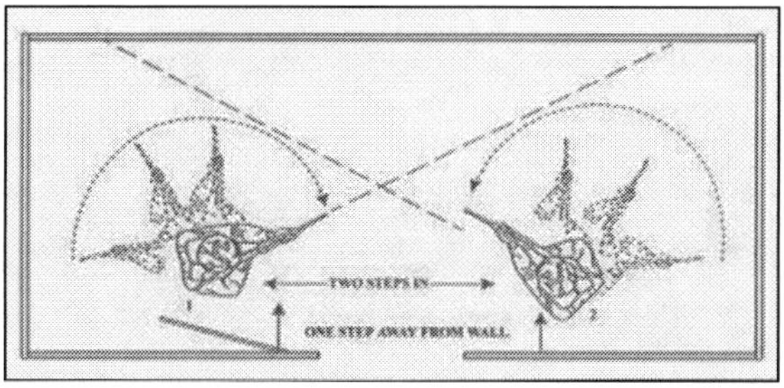

Figure A-35. Clearing a Room (Continued)

g. Another option for entering a room (situation dependent): If Shooter Number One of the clearing team enters a room without Shooter Number Two immediately following him (this could be due to room size, entering the room through a narrow opening, or a number of factors caused by the situation), then Shooter Number One will position himself inside the room to the left or right of the door and quickly scan the entire room. He then gives the command to Shooter Number Two, "**NEXT MAN IN, LEFT (RIGHT)**." Shooter Number Two shouts, "**COMING IN, LEFT (RIGHT)**," enters the room, positions himself up against the wall to the left (right) of the entrance as designated, and scans the room. Once in position, the clearing team leader can call in additional members with the "**NEXT MAN IN**" command, as the situation dictates. It is critical that all assault element members tell each other where they are to avoid fratricide.

h. When the clearing team has cleared the room and is ready to exit the room, it uses the following voice alerts. Once a room has been cleared, the clearing team yells, "**CLEAR**," to inform the covering team. Before leaving the room and rejoining the covering team, the clearing team yells, "**COMING OUT**" and waits to hear the covering team yell, "**ALL CLEAR**." The clearing team then executes its movement out of the room. The assault element then marks the room according to unit SOP.

When moving up or down a staircase, the appropriate team yells, "**COMING UP**" or "**COMING DOWN**." As before, it waits for the response "**ALL CLEAR**" before executing movement and rejoining its element. When exiting a building, the clearing team or assault element yells, "**COMING OUT**" and waits for the support force or covering team to respond with "**ALL CLEAR**."

20. Clearing a Room, Door Closed, Split Positions

a. Before opening a door, Shooter Number Two of the clearing team positions himself opposite Shooter Number One on the other side of the door and away from the wall in a safe position that allows Shooter Number One to shoot the door-opening mechanism (Figure A-36). This positioning allows Shooter Number Two to be in position to immediately move to a wall position opposite Shooter Number One. No matter what method is used to open the door, Shooter Number Two should get in a position on the side of the door opposite Shooter Number One. If the door is already open, the stacked positions in Paragraph 19.a. of this section are used.

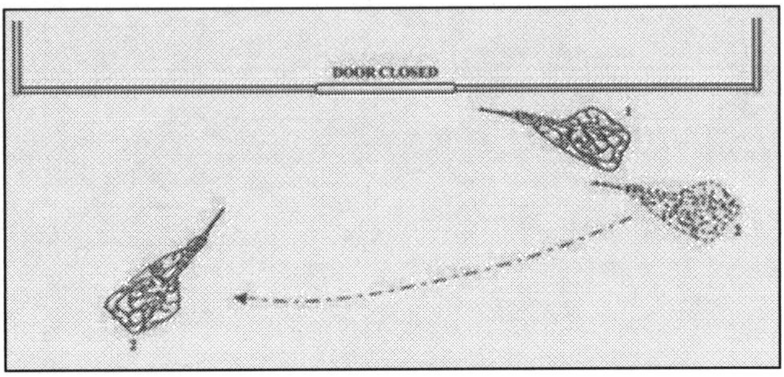

Figure A-36. Positioning To Open a Door

b. After Shooter Number One shoots the door open, Shooter Number Two immediately moves to a kneeling position against the wall (Figure A-37). Shooter Number One kicks the door open from a standing position. If the Marines have grenades, then Shooter Number Two immediately throws a grenade into the room as discussed in Paragraph 19.c. of this section. If the Marines do not have any grenades, the stand/kneel positioning of shooters places the shooters' weapons at different levels, thus allowing each shooter to engage targets inside the room while staying out of the other shooter's line of fire. The shooters scan the room for targets.

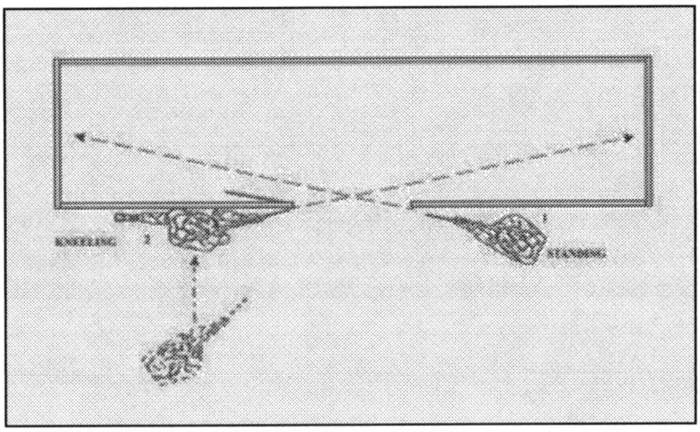

Figure A-37. Positioning To Enter a Room

c. If no grenade was thrown: Once the initial scan of the room is complete, Shooter Number One yells, "**MOVE**," and both shooters pivot from their respective positions toward each other, pieing (see Paragraph 3.a. of this appendix) while maintaining eye-muzzle-target contact. Shooter Number Two is kneeling, and Shooter Number One is still standing (Figure A-38). Both shooters clear their respective sector of fire.

If a grenade was thrown: After the explosion, Shooter Number One yells, "**MOVE**," and both shooters pivot from their respective positions toward each other, immediately enter the room, and buttonhook (Figures A-38, A-39A, and A-39B). Shooter Number Two goes from kneeling to standing as he pivots. Both shooters clear their respective sector of fire.

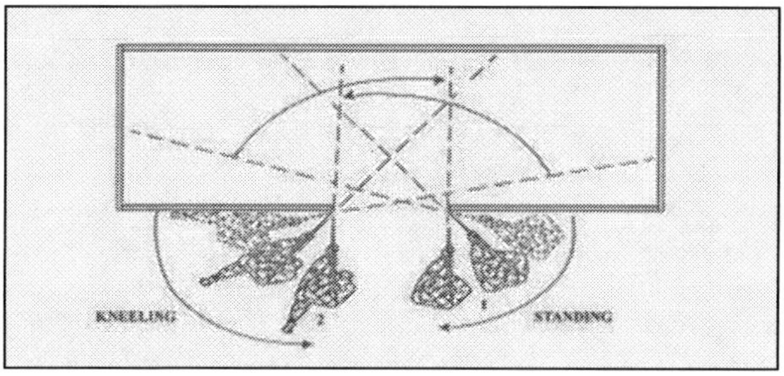

Figure A-38. Shooters Clear Sectors of Fire From Doorway

d. If no grenade was thrown: After scanning is completed, Shooter Number Two stands up and yells, "**READY**," and Shooter Number One Yells, "**MOVE**." Both shooters enter the room together (Figure A-39A).

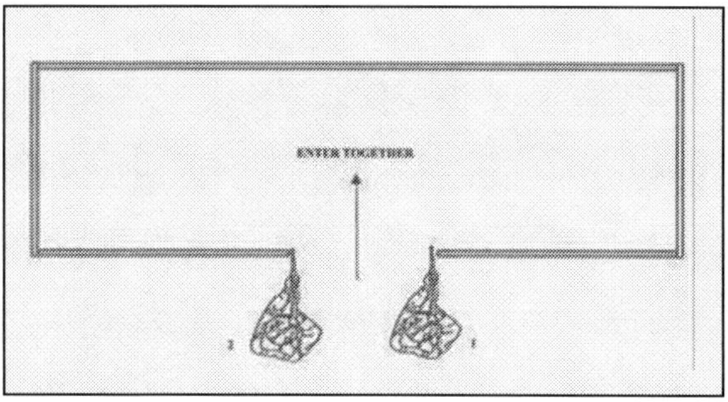

Figure A-39A. Shooters Enter the Room Together

e. Both shooters pass through the doorway together and immediately face respective corners, using the adjacent wall for cover. Both shooters buttonhook their respective areas (Figure A-39B). At this point, clearing the room is the same as discussed in Paragraphs 19.e. - 19.h.

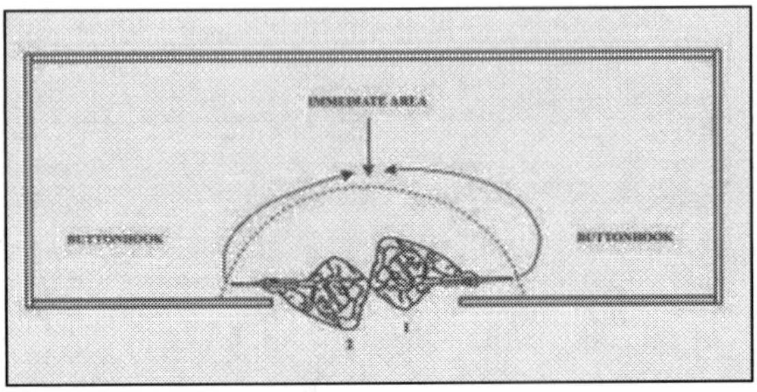

Figure A-39B. Shooters Clear Immediate Area

f. Cross Method. When employing the cross method, two Marines position themselves on either side of the entryway. Each Marine faces into the room covering the corner of the room opposite his position. On a prearranged signal, each Marine alternately enters the room. Each Marine crosses quickly to the opposite corner while covering the half of the room toward which he is moving. Once in the near corner, he assumes an outboard kneeling position to reduce his silhouette and continues to maintain coverage of his half of the room. He may change the position of his weapon to best cover the room. (Figure A-40)

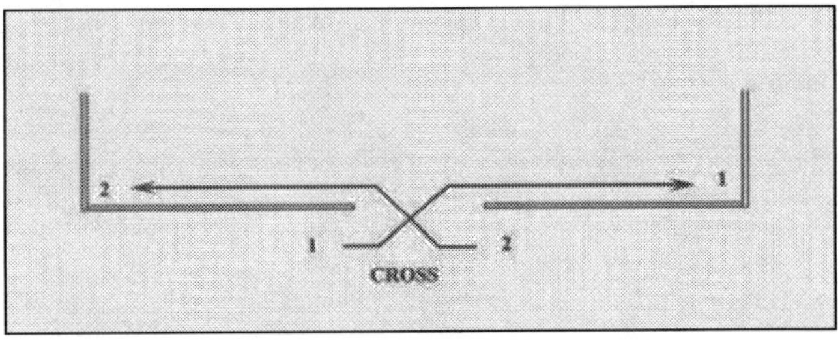

Figure A-40. Cross Method of Entering and Clearing Rooms

21. Covering Team Members Used To Support Clearing of a Room

a. Three Marines Clearing a Room. Clearing a room with three men involves the same procedures used by the aforementioned two-man clearing team, with a third Marine (from the covering team) following immediately (stacked) behind Shooters Number One and Two. Shooter Number Three moves to one side of the door and establishes a center sector of fire coverage (Figure A-41). The remaining covering team Marine provides outside security. If the split position is used, Shooters Number Two and Three are placed on the opposite side of the door.

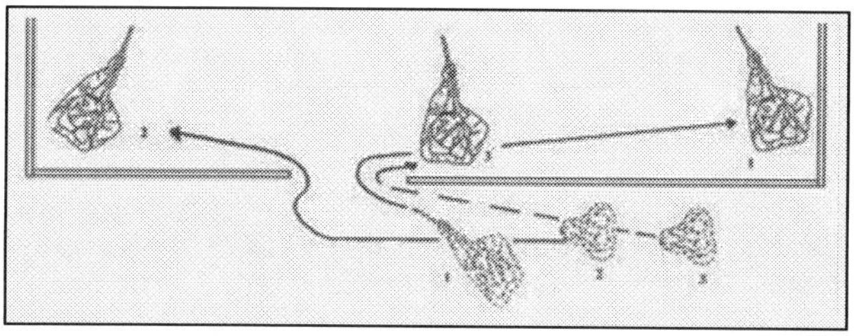

Figure A-41. Three-Marine Clearing Team (Stacked)

b. Four Marines Clearing a Room. When four Marines (entire assault element) are used to clear a room, outside security should be established by another assault element. (If outside security cannot be established, then the two- or three-Marine clearing team should be used.) The two-Marine clearing team procedures are used for initial entry into the room. Shooters Number Three and Four (split position example) stack up opposite Shooters Number One and Two. Shooters Number Three and Four follow Shooter Number Two into the room. Shooter Number Three crosses the threshold and establishes a center sector of fire coverage. Shooter Number Four follows Shooter Number Three and establishes a center sector of fire coverage opposite Shooter Number Three (Figure A-42). If the stacked position is used to enter a room, all shooters stack up on the same side of the door and follow each other in with the same positioning.

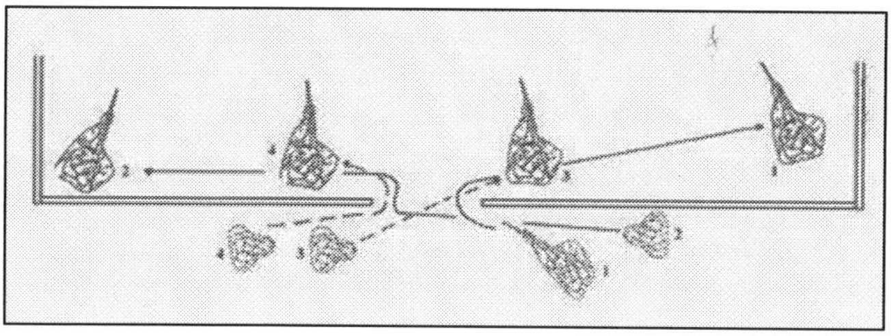

Figure A-42. Four-Marine Clearing Team (Split)

c. Single Marine Clearing a Room. This is the *least preferred* method of clearing a room. However, there may be times when this is the only method available, for example, when gaining access to an upper story window. When this is the case, the Marine gains entry to the room by shooting, kicking, ramming, or breaking a window. He then throws a grenade into the room if one is available and no restrictions apply and takes cover. Immediately after the grenade explodes, he crosses the threshold using the buttonhook technique depicted in earlier figures, scans the room for targets, and engages targets with two shots as they appear.

22. Clearing a Room, Entering Through Mousehole. Mouseholes measure about 2 feet wide and are blown or cut through a wall so that Marines can enter a room (Figure A-43). Mouseholes are safer entrances than doors, which can be easily boobytrapped and should be avoided. Procedures discussed in Paragraph 19 are used. The assault element leader positions the covering team to provide security.

a. In this situation, Shooter Number One would enter the room through a mousehole after throwing a grenade and waiting for the detonation.

b. Shooter Number One will position himself inside the room to the left or right of the mousehole and quickly scan the entire room. He then gives the command to Shooter Number Two, "**NEXT MAN IN, LEFT (RIGHT)**." Shooter Number Two shouts, "**COMING IN, LEFT (RIGHT)**," enters the room, positions himself against the wall to the left (right) of the entrance, and scans the room.

c. The clearing team continues clearing the room as discussed in Paragraph 19.

Figure A-43. Entering Through a Mousehole

23. Clearing an L-Shaped Hallway. After entering a room or building, the clearing team may encounter an L-shaped hallway that should be cleared. The following procedures are used:

a. The clearing team finds an L-shaped hallway.

b. Each shooter takes a dominant position.

c. Shooter Number Two aims his weapon to cover the dead space (Figure A-44).

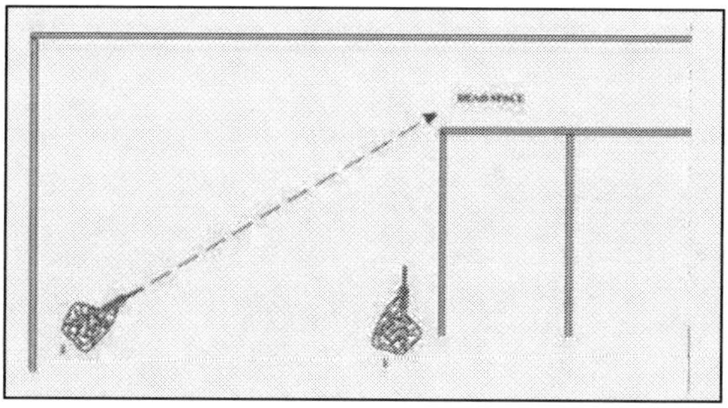

Figure A-44. Positioning To Clear an L-Shaped Hallway

d. Shooter Number Two, with his weapon covering the dead space, pies as far as possible and moves forward to increase his angle of fire further into the dead space. Shooter Number One moves accordingly with Shooter Number Two (Figure A-45).

e. Shooter Number Two stops pieing just before he reaches the position where he cannot cover the dead space. Shooter Number One stops short of the hallway and goes to a kneeling position.

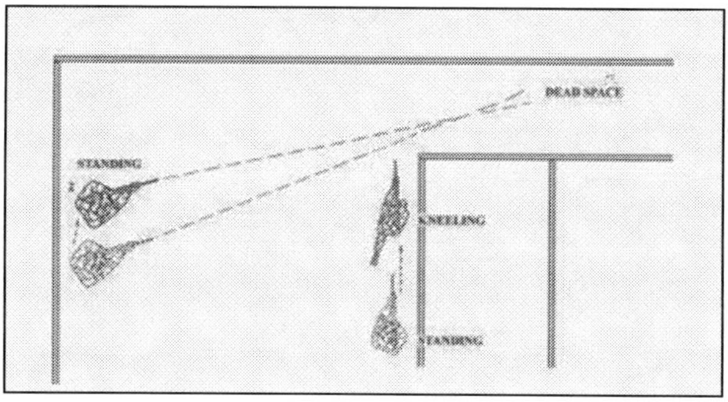

Figure A-45. Clearing an L-Shaped Hallway

f. Shooter Number One says, "**READY!**"

g. Shooter Number Two then says, "**MOVE!**" and both shooters move simultaneously to clear the hallway (Figure A-46). Shooter Number Two remains standing, and Shooter Number One remains kneeling as he pivots around the corner of the wall into the hallway.

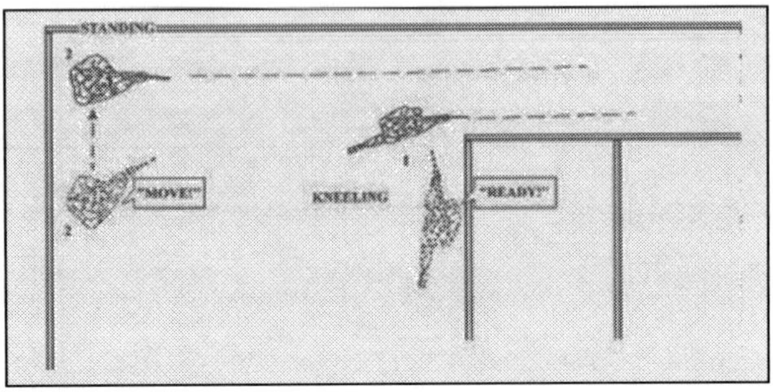

Figure A-46. Clearing an L-Shaped Hallway (Continued)

24. Clearing a T-Shaped Hallway. After entering a room or building, the clearing team may find a T-shaped hallway that should be cleared. The following procedures are used:

a. Both shooters approach the T-shaped intersection together, pieing their individual sectors while maintaining eye-muzzle-target contact (Figure A-47).

b. Both shooters move to dominant positions without entering the intersection.

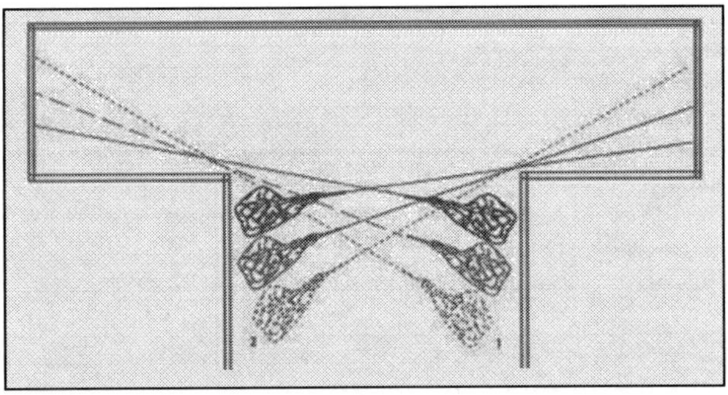

Figure A-47. Positioning To Clear a T-Shaped Hallway

c. Shooter Number One clears his sector of fire and then says, "**READY!**" (Figure A-48).

d. Shooter Number Two clears his sector of fire and then says, "**MOVE!**"

e. Both shooters buttonhook into their next positions.

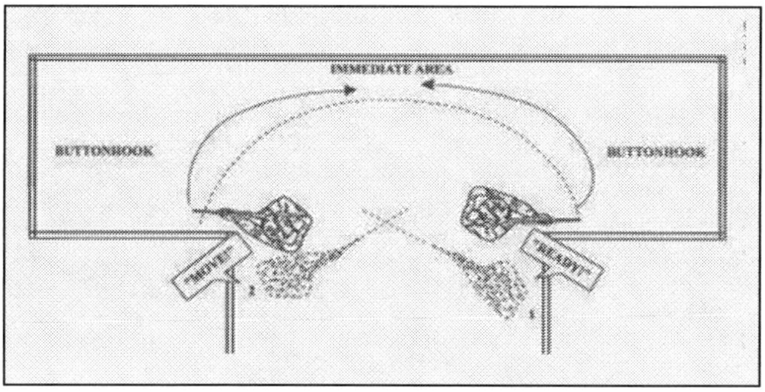

Figure A-48. Clearing a T-Shaped Hallway

f. Both shooters clear their immediate areas along their respective walls, starting from the nearest respective corner and continuing to the farthest respective corner (Figure A-49).

g. Both shooters establish dominant positions that give them control of the hallway and the doorways leading into hallways. The covering team can then be called forward to provide security while the clearing team clears one end of the hallway and associated rooms. The clearing team then clears the other end of the hallway.

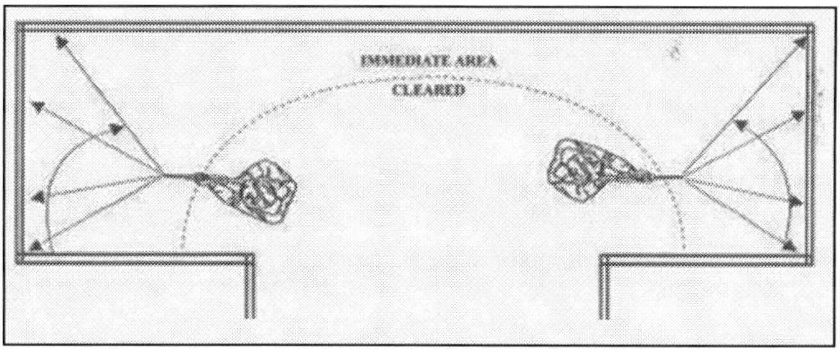

Figure A-49. Clearing a T-Shaped Hallway (Continued)

25. Clearing a Stairwell. After entering a building, the assault element may encounter stairwells. The following procedures are used:

a. Shooter Number One leads upstairs, one step ahead of Shooter Number Two (Figure A-50). Upon reaching the point just before he can be engaged from above, Shooter Number One turns around and covers overhead. From this point, Shooter Number One ascends the stairs, moving backward while covering behind and above.

b. Shooter Number Two follows Shooter Number One upstairs, one step behind and to the side of Shooter Number One. When Shooter Number One turns to cover overhead, Shooter Number Two remains oriented to the front, covering directly up the stairwell.

The clearing team's speed of movement is determined by Shooter Number One. Marines pie as much of an area as possible before ascending each step.

Figure A-50. Clearing a Stairwell

Section VI

Firing Positions

Whether a unit is attacking, defending, or conducting retrograde operations, its success or failure depends on the ability of the individual Marine to place accurate fire on the enemy while providing the least exposure to return fire. Consequently, Marines should immediately seek and properly use firing positions.

26. Hasty Firing Position. A hasty firing position is one that is normally occupied in the attack or the early stages of the defense. It is a position from which a Marine can place fire on the enemy while using available cover for protection from return fire. The Marine may occupy this position voluntarily, or he may be forced to occupy it because of enemy fire. In either case, the position lacks preparation before occupation. Some of the more common hasty firing positions in a built-up area involve firing around corners of buildings, firing from behind walls, firing from windows, firing from unprepared loopholes, and firing from the peak of a roof.

a. Corners of Buildings. The corner of a building provides cover for a hasty firing position if used properly.

(1) The shooter should be capable of both right-handed and left-handed firing of his weapon (using the "short stocking" technique) to be effective around corners. Short stocking the weapon will prevent the muzzle from protruding and keep the weapon ready to fire the instant visual contact with the enemy is made. Furthermore, it reduces the Marine's exposure as a target. (Figure A-51)

(2) A common mistake made when firing around corners is firing from the standing position. The shooter exposes himself at the height the enemy would expect a target to appear and risks exposing the entire length of his body as a target for the enemy.

Figure A-51. Firing, Using Either Hand, Around the Corner of a Building

b. Walls. When firing from behind walls, a Marine should fire around cover—not over it (Figure A-52). Marines in this situation should be able to fire from the right or left shoulder. When firing from behind cover, a Marine should remain far enough back from that cover so that the weapon does not extend beyond it. This aids in weapon retention by ensuring that no enemy on the other side of the wall can get control of the barrel.

Figure A-52. Firing Around Cover

c. Windows. In a built-up area, windows provide convenient firing ports. A Marine should avoid firing from the standing position because it exposes most of his body to return fire from the enemy and could silhouette him against a light-colored interior beyond the window. This is an obvious indicator of the firer's position, especially at night when the muzzle flash can easily be observed. When using the proper method of firing from a window (Figure A-53), the Marine is well back into the room to prevent the muzzle flash from being seen, and he is kneeling to limit exposure and avoid silhouetting himself.

Figure A-53. Firing From a Window

d. Loopholes. A Marine may fire through a hole torn in the wall and avoid windows (Figure A-54). He stays well back from the loophole so the muzzle of the weapon does not protrude beyond the wall and the muzzle flash is concealed.

Figure A-54. Firing From a Loophole

e. Roof. The peak of a roof provides a vantage point for snipers that increases their field of vision and the ranges at which they can engage targets (Figure A-55). A chimney, a smokestack, or any other object protruding from the roof of a building can reduce the size of the exposed target and should be used.

Figure A-55. Firing From the Peak of a Roof

f. No Position Available. When a Marine is subjected to enemy fire and none of the positions mentioned are available, he should try to expose as little of himself as possible. When a Marine is in an open area between buildings (a street or alley) with no cover available and he is fired upon by an enemy from one of the buildings to his front, he should lie prone as close as possible to a building on the same side of the open area as the enemy. To engage the Marine, the enemy must then lean out of the window and expose himself to return fire.

g. No Cover Available. When no cover is available, target exposure can be reduced by firing from the prone position, by firing from shadows, and by presenting no silhouette against buildings.

27. Prepared Firing Position. A prepared firing position is one built or improved to allow the shooter to engage a particular area, avenue of approach, or enemy position, while reducing his exposure to return fire. Examples of prepared positions include: barricaded windows, fortified loopholes, sniper positions, antiarmor positions, and machine gun positions.

a. The natural firing port provided by windows can be improved by barricading the window and leaving a small hole for the firer's use (Figure A-56). The barricading may be accomplished by using materials torn from the interior walls of the building or any other available material. When barricading windows, a Marine should avoid:

(1) Barricading only the windows that will be used as firing ports. The enemy will soon determine that the barricaded windows are firing positions.

(2) Neat square or rectangular holes that are easily identified by the enemy. A barricaded window should not have a neat, regular firing port. The window should keep its original shape so that the position of the firer is hard to detect. Firing from the bottom of the window gives the firer the cover advantage of the wall because the firing port is less obvious to the enemy. Sandbags are used to reinforce the wall below the window and to increase protection for the firer. All glass should be removed from the window to prevent injury to the firer. Lace curtains permit the firer to see out and prevent the enemy from seeing in. Wet blankets should be placed under weapons to reduce dust that may create a signature. Wire mesh placed over the window keeps the enemy from throwing in hand grenades.

b. Although windows are usually good firing positions, they do not always allow the firer to engage targets throughout his sector.

(1) To avoid establishing a pattern of always firing from windows, an alternate position is required such as the prepared loophole (Figure A-57). This involves cutting or blowing a small hole in the wall to allow the shooter to observe and engage targets in his sector.

(2) Sandbags are used to reinforce the walls below, around, and above the loophole. If the position is on the second floor or higher, two layers of sandbags are placed on the floor under the firer to protect him from an explosion on a lower floor. A wall of sandbags, rubble, furniture, and so on should be constructed to the rear of the position to protect the firer from explosions in the room.

Figure A-56. Window Firing Position

(3) A table, a bedstead, or other available material provides overhead cover for the position. This prevents injury to the shooter from falling debris or explosions above his position.

(4) The position should be camouflaged by knocking other holes in the wall. This makes it difficult for the enemy to determine which hole the fire is coming from. Siding material should be removed from the building in several places to make loopholes less noticeable.

Figure A-57. Prepared Loopholes

c. A chimney or other protruding structure provides a base from which a sniper position can be prepared. Part of the roofing material is removed to allow the sniper to fire around the chimney. He should stand inside the building on the beams or on a platform with only his head and shoulders above the roof (behind the chimney). Sandbags placed on the sides of the position protect the sniper's flanks.

d. When the roof has no protruding structure to provide protection, the sniper position should be prepared from underneath on the enemy side of the roof (Figure A-58). The position is reinforced with sandbags, and a small piece of roofing material should be removed to allow the sniper to engage targets in his sector. The missing piece of roofing material should be the only sign that a position exists. Other pieces of roofing should be removed to deceive the enemy as to the true sniper position. The sniper should be invisible from outside the building, and the muzzle flash should be hidden from view.

Figure A-58. Sniper Position

e. Some rules and considerations for selecting and occupying individual firing positions are:

(1) Make maximum use of available cover and concealment.

(2) Avoid firing over cover; when possible, fire around it.

(3) Avoid silhouetting against light-colored buildings, the skyline, and so on.

(4) Carefully select a new firing position before leaving an old one.

(5) Avoid setting a pattern; fire from both barricaded and unbarricaded windows.

(6) Keep exposure time to a minimum.

(7) Begin improving a hasty position immediately after occupation.

(8) Use construction material that is readily available in the built-up area for prepared positions.

(9) Remember that positions that provide cover at ground level may not provide cover on higher floors.

f. In attacking a built-up area, recoilless weapon and ATGM crews may be hampered in choosing firing positions by the backblast of their weapons. They may not have enough time to knock out walls in buildings and clear backblast areas. They should select positions that allow the backblast to escape, such as corner windows where the round fired goes out one window and the backblast escapes from another. The corner of a building can be improved with sandbags to create a firing position (Figure A-59).

Figure A-59. Corner Firing Position

g. The rifle squad is often reinforced with attached antitank weapons during an attack on and in defense of a built-up area. Therefore, the rifle squad leader should be able to choose good firing positions for the antitank weapons under his control.

h. Various principles for employing antitank weapons have universal applications, such as making maximum use of available cover, trying to achieve mutual support, and allowing for the backblast when positioning recoilless weapons, TOWs, Dragons, and AT4s.

i. Operating in a built-up area presents new considerations. Marines should select numerous alternate positions, particularly when the structure does not provide cover from small-arms fire. They should position their weapons in the shadows and within the building.

j. Recoilless weapons and ATGMs firing from the top of a building can use the chimney for cover (Figure A-60). The rear of this position should be reinforced with sandbags.

Figure A-60. Antitank Weapon Firing From a Rooftop

k. When selecting firing positions for recoilless weapons and ATGMs, Marines should make maximum use of rubble, corners of buildings, and destroyed vehicles to provide cover for the crew. Recoilless weapons and ATGMs can also be moved along rooftops to obtain a better firing angle from which to engage enemy armor. When buildings are elevated off the ground, positions can be prepared using a building for overhead cover (Figure A-61). The backblast under the building must not damage or collapse the building or injure the crew.

Note: When firing from a slope, ensure that the angle of the launcher relative to the ground or firing platform is not greater than 20 degrees. When firing within a building, ensure that the enclosure is at least 10 feet by 15 feet, is clear of debris and loose objects, and has windows, doors, or holes in the walls to allow the backblast to escape.

l. The machine gun can be emplaced almost anywhere. In the attack, windows and doors offer ready-made firing ports (Figure A-62). For this reason, the enemy normally has windows and doors under observation and fire; they should therefore be avoided. Any openings in walls that were created during the fighting may be used. When other holes are not present, small explosive charges can create loopholes (Figure A-63). Regardless of what openings are used, machine guns should be sited within the building and in the shadows.

Figure A-61. Prepared Positions Using a Building for Overhead Cover

Figure A-62. Emplacement of a Machine Gun in a Doorway

m. Upon occupying a building, Marines board up all windows and doors. By leaving small gaps between the slots, Marines can use windows and doors as good alternate firing positions.

n. Loopholes should be used extensively in the defense. They should not be constructed in any logical pattern, nor should they all be at floor or tabletop level. Varying their height and location makes them hard to pinpoint and identify. Dummy loopholes, knocked-off shingles, or holes that are cut but not intended to be used as firing positions aid in deception. Loopholes located behind shrubbery, under doorjambs, and under the eaves of a building are hard to detect.

Figure A-63. Use of a Loophole With a Machine Gun

o. Increased fields of fire can be obtained by locating the machine gun in the corner of the building or sandbagging it under a building (Figure A-64). Available materials, such as desks, overstuffed chairs, couches, and other items of furniture, should be integrated into the construction of bunkers to add both cover and concealment (Figure A-65).

Figure A-64. Sandbagged Machine Gun Emplacement Under a Building

p. Although grazing fire is desirable when employing the machine gun, it may not always be practical or possible. Where destroyed vehicles, rubble, and other obstructions restrict the fields of grazing fire, the gun can be elevated to a position from which it can deliver plunging fire over obstacles. Firing from loopholes on the second or third story may be necessary. A firing platform can be built under the roof (Figure A-66) and a loophole constructed. Again, the exact location of the position should be concealed by knocking off shingles in isolated patches over the entire roof.

Figure A-65. Corner Machine Gun Bunker

Figure A-66. Firing Platform Built Under a Roof

28. Target Acquisition. Built-up areas provide unique challenges to acquiring targets. Buildings mask movement and the effects of direct and indirect fires. Also, the rubble from destroyed buildings provides concealment and protection for attackers and defenders alike, making target acquisition difficult.

a. The techniques of patrolling and using OPs apply in the city as well as in wooded terrain. These techniques enable units to locate the enemy, to develop targets for direct and indirect fires in the defense, and to find uncovered avenues of approach in the offense.

b. Most weapons and vehicles produce recognizable signatures. These come from design features or from the environment in which the equipment is used. For example, firing a tank main gun in dry, dusty, and debris-covered streets raises a dust cloud; a tank being driven in built-up areas produces more noise than one moving through an open field; people moving through rubble on a street or in the halls of a damaged building create more noise than in a wooded area. Marines should learn to recognize signatures so they can locate and identify targets. Seeing and hearing assist in detecting and identifying signatures that lead to target location, identification, and rapid engagement. Marines should look for targets in areas where they are most likely to be employed.

c. Target acquisition should be continuous, whether halted or moving. Built-up areas provide both the attacker and defender with good cover and concealment but usually favor the defender. This makes rapid and accurate target acquisition extremely important.

d. When a unit is moving and enemy contact is likely, the unit should employ an overwatching element. This principle applies in built-up areas as it does in other kinds of terrain except that in built-up areas the overwatching element should observe both the upper floors of buildings and the street level.

e. Stealth should be used when moving in built-up areas because little distance separates attackers from defenders. Only hand and arm signals should be used until contact is made. The unit should stop periodically to listen and observe to ensure that it is not being followed or that the enemy is not moving parallel to the unit's flank for an ambush. Routes should be carefully chosen so that buildings and piles of rubble can be used to mask the unit's movement.

f. Observation duties should be clearly issued to squad members to ensure all-around security as they move. This security continues at the halt. All of the senses should be used to acquire targets, especially hearing and smelling. Marines soon recognize the sounds of vehicles and people moving through streets that are littered with rubble. The smell of fuel, cologne, or food cooking can disclose enemy positions.

g. OPs are "positions from which military observations are made, or fire directed and adjusted, and which possess appropriate communications" (Joint Pub 1-02). They are positions from which Marines can watch and listen to enemy activity in a specific sector. They warn the unit of enemy approach and should be positioned in the upper floors of buildings to give Marines a better vantage point than they would have at street level.

h. In the defense, a platoon leader positions OPs for local security as ordered by the company commander. The platoon leader selects the general location, but the squad leader establishes the OP (Figure A-67). Normally, there is at least one OP per platoon. An OP consists of two to four men and is within small-arms supporting range of the platoon. Leaders look for positions that allow good observation of the target sector. Ideally, an OP has a field of observation that overlaps those of adjacent OPs. The position selected for the OP should provide cover and concealment for units moving to and from the OP. The upper floors of houses or other buildings should be used. The squad leader should not routinely select obvious positions such as water towers or church steeples because those positions naturally attract the enemy's attention.

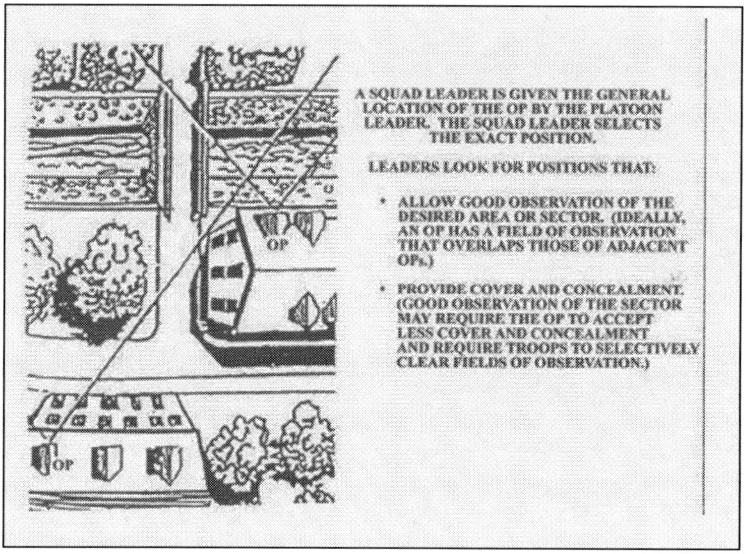

Figure A-67. Selection of an OP Location

i. The Marine should be taught how to scan a target area from OPs or from fighting positions. Use of proper scanning techniques enables observers to quickly locate and identify targets. The Marine searches quickly without optics for obvious targets, using all of his senses to detect target signatures. If no targets are found and time permits, he makes a more detailed search (using binoculars, if available) of the terrain in the assigned sector using the 50-meter method. First, he searches a strip 50 meters deep from right to left. He then searches a strip from left to right that is farther out, overlapping the first strip. This process is continued until the entire sector has been searched. In the city core or core periphery where the observer is faced with multistory buildings, the overlapping sectors may extend up rather than out.

j. Marines who man OPs and other positions should employ target acquisition devices. These devices include binoculars, image intensification devices, and thermal sights. All of these devices can enhance the unit's ability to detect and engage targets, especially at night or during periods of reduced visibility.

k. Target acquisition techniques used at night are similar to those used during the day. At night, whether using daylight optics or the unaided eye, the observer should not look directly at an object but a few degrees off to the side. The side of the eye is more sensitive to dim light. When scanning with off-center vision, the Marine moves his eyes in short, abrupt, irregular movements. At each likely target area, he pauses for a few seconds to detect any motion.

l. At night, sounds and smells can aid in acquiring targets. They transmit better in the cooler, damper night air. Running engines, vehicles, and people moving through rubble- covered streets can be heard for great distances. Odors from diesel fuel, gasoline, cooking food, burning tobacco and aftershave lotion can reveal enemy and friendly locations.

29. Firefighting Planning and Operations. Incendiary ammunition, special weapons, and the ease with which incendiary devices can be constructed from gasoline and other flammables make fire a true threat in operations in built-up areas. During defensive operations, firefighting should be a primary concern. Proper steps should be taken to reduce the risk of a fire that could make a chosen position indefensible.

a. Marines choose or create positions that do not have large openings. These positions provide as much built-in cover as possible to prevent penetration by incendiary ammunition. All unnecessary flammable materials are removed, including ammunition boxes, furniture, rugs, newspapers, curtains, and so on. The electricity and gas coming into the building should be shut off.

b. A building of concrete block construction with concrete floors and a tin roof is an ideal place for a fighting position. However, most buildings have wooden floors or subfloors, wooden rafters, and wooden inner walls that may require improvement. Inner walls are removed and replaced with blankets to resemble walls from the outside. Sand is spread 2 inches deep on floors and in attics to retard fire.

c. All available firefighting gear is prepositioned so that it can be used during combat. For the individual Marine, such gear includes entrenching tools, helmets, sand, and blankets. These items are supplemented with fire extinguishers.

d. Fire is so destructive that it can easily overwhelm personnel regardless of extraordinary precautions. Marines should plan routes of withdrawal so that evacuation can be accomplished from their fighting positions. This allows Marines to exit through areas that are free of combustible material but that still provide cover from enemy direct fire.

e. The confined space and large amounts of combustible material found in built-up areas may influence the enemy to use incendiary devices. First-aid problems that are expected include burns and smoke/flame inhalation. These can easily occur in buildings, thereby rendering the victims combat ineffective. Although there is little defense against flame inhalation and lack of oxygen, smoke inhalation may be greatly reduced by wearing the individual protective mask. Regardless of the fire hazard, defensive planning for combat in built-up areas should include location of corpsmen. Corpsmen must reach victims and their equipment and should stock extra supplies for the treatment of burns and inhalation injuries.

f. Offensive operations also require plans for firefighting because the success of the mission can easily be threatened by fire. Poorly planned use of incendiary munitions can make fires so extensive that they become obstacles to offensive operations. The enemy may use fire to cover his withdrawal and to create obstacles and barriers to the attacker.

g. Every Marine participating in an attack should be ready to deal with the problems of fire. The normal firefighting equipment available includes the entrenching tool, helmet (for carrying sand or water), and blankets (for snuffing out small fires). Fire extinguishers are normally available on vehicles.

30. Employment of Snipers. The value of the sniper to a unit operating in a built-up area depends on several factors. These factors include the type of operation, the level of conflict, and the ROE. Where ROE allow destruction, snipers may not be needed because other weapons systems available to the infantry commander have a greater destructive effect. However, snipers can contribute to the fight. Where the ROE prohibit collateral damage, snipers may be the most valuable tool the commander has. (See MCWP 3-15.3 (under development).)

a. Sniper effectiveness depends in part on the terrain. Control is decentralized by the characteristics of an urban area. To provide timely and effective support, the sniper should have a clear picture of the commander's mission and intent.

b. Where possible, snipers should be positioned in masonry buildings. These buildings should also offer long-range fields of fire and all-around observation. The sniper has an advantage because he does not have to move with or be positioned with lead elements. He may occupy a higher position to the rear or flanks of, and some distance away from, the element that he is supporting. By operating far from the other elements, a sniper avoids decisive engagement but remains close enough to kill distant targets that threaten the unit. Snipers should not be placed in obvious positions such as church steeples and rooftops because the enemy often watches these and targets them for destruction. Also, snipers should not be positioned where there is heavy traffic; these areas also invite enemy observation.

c. Snipers should operate throughout the AO, moving with and supporting the companies as necessary. Some teams may be designated to operate independently of other forces. They may be tasked to search for targets of opportunity, especially for enemy snipers. The sniper team may occupy multiple positions. A single position may not afford adequate observation for the entire team without increasing the risk of detection by the enemy. Separate positions should maintain mutual support. Alternate and supplementary positions should also be established in urban areas.

d. Snipers may be assigned the following tasks:

- Killing enemy snipers (countersniper fire)

- Killing targets of opportunity (These targets should be prioritized by the commander. Types of targets might include enemy snipers, leaders, vehicle commanders, radio men, sappers, and machine gun crews.)

- Denying enemy access to certain areas or avenues of approach (controlling key terrain)

- Providing fire support for covering barricades and other obstacles

- Maintaining surveillance of flank and rear avenues of approach (screening)

- Supporting local counterattacks with precision fire.

Section VII

Advancing/Patrolling Along City Streets

31. Advancing/Patrolling Along City Streets. During MOUT operations, skills for advancing/ patrolling along city streets will be required. The platoon, squad, or fire team mission may be such that building-to-building clearing is neither required nor desired until contact with the enemy has been made. This section will discuss advancing/patrolling along urban streets. The standard tactics, techniques, and procedures for the conduct of a patrol, as outlined in MCWP 3-11.6 *Scouting and Patrolling* (draft), apply except as otherwise discussed within this section.

a. Organization. The platoon's three-squad organization or the squad's three-fire team organization is already structured to conduct an advance/patrol down an urban street. The typical patrol is organized into assault, support, security, and command and control elements. Security is divided into front, flank, and rear security responsibilities. The platoon or squad can move down the street in a double or single column.

b. Formation and Movement

(1) Double Column. The double column is the *preferred method* for moving along an urban street (Figures A-68 and A-69). The double column provides 360 degrees of security and allows for mutual support and interlocking sectors of fire. The double column is flexible but can make control more difficult. There should be at least two squads or fire teams to conduct a double column. The two squads or fire teams move down the sides of the street using the walls of the buildings for cover and concealment. The pointmen for the two columns should always stay abreast of each other and not move in front of or behind the opposite column. The Marines following in trace of the pointmen stagger themselves accordingly. The platoon or squad leaders move where they can best control their units. If a squad is conducting the advance/patrol along the street, the trail fire team is never split.

(a) Sectors of Responsibility. Every Marine moving in a column has a particular sector for whose security he is responsible. The pointmen are responsible for security to the direct front and for pieing off windows and doorways. They are also responsible for staying in line with one another.

The Marines immediately behind the pointmen are designated as the covermen. They are responsible for providing guidance to their respective pointmen from the leader of the unit and for covering the 45-degree oblique angle across from one another. The second coverman is also responsible for covering the direct front when the pointman is pieing and checking for the enemy. The third and fourth covermen in the column are responsible for the far-side flank security; one covers the upper level and one covers the lower level building windows, doors, and stairwells. Marines will alternate providing security on upper level and lower level windows, doors, and stairwells. The fourth Marine also provides rear security if he is the last man in the column or interlocks his sector of fire with that of the lead Marine in the next fire team in the column. Should the second coverman have to take the point while the pointman pies a window or door, the third coverman would automatically shift his sector of fire to the 45-degree oblique angle. If the second coverman has to assist the pointman in pieing a large opening, the third coverman would provide security to the direct front while the fourth coverman would take the 45-degree oblique angle. Marines following in the column would adjust their sectors of fire accordingly.

The two columns provide mutual support for each other and the greatest all-around security. Teamwork and coordination between the two columns are vital.

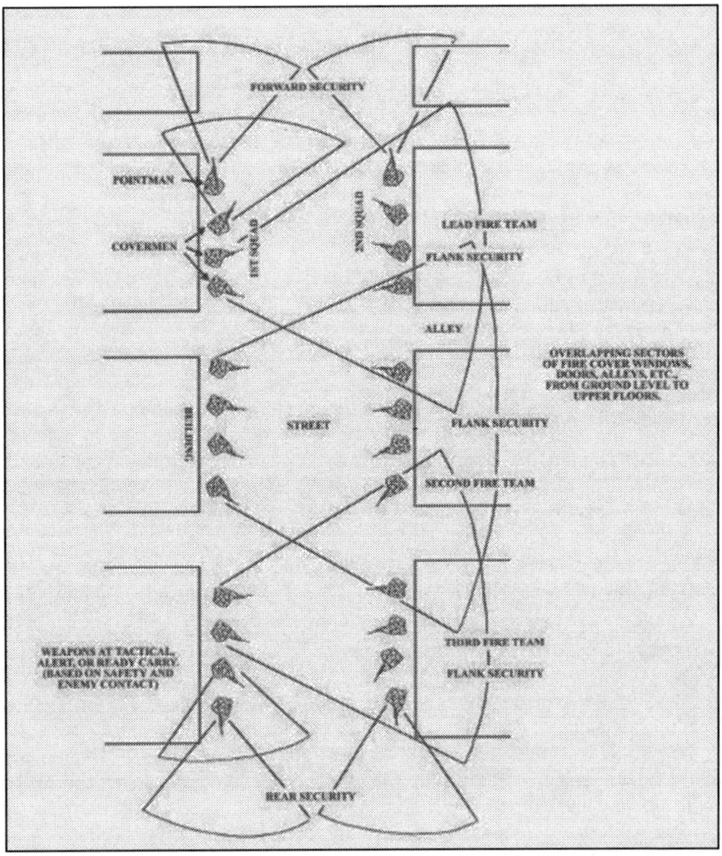

Figure A-68. Advancing/Patrolling Along an Urban Street (Squad, Double Column)

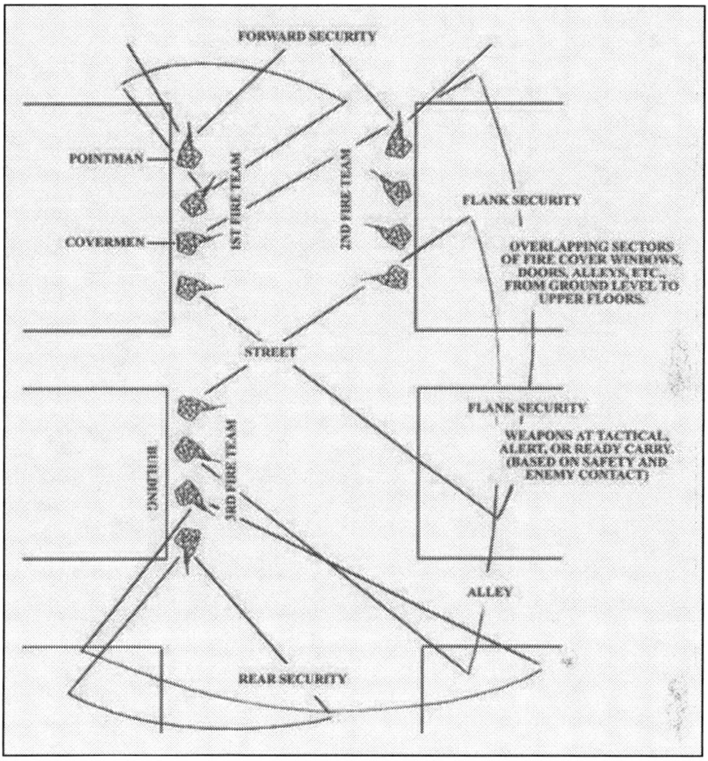

Figure A-69. Advancing/Patrolling Along an Urban Street (Fire Team, Double Column)

(b) Rear Security. The last two Marines in the squad of each column are responsible for the rear security of each of their formations. They will move in bounds in order to allow a Marine to always provide active security to the squad's rear (Figures A-70 and A-71). Two methods of bounding may be used: *alternate* or *successive bounding*.

1 **Alternate Bounding.** In this method, the two rear-security Marines alternate being the last Marine in the squad. The last Marine in the formation will have his weapon at the ready and will be in the kneeling position along the wall of the building. The second to last Marine will move in trace of the unit until he has moved approximately 10 to 15 meters. At this point, he will step out one pace from the wall. He will then assume a good kneeling firing position and announce, "**SET**." The last Marine will rise from his position, turn around, and move next to the wall to catch up with the column. It is important that the Marine moving back

does not step out and mask the other Marine's field of fire. At this point, the Marine covering the rearward movement of the last Marine can assume the rear guard position and move back to the wall as the Marine passes by his position (Figure A-70). The process is repeated continuously. The rear fire team leaders coordinate the movements of the rear guard to keep them abreast of each other and ensure that the column does not move too fast and leave the rear guard behind.

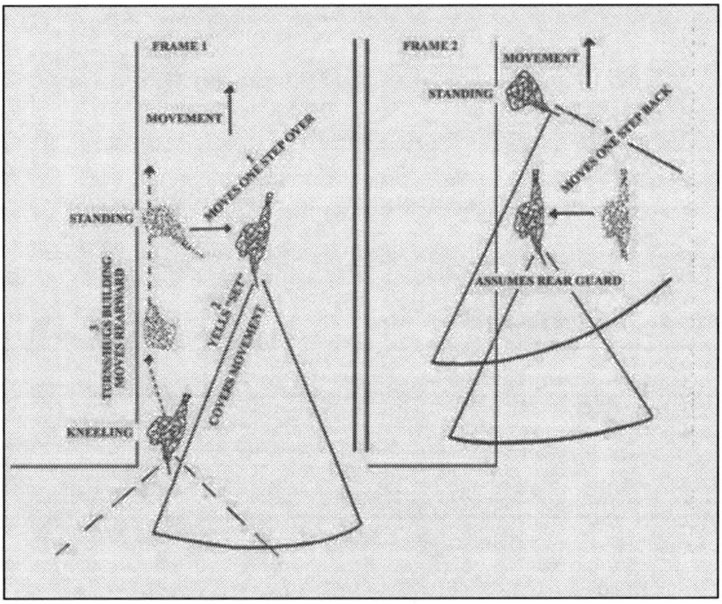

Figure A-70. Rear Security (Alternate Bounds)

2 **Successive Bounding.** In this method, the last Marine in the squad continues to be the rear guard after bounding. The same procedures are employed as in alternate bounding; however, the rear-guard Marine bounds abreast of the Marine providing cover, turns around, drops to a good kneeling position, provides rear-guard security, and announces, "**SET**." The covering Marine moves behind the rear-guard Marine and continues to move 10 to 15 meters before repeating the process (Figure A-71). The rear fire team leaders coordinate the movements of the rear guard as in alternate bounding.

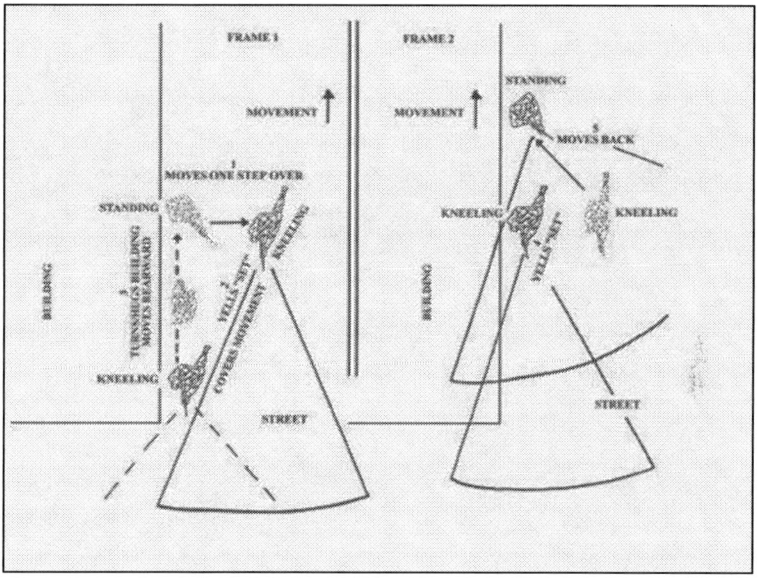

Figure A-71. Rear Security (Successive Bounds)

(2) Single Column. The single column is used to move rapidly along a street or when streets are restricted to a single column. The single column is the least preferred advance/patrol method of moving along an urban street. The same principles apply as in the double column, except the single column does not provide 360-degree security. The single column is normally used when speed is desired and contact with the enemy is considered to be unlikely.

32. Crossing a Street Intersection. Four-way intersections are very dangerous areas that should be crossed quickly. Security at an intersection is extremely important for the protection of crossing units. When moving, the pointmen notify the unit leaders, through the covermen, of an approaching intersection. Crossing the intersection will force the three elements of a unit moving along a city street to rotate positions and assume appropriate responsibilities.

a. First Fire Team To Cross. The lead (1st) fire teams in the two columns will simultaneously set security around the corners of the building. They will provide mutually supporting security to the direct front, security to the building on the far side of the intersection, and intersection security (Figure A-72). The middle (2nd) fire teams provide mutually supporting flank and rear security. On order, the rear (3rd) fire teams move forward, cross the intersection, and assume the lead (1st) fire teams' forward security mission.

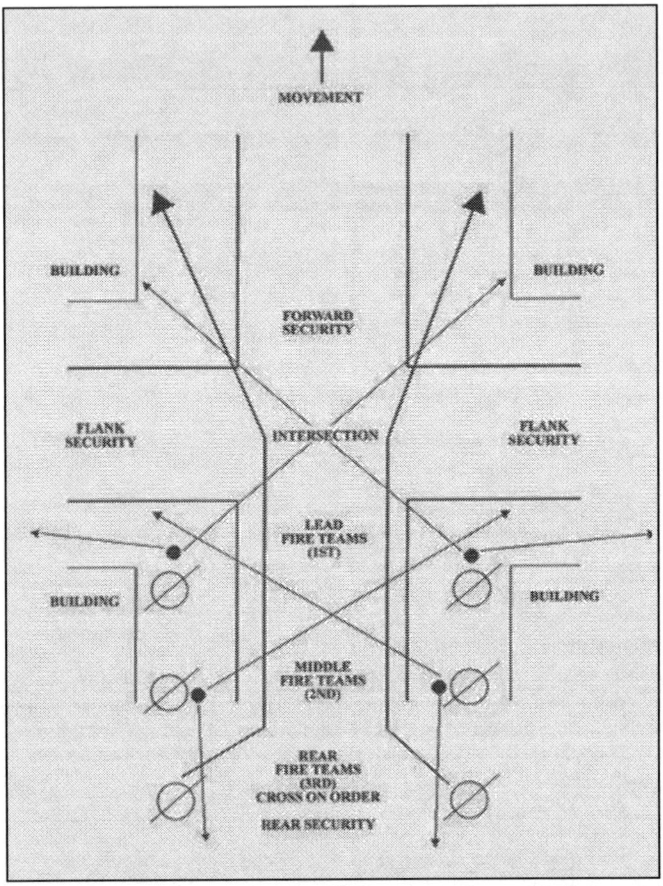

Figure A-72. Crossing an Intersection (Frame 1)

b. Second Fire Team to Cross. The 1st fire teams (now the middle fire teams) continue to provide intersection security and assume rear area security. The 2nd fire teams (now rear fire teams) cross the intersection on order. The second fire teams position themselves to provide rear and intersection security to support the covering of the last fire teams (Figure A-73).

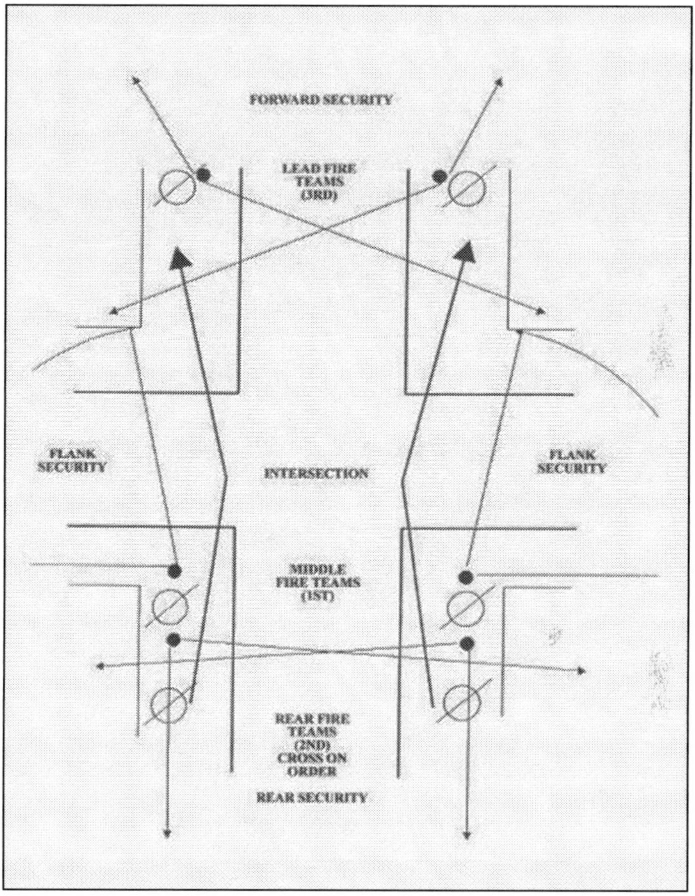

Figure A-73. Crossing an Intersection (Frame 2)

c. Third Fire Team To Cross. The 1st fire teams (now the rear fire teams) provide rear security and cross the intersection on order. The 1st fire teams bypass the 2nd fire teams to become the middle fire teams and provide mutually supporting forward and flank security. The 2nd fire teams become the rear security (Figure A-74).

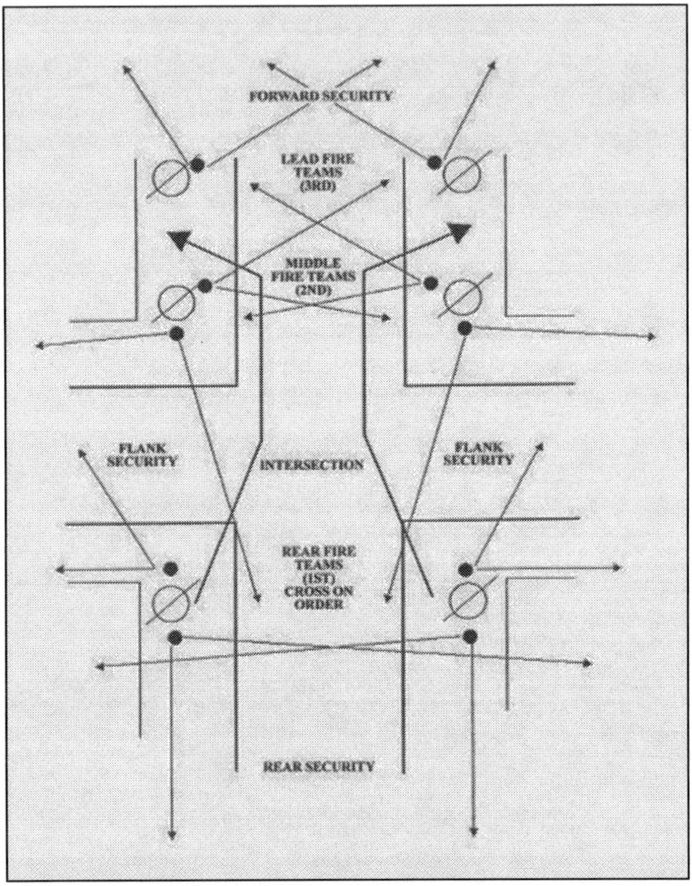

Figure A-74. Crossing an Intersection (Frame 3)

d. The squads and three fire teams have now crossed the intersection, and the fire teams have rotated (Figure A-75). This rotation is convenient because it alternates the lead units, who may quickly become fatigued during MOUT operations. The same procedures are used to cross an intersection with a single column.

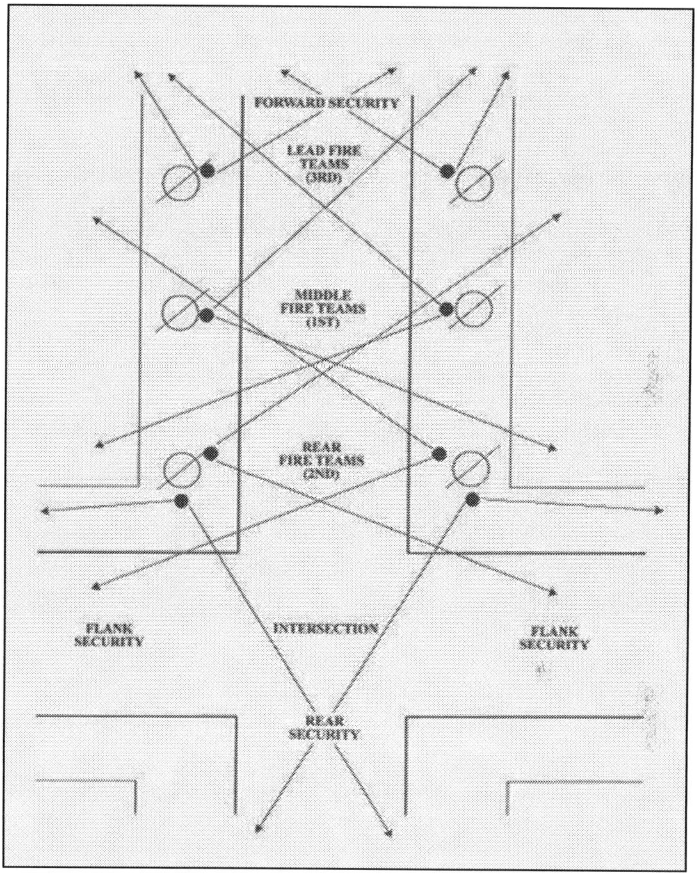

Figure A-75. Crossing an Intersection (Frame 4)

33. Reacting to Enemy Contact. When a platoon or squad is moving along a city street, contact with the enemy could happen at any time, even if the area has been declared secure. If a unit engages or is engaged by the enemy, there are two basic options that Marines may follow. The first and most preferred option is to immediately return fire and conduct a hasty clearing of a structure to seek cover inside of the building to which the unit was adjacent. The second and least preferred method is to remain outside of the building(s) and fight from the street.

a. Hasty Clearing. To conduct a hasty clearing of a structure in order to seek cover, Marines immediately return fire and enter buildings as quickly as possible. The basic principles of entry are applied as discussed earlier in the appendix; however, speed of action is essential. Once inside, Marines quickly scan the area and engage any threat. If no threat is present, the Marines acquire the enemy's positions and deliver carefully aimed shots to achieve fire superiority. At this point, the situation is developed as rapidly as possible and an attack to clear the enemy may be ordered.

b. Remain Outside and Fight. This option should not be conducted unless strict ROE do not allow the occupation of buildings without being fired on from them first or unless buildings cannot be entered because of obstruction, obstacles, or boobytraps. In such cases, the Marines should quickly acquire the target(s) and return fire to gain fire superiority. They should also seek whatever cover may exist. Once fire superiority is achieved, an assault to clear the enemy may be conducted.

Section VIII

Navigation in Built-Up Areas

Navigation in built-up areas present a unique set of challenges. Deep in the city core, the normal terrain features depicted on maps may not apply—buildings become the major terrain features, and units become tied to streets. Fighting in the city destroys buildings, whose rubble then blocks streets. Street and road signs are destroyed during the fighting if they are not switched or removed by the defenders. Operations in subways and sewers present other unique challenges. However, maps and photographs are available to help units overcome these problems. Additionally, the global positioning system (GPS) can help supplement navigational abilities in built-up areas.

34. Military Maps. The military city map is a topographical map of a city that delineates streets and shows street names, important buildings, and other urban elements. The scale of a city map can vary from 1:25,000 to 1:50,000 to 1:100,000 depending on the importance and size of the city, the density of detail, and intelligence information available.

a. Special maps prepared by supporting topographical engineers can assist units in navigating in built-up areas. These maps are designed or modified to provide information that is not covered on a standard military map. This may include maps of road and bridge networks, railroads, built-up areas, and electric power fields. Special maps can be used to supplement military city maps and topographical maps.

b. Once in the built-up area, Marines may use street intersections as reference points as they would use hills and streams in rural terrain. City maps supplement or replace topographical maps as the basis of navigation. These maps enable units moving in the built-up area to know where they are and to move to new locations even though streets have been blocked or a key building has been destroyed.

c. The techniques of compass reading and pace counting can still be used, especially in a blacked-out city where street signs and buildings are not visible. The presence of steel and iron in the MOUT environment may cause inaccurate compass readings. Sewers may be navigated in much the same way. Maps providing the basic layout of the sewer system are maintained by city sewer departments. This information includes directions in which the sewer lines run and distances between manhole covers. Along with basic compass and pace-count techniques, such information enables a unit to move through the city sewers with accuracy.

d. Operations in a built-up area adversely affect the performance of some types of communications-electronic devices such as the GPS, the Position Location Reporting System (PLRS), or other data-distribution systems. These systems function the same as some communications equipment—by LOS. They cannot determine underground locations or positions within a building. These systems should be employed on the tops of buildings, in open areas, and down streets where obstacles will not affect LOS readings.

e. City utility workers are assets to units fighting in built-up areas. They can provide maps of sewers and electrical fields and information about the city. This is especially important with regard to the use of the sewers. Sewers can contain pockets of methane gas that are highly toxic to humans. City sewer workers know the locations of these danger areas and can advise a unit on how to avoid them.

35. Aerial Photographs. Current aerial photographs are also excellent supplements to military city maps and can be substituted for a map. A topographical map or military city map could be obsolete if compiled many years ago. A recent aerial photograph shows changes that have taken place since the map was made. This could include destroyed buildings and streets that have been blocked by rubble as well as enemy defensive preparations. More information can be gained by using aerial photographs and maps together than by using either one alone.

Section IX

Camouflage

To survive and win in combat in built-up areas, a unit should supplement cover and concealment with camouflage. To properly camouflage men, carriers, and equipment, Marines should study the surrounding area and make their fighting positions blend with the local terrain.

36. Application. Only the proper amount of material needed for camouflaging a position should be used. Excess material could reveal the position. Material should be obtained from a wide area around the position. For example, Marines defending a cinder-block building do not strip the front, sides, or rear of the building to camouflage a position.

a. Buildings provide numerous opportunities for establishing concealed positions. Armored vehicles can often find prospective positions under archways or inside small industrial or commercial structures. Thick masonry, stone, or brick walls offer excellent protection from direct fire and may provide concealed routes into or out of the position.

b. After camouflage is completed, the Marine inspects his position from the enemy's viewpoint. He makes routine checks to see whether the camouflage remains natural looking and actually conceals the position. If it does not look natural, the Marine should rearrange or replace it until a natural look is achieved.

c. Positions should be progressively camouflaged as they are prepared. Work should continue until all camouflage is complete. If the enemy has air superiority, work may be possible only at night. Shiny or light-colored objects that attract attention from the air should be hidden.

d. Camouflage face paint is issued in three standard, two-tone sticks. When issue-type face-paint sticks are not available, burnt cork, charcoal, or lampblack can be used to tone down exposed skin. Mud may be used as a last resort because it dries and may peel off, leaving the skin exposed, and may also contain harmful bacteria.

37. Use of Shadows. Buildings throw sharp shadows that can be used to conceal vehicles and equipment (Figure A-76). Marines should avoid movement or positions in areas that are not in shadows. Vehicles may have to be moved periodically as shadows shift during the day. Emplacements inside buildings provide better concealment and cover as well.

38. Color and Texture. Standard camouflage-pattern painting of equipment is not as effective in built-up areas as a solid, dull, dark color hidden in shadows. Because repainting vehicles before entering a built-up area is not always practical, the lighter sand-colored patterns should be subdued with mud or dirt.

Figure A-76. Use of Shadows for Concealment

a. The need to break up the silhouette of helmets and individual equipment exists in built-up areas the same as it does elsewhere. However, burlap or canvas strips are a more effective camouflage than foliage (Figure A-77). Predominant colors are normally browns, tans, and sometimes grays rather than greens, but each camouflage location should be evaluated to determine ideal colors and patterns.

Figure A-77. Helmet Camouflaged With Burlap Strips

b. CPs and logistics emplacements are easier to camouflage and better protected if located underground. Antennas can be remoted to upper stories or to higher buildings based on remote capabilities. Field telephone wire should be laid in conduits, in sewers, or through buildings.

c. Marines should consider the background to ensure that they are not silhouetted or skylined, but rather blend into their surroundings. To defeat enemy urban camouflage, Marines should be alert for common camouflage errors such as:

- Tracks or other evidence of activity
- Shine or shadows
- Unnatural color or texture
- Muzzle flash, smoke, or dust
- Unnatural sounds and smells
- Movement.

d. Dummy positions can be used effectively to distract the enemy and make him reveal his position by firing on them.

e. Built-up areas afford cover, resources for camouflage, and locations for concealment. Following are the basic rules of cover, camouflage, and concealment to which Marines should adhere:

(1) Use the terrain and alter camouflage habits to suit your surroundings.

(2) Employ deceptive camouflage of buildings.

(3) Continue to improve positions. Reinforce fighting positions with sandbags or other fragment- and blast-absorbent material.

(4) Maintain the natural look of the area.

(5) Keep positions hidden by clearing away minimal debris for fields of fire.

(6) Choose firing ports in inconspicuous spots when available.

Note: Remember that a force that *covers* and *conceals* itself has a significant advantage over a force that does not.

Appendix B

Employment and Effects of Weapons

Success against the enemy in urban operations to a large extent depends on the effective employment of weapons. Marines at all levels must understand the effects that their weapons will produce in urban combat.

1. Effectiveness of Weapons and Demolitions. The characteristics and nature of combat in built-up areas affect the results and employment of weapons. Leaders at all levels must consider the following factors in various combinations when choosing weapons to employ.

a. Hard, smooth, flat surfaces are characteristic of urban targets. Rounds rarely impact perpendicular to these flat surfaces, but rather tend to impact at some angle of obliquity. This reduces the effect of a round and increases the threat of ricochets. The tendency of rounds to strike glancing blows against hard surfaces means that up to 25 percent of impact-fuzed explosive rounds may not detonate when fired into rubbled areas.

b. Engagement ranges in MOUT are close. Studies and historical analyses have shown that only 5 percent of all targets are more than 100 meters away. About 90 percent of all targets are located 50 meters or less from the identifying Marine. Few personnel targets will be visible beyond 50 meters, and most occur at 35 meters or less. Minimum arming ranges and unit safety from backblast or fragmentation effects must be considered.

c. Engagement times are short because of the close fight. Enemy personnel present only fleeting targets. Enemy-held buildings or structures are normally covered by mutually supporting fires and often cannot be engaged with deliberate, well-aimed shots without prolonging exposure to their observation and weapons.

d. Depression and elevation limits for some weapons create dead space. Tall buildings form deep canyons that are often safe from indirect fires. Some weapons, such as the M203, can fire rounds to ricochet behind cover and inflict casualties. Target engagement from oblique angles, both horizontal and vertical, demands superior marksmanship skills.

e. Smoke from burning buildings, dust from explosions, shadows from tall buildings, and the lack of light penetrating inner rooms all combine to reduce visibility and to increase a sense of isolation. Added to this is the masking of fires caused by rubble and manmade structures. Targets, even at close range, tend to be indistinct.

f. Urban fighting often leads to confused melees with several small units attacking on converging axes. The risks from friendly fires, ricochets, and fratricide must be considered during the planning phase of operations, and control measures must be continually adjusted to lower these risks. Marines and leaders must maintain a sense of situational awareness and clearly mark their progress in accordance with unit SOPs to avoid fratricide.

g. The firer and target may each be inside or outside the same or separate buildings. The enclosed nature of combat in built-up areas means that a weapon's signature effect, such as a muzzle blast or backblast, must be considered, as well as the round's impact on the target.

h. Usually the manmade structure that is protecting the enemy must be attacked before enemy personnel inside can be engaged. Therefore, weapons and demolitions can be chosen for employment based on their effects against masonry and concrete rather than against enemy personnel.

i. Modern engineering and design improvements mean that most large buildings constructed since World War II are resilient to the blast effects of bomb and artillery attack. Even though modern buildings may burn easily, they often retain their structural integrity and remain standing. Once high-rise buildings burn out, they are still useful to the military and are almost impossible to damage further. A large structure can take 24 to 48 hours to burn out and become cool enough for people to enter.

j. The most common worldwide building type is the 12- to 24-inch brick building. Table B-1 lists the frequency of occurrence of building types worldwide.

Type of Building	Frequency of Occurrence (Percentage)
30-inch stone	1
8- to 10-inch reinforced concrete	6
12- to 24-inch brick	63
6-inch wood	16
14-inch steel and concrete (heavy clad)	2
7-inch steel and concrete (light clad)	12

Table B-1. Types of Buildings and Frequency of Occurrence

2. M16 Rifle and M249 Squad Automatic Weapon. The M16A2 rifle and the M249 squad automatic weapon are the most common weapons used in combat in built-up areas. The M16A2 rifle and the M249 are used to kill enemy personnel, to suppress enemy fire and observation, and to penetrate light cover. Leaders can use 5.56-mm tracer fire to designate targets for other weapons.

a. Employment. Close combat is the predominant characteristic of urban engagements. Marine riflemen must be able to hit small, fleeting targets from bunker apertures, windows, and loopholes. This requires pinpoint accuracy with weapons fired in the semiautomatic mode. Killing an enemy through an 8-inch loophole at a range of 50 meters is a challenge, but one that may be common in combat in built-up areas.

(1) When fighting and engaging targets inside buildings, the weapons handling and firing techniques discussed in Appendix A are used.

(2) Within built-up areas, burning debris, reduced ambient light, strong shadow patterns of varying density, and smoke all limit the effectiveness of night vision and sighting devices. Aiming stakes and noise/light boobytraps coupled with night vision goggles and firing techniques can be used in the defense. During the offense, night vision goggles and illumination munitions can be used. Any Marine using night vision goggles should be teamed with at least one Marine not wearing them.

b. Weapon Penetration. The penetration that can be achieved with a 5.56-mm round depends on the range to the target and the type of material being fired against. Single 5.56-mm rounds are not effective against structural materials (as opposed to partitions) when fired at close range—the closer the range, the less the penetration.

(1) For the 5.56-mm round, maximum penetration occurs at 200 meters. At ranges of less than 25 meters, penetration is greatly reduced. At 10 meters, penetration by the M16 round is poor as a result of the tremendous stress placed on this high-speed round, which causes it to yaw upon striking a target. Stress causes the projectile to break up, and the resulting fragments are often too small to penetrate.

(2) Even with reduced penetration at short ranges, interior walls made of thin wood paneling, sheetrock, or plaster offer no protection against 5.56-mm rounds. Common office furniture such as desks and chairs cannot stop these rounds, but a layer of books 18 to 24 inches thick can.

(3) Wooden-framed buildings and single cinder-block walls offer little protection from 5.56-mm rounds. When clearing such structures, Marines must ensure that friendly casualties do not result from rounds passing through walls, floors, or ceilings.

(4) Armor-piercing rounds are slightly more effective than ball ammunition in penetrating urban targets at all ranges. They are also more likely to ricochet than ball ammunition, especially when the target presents a high degree of obliquity.

c. Protection. The following common barriers in built-up areas stop a 5.56-mm round fired at less than 50 meters:

- One thickness of sandbags
- A 2-inch concrete wall (unreinforced)
- A 55-gallon drum filled with water or sand
- A small ammunition can filled with sand

- A cinder block filled with sand (block will probably shatter)
- A plate-glass windowpane at a 45-degree angle (glass fragments will be thrown behind the glass)
- A brick veneer
- A car body (5.56-mm round will penetrate but normally not exit).

d. Wall Penetration. Although most structural materials repel single 5.56-mm rounds, continued and concentrated firing can breach some typical urban structures (See Table B-2).

Type	Penetration	Rounds Required
8-inch reinforced concrete	Initial Loophole	35 250
14-inch triple brick	Initial Loophole	90 160
12-inch cinder block with single-brick veneer	Loophole Breach hole	60 250
9-inch double brick	Initial Loophole	70 120
16-inch tree trunk or log wall	Initial[1]	1 to 3
12-inch cinder block (filled w/sand)	Loophole	35
24-inch double sandbag wall	Initial[1]	220
3/8-inch mild steel door	Initial[1]	1

[1]Penetration only, no loophole

Table B-2. Structure Penetration Capabilities of the 5.56-mm Round Against Typical Urban Targets (Range 25 to 100 Meters)

(1) The best method for breaching a masonry wall is by firing short bursts (3 - 5 rounds) in a U-shaped pattern. The distance from the gunner to the wall should be minimized for best results—ranges as close as 25 meters are relatively safe from ricochet. Ballistic eye protection, a protective vest, and a helmet should be worn.

(2) Ball ammunition and armor-piercing rounds produce almost the same results, but armor-piercing rounds are more likely to fly back at the firer. The 5.56-mm round can be used to create either a loophole (about 7 inches in diameter) or a breach hole (large enough for a man to enter). When used against reinforced concrete, the M16 rifle and M249 cannot cut the reinforcing bars.

3. Medium/Heavy Machine Guns and Sniper Rifles (7.62 mm and .50 caliber)

a. Machine Guns. In the urban environment, the .50-caliber machine gun and the 7.62-mm M240G machine gun provide high-volume, long-range, automatic fires for the suppression or destruction of targets. They provide final protective fires along fixed lines and can be used to penetrate light structures; the .50-caliber machine gun is most effective in this role. Tracers from both machine guns are likely to start fires, but the .50-caliber tracer is more apt to do so.

(1) Employment. The primary consideration affecting the employment of machine guns within built-up areas is the limited availability of long-range fields of fire. Although machine guns should be emplaced at the lowest level possible, grazing fire at ground level is often obstructed by rubble.

(a) The .50-caliber machine gun is often employed on its vehicular mount during both offensive and defensive operations. If necessary, it can be mounted on the M3 tripod for use in the ground role or in the upper level of buildings. When mounted on a tripod, the .50-caliber machine gun can be used as an accurate, long-range weapon and can supplement sniper fires.

(b) The M240G machine gun is cumbersome, making it difficult to use inside while clearing a building. However, it is useful outside to suppress and isolate enemy defenders. The M240G can be fired from either the shoulder or the hip to provide a high volume of assault and suppressive fires. The use of the long sling to support the weapon and ammunition is preferred.

(c) Because of their reduced penetration power, M240G machine guns are less effective against masonry targets than .50-caliber machine guns. However, their availability and light weight make them well suited to augment heavy machine gun fire, to be used in areas where .50-caliber machine guns cannot be positioned, or as a substitute when heavy machine guns are not available. The M240G machine gun can be employed on its tripod to deliver accurate fire along fixed lines and can then quickly be converted to bipod fire to cover alternate fields of fire.

(2) Penetration. The ability of the 7.62-mm and .50-caliber rounds to penetrate is also affected by the range to the target and type of material against which the rounds are fired. Close ranges affect penetration of the 7.62-mm round less than the 5.56-mm round; the .50-caliber round's penetration is reduced the least.

(a) At 50 meters, the 7.62-mm ball round cannot penetrate a single layer of sandbags. It can penetrate a single layer at 200 meters, but not a double layer. The armor-piercing round does only slightly better against sandbags. It cannot penetrate a double layer but can penetrate up to 10 inches of sand at 600 meters.

(b) The penetration of the 7.62-mm round is best at 600 meters, but most urban targets are closer. The longest engagement range is usually 200 meters or less. Table B-3 lists the penetration capabilities of a single 7.62-mm (ball) round at closer ranges.

Range (meters)	Pine Board (inches)	Dry, Loose Sand (inches)	Cinder Block (inches)	Concrete (inches)
25	13	5	8	2
100	18	4.5	10	2
200	41	7	8	2

Table B-3. Penetration Capabilities of a Single 7.62-mm (Ball) Round

(c) The .50-caliber round is also optimized for penetration at long ranges (about 800 meters). For hard targets, .50-caliber penetration is affected by obliquity and range. Both armor-piercing and ball ammunition penetrate 14 inches of sand or 28 inches of packed earth at 200 meters if the rounds impact perpendicular to the flat face of the target. Table B-4 depicts the effect of a 25-degree obliquity on a .50-caliber penetration.

Thickness (feet)	100 Meters (rounds)	200 Meters (rounds)
2	300	1,200
3	450	1,800
4	600	2,400

Table B-4. Number of Rounds Needed To Penetrate a Reinforced Concrete Wall at a 25-Degree Obliquity

(3) Protection. Barriers that offer protection against 5.56-mm rounds are also effective against 7.62-mm rounds with some exceptions. The 7.62-mm round can penetrate a windowpane at a 45-degree obliquity, a hollow cinder block, or both sides of a car body. It can also easily penetrate wooden-framed buildings. The .50-caliber round can penetrate all of the commonly found urban barriers except a sand-filled 55-gallon drum.

(4) Wall Penetration. Continued and concentrated machine gun fire can breach most typical urban walls. Such fire cannot breach thick reinforced concrete structures or dense natural stone walls. Internal walls, partitions, plaster, floors, ceilings, common office furniture, home appliances, and bedding can be easily penetrated by both 7.62-mm and .50-caliber rounds (Tables B-5 and B-6).

(a) The M240G machine gun can be hard to hold steady enough to repeatedly hit the same point on a wall. The dust created by the bullet strikes also makes precise aiming difficult. Firing with a tripod is usually more effective than without, especially if sandbags are used to steady the weapon. Short bursts of three to five rounds fired in a U-type pattern are best.

(b) Breaching a brick veneer over cinder block presents a special problem for the M240G machine gun. Rounds penetrate the cinder block but leave a net-like structure of unbroken block. Excessive ammunition is required to destroy the net because most rounds only pass through a previously eroded hole. One or two minutes of work with an entrenching tool, crowbar, or ax can remove this web and allow entry through the breach hole.

(c) The .50-caliber machine gun can be fired accurately from the tripod using the single-shot mode. This is the most efficient method for producing a loophole. Automatic fire in three- to five-round bursts in a U-type pattern is more effective in producing a breach.

Type	Thickness (Inches)	Hole Diameter (Inches)	Rounds Required
Reinforced concrete	8	7	100
Triple brick wall	14	7	170
Concrete block with single brick veneer	12	6 and 24	30 and 200
Cinder block (filled)	12	N/A[1]	18
Double brick wall	9	N/A[1]	45
Double sandbag wall	24	N/A[1]	110
Log wall	16	N/A[1]	1
Mild steel door	3/8	N/A[1]	1

[1]Penetration only, no loophole

Table B-5. Structure Penetrating Capabilities of 7.62-mm Round (NATO) Against Typical Urban Targets (Range 25 Meters)

Type	Thickness (Inches)	Hole Diameter (Inches)	Rounds Required
Reinforced concrete	10	12	50
		24	100
	18	7	140
Triple brick wall	12	8 and 26	15 and 50
Concrete block with single brick veneer	12	10 and 33	25 and 45
Armor plate	1	N/A[1]	1
Double sandbag wall	24	N/A[1]	5
Log wall	16	N/A[1]	1

[1]Penetration only, no loophole

Table B-6. Structure Penetrating Capabilities of .50-caliber Ball Against Typical Urban Targets (Range 35 Meters)

b. Sniper Rifles. The 7.62-mm match ball ammunition fired from the M-40A1 sniper rifle will perform similarly to the 7.62-mm ball round. The M-82A1A special application sniper rifle (SASR) can fire the .50-caliber armor-piercing incendiary (API) ammunition, which will have the same performance as when fired from the M-2 HB. The M-82A1A also fires the .50-caliber RAUFOSS ammunition, which contains a tungsten penetrator and a more powerful explosive charge than the API ammunition. When the RAUFOSS penetrator punches through the metal, the explosive detonates inside, acting as an incendiary round. Although the SASR's exact penetration capabilities have yet to be defined, it has penetrated an inch of steel at 2000 yards. Both sniper rifles are organic to the Marine infantry battalion.

4. Grenade Launchers, 40-mm (M203 and MK19). Both the M203 dual-purpose weapon and the MK19 grenade machine gun fire 40-mm HE and HE dual-purpose (HEDP) ammunition. Ammunition for these weapons is not interchangeable, but the grenade and fuze assembly that actually hits the target is identical. Both weapons provide point and area destructive fires as well as suppression. The MK19 has a much higher rate of fire and a longer range. The M203 is much lighter and more maneuverable.

a. Employment. The main consideration affecting the employment of 40-mm grenades within built-up areas is the typically short engagement range. The 40-mm grenade has a *minimum arming range of 14 to 28 meters*. If the round strikes an object before it is armed, it will not detonate. Both the HE and HEDP rounds have 5-meter burst radii against exposed forces, which means that the *minimum safe firing range for combat is 33 meters*. The 40-mm grenades can be used to suppress the enemy in a building or inflict casualties by firing through apertures or windows. The MK19 can use its high rate of fire to concentrate rounds against light structures. This concentrated fire can create extensive damage. The 40-mm HEDP round can penetrate the armor on the flank, rear, and top of Soviet-made BMPs and BTRs. Marines can use the M203 from upper stories to deliver accurate fire against the top decks of armored vehicles. Multiple hits are normally required to achieve a kill.

b. Weapon Penetration. The 40-mm HEDP grenade has a small shaped charge that penetrates better than the HE round. It also has a thin wire wrapping that bursts into a dense fragmentation pattern, creating casualties out to 5 meters. Because they explode on contact, 40-mm rounds achieve the same penetration regardless of range. Table B-7 lists the penetration capabilities of the HEDP round.

Target	Penetration (Inches)
Sandbags	20
Sand-filled cinder block	16
Pine logs	12
Armor plate	2

Table B-7. Penetration Capabilities of the HEDP Round

(1) If projected into an interior room, the 40-mm HEDP can penetrate all interior partition-type walls. It splinters plywood and plaster walls, making a hole large enough to fire a rifle through. It is better to have HEDP rounds pass into a room and explode on a far wall, even though much of the round's energy is wasted penetrating the back wall (Figure B-1). The fragmentation produced in the room causes more casualties than the high-explosive jet formed by the shaped charge.

(2) The fragments from the HEDP round do not reliably penetrate interior walls. They are also stopped by office furniture, sandbags, helmets, and protective vests (flak jackets). The M203 dual-purpose weapon has the inherent accuracy to place grenades into windows at 125 meters and bunker apertures at 50 meters. These ranges are significantly reduced as the angle of obliquity increases. Combat experience shows that M203 gunners cannot consistently hit windows at 50 meters when forced to aim and fire quickly.

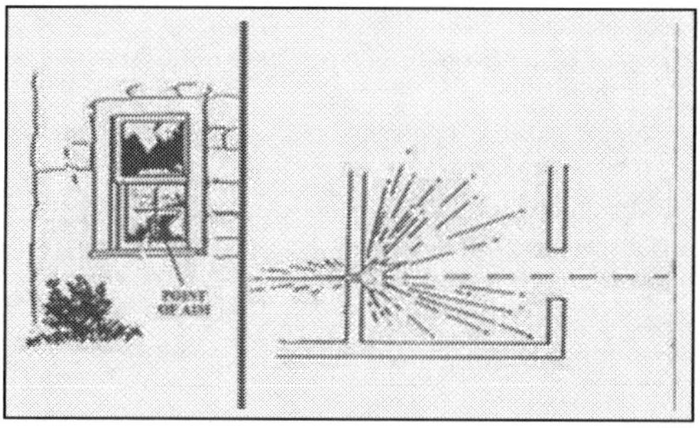

Figure B-1. Aim Point for 40-mm HEDP

c. Wall Penetration. The M203 cannot reasonably deliver the rounds needed to breach a typical exterior wall. The MK19 can concentrate its fire and achieve wall penetration. Firing from a tripod using a locked-down traversing and elevating mechanism is best for this role. Brick, cinder block, and concrete can be breached by using the MK19; individual HEDP rounds can penetrate 6 to 8 inches of brick. The only material that has proven resistant to concentrated 40-mm fire is dense stone such as that used in some European building construction. No precise data exist as to the number of rounds required to produce loopholes or breach holes with the MK19. However, the rounds' explosive effects are dramatic and should exceed the performance of the .50-caliber machine gun.

5. Light and Medium Recoilless Weapons. Light and medium recoilless weapons are used to attack enemy personnel, field fortifications, and LAVs. They have limited capability against main battle tanks, especially those equipped with reactive armor, except when attacking from the top, flanks, or rear. This category of weapons includes the AT4, the M47 Dragon, and the SMAW.

a. Employment. Other than defeating LAVs, the most common task for which light recoilless weapons are used is to neutralize fortified firing positions. Because of the design of the warhead and the narrow blast effect, these weapons are not as effective in this role as heavier weapons such as a tank main gun round. Their light weight allows Marines to carry several AT4s. Light recoilless weapons can be fired from the tops of buildings or from areas with extensive ventilation.

(1) Light and medium recoilless weapons with the exception of the SMAW employ shaped-charge warheads. As a result, the hole they punch in walls is often too small to use as a loophole. The fragmentation and spall that these weapons produce are limited. Normally, shaped-charge warheads do not neutralize enemy forces behind walls unless they are located directly in line with the point of impact.

(2) Against structures, shaped-charge weapons should be aimed about 6 inches below or to the side of a firing aperture (Figure B-2). This enhances the probability of killing the enemy behind the wall. A round that passes through a window wastes much of its energy on the back wall. Because these shaped-charge rounds lack the wire wrapping of the 40-mm HEDP, they burst into few fragments and are often ineffective casualty producers.

(3) Sandbagged emplacements present a different problem (Figure B-3). Because sandbags absorb much of the energy from a shaped charge, the rounds should be aimed at the center of the firing aperture. Even if the round misses the aperture, the bunker wall area near it is usually easiest to penetrate.

Figure B-2. Point of Aim for a Shaped-Charge Weapon Against a Masonry Structure

Figure B-3. Point of Aim for Sandbagged Emplacement

(4) Light and medium recoilless weapons obtain their most effective short-range antiarmor shots by firing from upper stories or from the flanks and rear. When firing at main battle tanks, these weapons should always be employed against these weaker areas in volley or paired firing. They normally require multiple hits to achieve a kill on a tank. Flank, top, and rear shots hit the most vulnerable parts of armored vehicles. Firing from upper stories protects the firer from tank main gun and coaxial machine gun fire because tanks cannot sharply elevate their cannons. The BMP-2 *can* elevate its 30-mm cannon to engage targets in upper stories. The BTR-series armored vehicles *can* also fire into upper stories with their heavy machine gun.

(5) Modern IFVs such as the BMP-2 and the BTR-80 have significantly improved frontal protection against shaped-charge weapons. Many main battle tanks have some form of reactive armor in addition to their thick armor plate. Head-on, ground-level shots against these vehicles have little probability of obtaining a kill. Even without reactive armor, modern main battle tanks are hard to destroy with a light antiarmor weapon.

(6) The easiest technique to improve the probability of hitting and killing an armored vehicle is to increase the firing depression angle. A 45-degree downward firing angle doubles the probability of a first-round hit as compared to a ground-level shot.

b. Backblast. Backblast characteristics must be considered when employing all recoilless weapons. During combat in built-up areas, the backblast area in the open is more hazardous because of the loose rubble and the channeling effect of the narrow streets and alleys. Figure B-4 shows the backblast areas of Marine light recoilless weapons in the open.

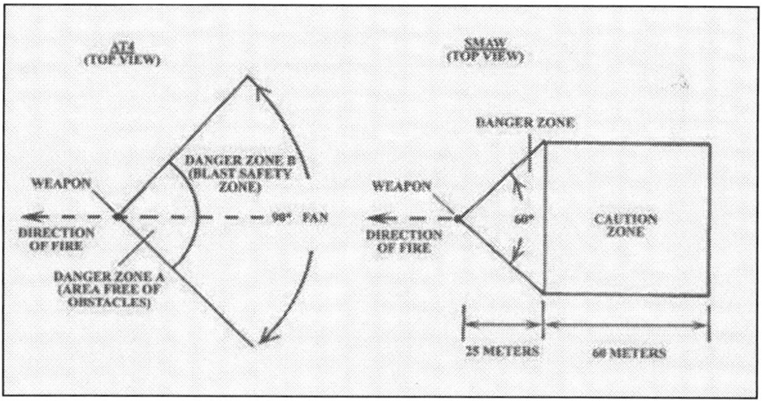

Figure B-4. Backblast Areas of Light Recoilless Weapons in the Open

(1) When firing recoilless weapons in the open, Marines should protect themselves from blast and burn injuries caused by the backblast. All personnel should be out of the danger zone. Anyone not able to vacate the caution zone should be behind cover. Marines in the caution zone should wear helmets, protective vests, and eye protection. The firer and all Marines in the area should wear earplugs.

(2) Extensive testing on the effects of firing recoilless weapons from within enclosures has been conducted since 1948. These tests showed that firing these weapons from enclosures presented no serious hazards, even when the overpressure was enough to produce structural damage to the building. The following were other findings of these tests.

(a) Little hazard exists to the gunnery or crew from any type of flying debris. Loose items were not hurled around the room.

(b) No substantial degradation occurs to the operator's tracking performance as a result of obscuration or blast overpressure.

(c) The most serious hazard that can be expected is hearing loss. This must be evaluated against the advantage gained in combat from firing from cover. To place this hazard in perspective, a gunner wearing earplugs and firing the loudest combination (the Dragon from within a masonry building) is exposed to less noise hazard than if he fired a LAW or AT4 in the open without earplugs.

(d) The safest place for other Marines in the room with the firer is against the wall from which the weapon is fired. Plastic ignition plugs are a hazard to anyone standing directly behind a LAW or TOW when it is fired.

(e) Firers should take advantage of all available sources of ventilation by opening doors and windows. Ventilation does not reduce the noise hazard, but it helps clear the room of smoke and dust and reduces the effective duration of the overpressure.

(f) The only difference between firing these weapons from enclosures and firing them in the open is the duration of the pressure fluctuation.

(g) Frame buildings, especially small ones, can suffer structural damage to the rear walls, windows, and doors. Large rooms suffer slight damage, if any.

(3) Recoilless weapons fired from within enclosures create some obscuration inside the room but almost none from the gunner's position looking out. Inside the room, obscuration can be intense, but the room remains inhabitable. Table B-8 shows the effects of smoke and obscuration.

Building	Weapon	From Gunner's Position Looking Out	Inside the Room	From Outside at a Distance
Masonry	LAW	None	Moderate	Slight smoke
	Dragon	Slight	Moderate	Small flash

Bunker	Dragon TOW	None None	Slight Slight	Moderate flash Moderate smoke
Small frame	LAW Dragon	None None	Moderate Severe	Moderate Moderate
Medium frame	LAW Dragon	None None	Slight Severe	Moderate Slight flash
Large frame	LAW Dragon TOW	None Slight None	Slight Severe Severe	None Slight flash Slight smoke

Table B-8. Smoke and Obscuration

(4) The Dragon causes the most structural damage, but only in frame buildings. There does not seem to be any threat of injury to the gunner because the damage is usually to the walls away from the gunner. The most damage and debris is from flying plaster chips and pieces of wood trim. Large chunks of plasterboard can be dislodged from ceilings. The backblast from a LAW, AT4, Dragon, or TOW rarely displaces furniture. Table B-9 shows the test results of structural damage and debris.

		Damage		
Building	**Weapon**	**Structure**	**Wall Covering**	**Debris Movement**
Masonry	LAW Dragon	None None	Slight Slight	Slight Slight
Bunker	Dragon TOW	None None	None None	None Leaves and dust disturbed
Small frame	LAW Dragon	None Severe	Slight Severe	None None
Medium frame	LAW Dragon	None Slight	None Slight	Slight Lamp and chair overturned
Large frame	LAW Dragon TOW	None Slight Slight	Slight Moderate Severe	Slight None None

Table B-9. Structural Damage and Debris Movement

(5) To fire a LAW, AT4, or SMAW from inside a room, the following safety precautions must be taken (Figure B-5).

(a) The building should be of a sturdy construction.

(b) The ceiling should be at least 7 feet high with loose plaster or ceiling boards removed.

(c) The floor size should be at least 15 feet by 12 feet. (The larger the room, the better.)

(d) At least 20 square feet of ventilation (room openings) should exist to the rear or side of the weapon. An open 7- by 3-foot door would provide minimum ventilation.

(e) All glass should be removed from windows, and small, loose objects should be removed from the room.

(f) Floors should be wet to prevent dust and dirt from blowing around and obscuring the gunner's vision.

(g) All personnel in the room should be forward of the rear of the weapon.

(h) All personnel in the room should wear helmets, protective vests, ballistic eye protection, and earplugs.

(i) If the gunner is firing from the prone position, his lower body must be perpendicular to the bore of the weapon, or the blast could cause injury to his legs.

Figure B-5. Firing an AT4 From Inside a Building

c. Weapon Penetration. The most important tasks to be performed against structures are the neutralization of fortified firing positions, personnel, and weapons behind barriers. Recoilless weapons can be used in this role; none, however, are as effective as heavy, direct-fire weapons or standard demolitions. Each recoilless weapon has different penetrating ability against various targets. Penetration does not always mean the destruction of the integrity of a position. Usually only those enemy soldiers directly in the path of the spall from a high-explosive antitank (HEAT) round become casualties. Other soldiers inside a fortification could be deafened, dazed, or shocked but could eventually return to action.

(1) AT4. The AT4 is heavier than the LAW and has a diameter of 84 millimeters, which gives the warhead much greater penetration. The AT4 can penetrate more than 17.5 inches (450 mm) of armor plate. Its warhead produces highly destructive results behind the armor. Tests against typical urban targets are still ongoing, but the AT4 should penetrate at least as well as the 90-mm recoilless rifle, if not better. The AT4 has a *minimum arming distance of 10 meters*, which allows it to be fired successfully against close targets. Firers should be well covered and protected when firing at close targets.

(2) MK153 83-mm SMAW. The SMAW is a lightweight assault weapon that is easily carried and placed into action by one man. It is used against fortified positions, but it is also effective against LAVs. The SMAW has a 9-mm spotting rifle and a 3.8-power telescope that ensure accuracy over ranges common to combat in built-up areas. The SMAW has excellent incapacitating effects for enemies behind walls and inside bunkers and can arm within 10 meters. Its fuze has the ability to distinguish between armor and soft earth, maximizing its capabilities against buildings, bunkers, or light armor. The warhead detonates immediately against hard targets but delays detonation against soft targets and burrows in to explode inside. The SMAW can destroy most bunkers with a single hit. The 83-mm HE warhead can create an 8-inch mousehole in reinforced concrete but will not cut the steel reinforcing bar. The SMAW is an outstanding urban support weapon that can be decisive during an urban assault. The SMAW allows for quick reduction of obstacles and bunkers as well as creation of entry points. SMAWs fired in a volley can be devastating to a building.

d. Wall Breaching. Wall breaching is a common combat task in built-up areas for which light recoilless weapons can be used. Breaching operations improve mobility by providing access to building interiors without using existing doors or windows. Breaching techniques can also be used to create loopholes for weapons positions or to allow hand grenades to be thrown into defended structures. Breaching holes for unit mobility should be about 24 inches (60 centimeters) in diameter. Loopholes should be about 8 inches (20 centimeters) in diameter (Figure B-6). Only the SMAW provides a one-shot wall breaching ability. However, in breaching walls a number of shots should be planned.

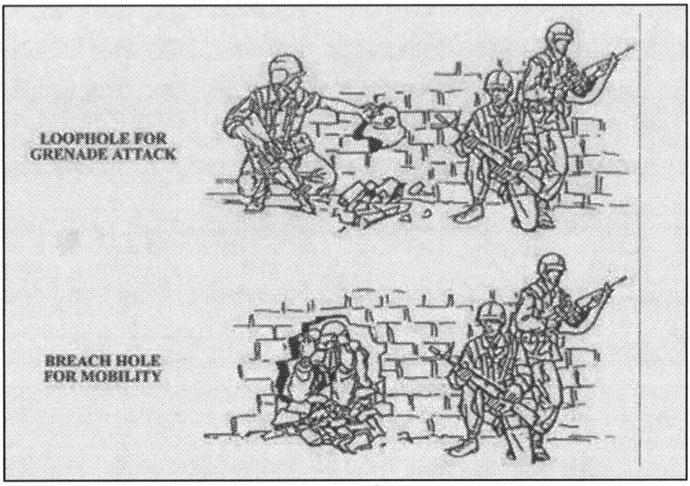

Figure B-6. Tactical Use of Holes in Masonry Walls

(1) Of all the common building materials, heavy stone is the most difficult to penetrate. The AT4 normally will not penetrate a heavy European-style stone wall. Surface cratering is usually the only effect it will have.

(2) Layered brick walls are also difficult to breach with light recoilless weapons. Some brick walls can be penetrated by multiple firings, especially if they are less than three bricks thick. Multiple AT4 rounds fired at the same spot on an 8-inch (double-brick) wall normally produce a loophole. The SMAW produces a hole in brick walls that is often large enough to be a breach hole.

(3) Wooden structural walls offer little resistance to light recoilless weapons. Even heavy timbered walls are penetrated and splintered. Multiple AT4 rounds fired at the same area of a wood-framed wall usually produce a man-sized hole. The SMAW has a devastating effect against wood-framed walls. A single round produces a breach hole as well as significant spall.

(4) Because of its high velocity, the AT4 may penetrate a soft target such as a car body or frame building before exploding.

(5) None of the light recoilless weapons are as effective against structural walls as demolitions or heavier weapons such as tank main guns or field artillery. Of all the light recoilless weapons, the SMAW is the most effective.

6. Antitank Guided Missiles. ATGMs are used mainly to defeat main battle tanks and other armored combat vehicles. They have a moderate capability against bunkers, buildings, and other fortified targets commonly found during combat in built-up areas. This category of weapons includes the TOW and Dragon missiles.

a. Employment. TOWs and Dragons provide overwatch antitank fires during the attack of a built-up area and an extended range capability for the engagement of armor during the defense. Within built-up areas, they are best employed along major thoroughfares and from the upper stories of buildings to attain long-range fields of fire. Their *minimum firing range of 65 meters* could limit firing opportunities in the confines of densely built-up areas.

(1) Obstacles. Rubble or other obstacles could interfere with the flight of missiles when fired from street level. At least 3.5 feet (1 meter) of vertical clearance over such obstacles must be maintained. Figure B-7 shows the most common obstacles to ATGM flight that are found in built-up areas. *Power lines are a special obstacle and present a unique threat to ATGM gunners. If the power in the lines has not been interrupted, the ATGM guidance wires could create a short circuit. This would allow extremely high voltage to pass to the gunner in the brief period before the guidance wires melted. This voltage could either damage the sight and guidance system or injure the gunner. Before any ATGM is fired over a power line, an attempt must be made to determine whether or not the power to that line has been interrupted.*

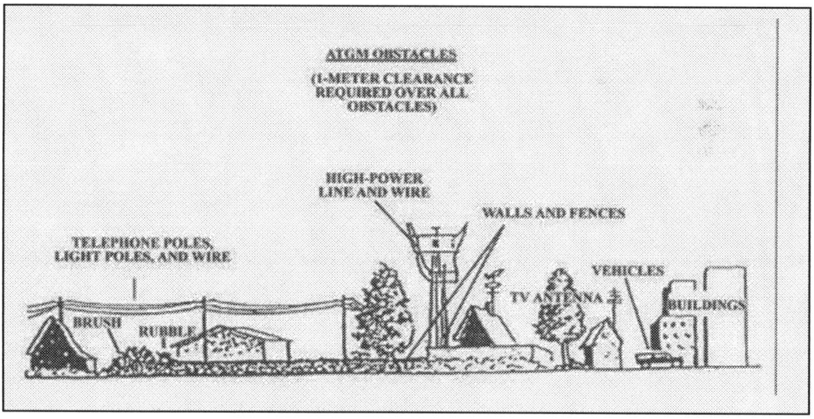

Figure B-7. Common Obstacles to ATGM Flights

(2) Dead Space. Three aspects of dead space that affect ATGM fires are arming distance, maximum depression, and maximum elevation.

(a) Both the Dragon and TOW missiles have a minimum arming distance of 65 meters, which severely limits their use in built-up areas. Few areas in the inner city permit fires much beyond the minimum arming distance. However, ground-level, long-range fires down streets or rail lines and across parks or plazas are possible. ATGMs may be used effectively from upper stories or roofs of buildings to fire into other buildings.

(b) The TOW is limited much more than the Dragon by its maximum depression and elevation. The maximum depression and elevation limits of the TOW mount could result in dead space and preclude the engagement of close targets (Figure B-8). A target located at the minimum arming range (65 meters) cannot be engaged by a TOW crew located any higher than the sixth floor of a building due to maximum depression limits. At 100 meters, the TOW crew can be located as high as the ninth floor and still engage the target.

(3) Backblast. Backblast for ATGMs is of more concern during combat in built-up areas than in open country. Any loose rubble in the caution zone could be picked up and thrown by the backblast, and the channeling effect of walls and narrow streets is even more pronounced due to the greater backblast. If the ATGM backblast strikes a wall at an angle, it can pick up debris or be deflected and cause injury to unprotected personnel (Figure B-9). Both types of ATGM can be fired from inside some buildings. In addition to the helmet and protective vest, eye protection and earplugs should be worn by all personnel in the room.

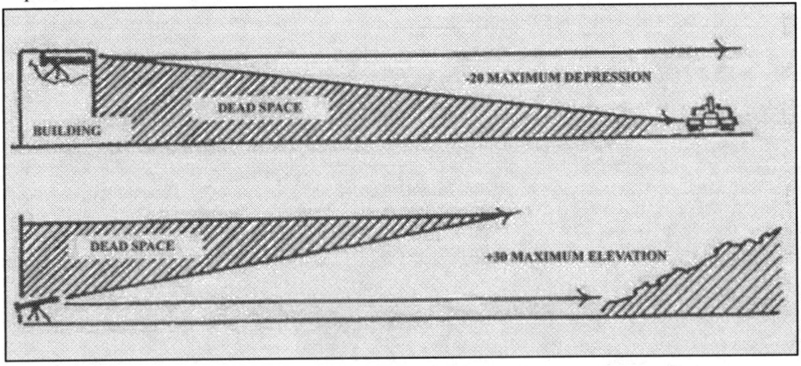

Figure B-8. TOW Maximum Elevation and Depression Limitations

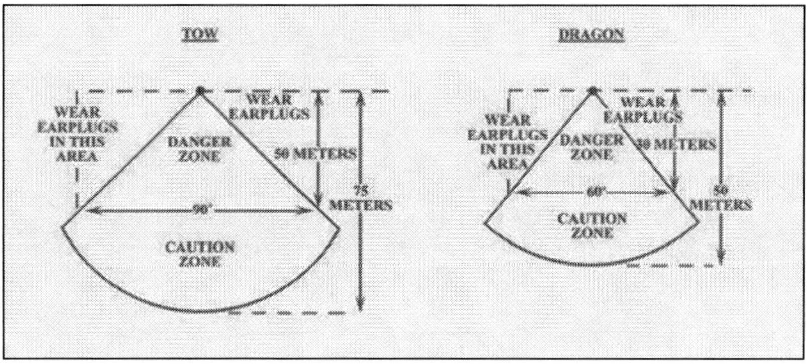

Figure B-9. ATGM Backblast in an Open Street

(a) To fire a TOW from inside a room, the following safety precautions must be taken (Figure B-10):

- The building must be of sturdy construction.
- The ceiling should be at least 7 feet high.
- The floor size of the room should be at least 15 by 15 feet, or larger if possible.
- At least 20 square feet of room ventilation should exist, preferably to the rear of the weapon. An open 7- by 3-foot door is sufficient. Additional ventilation can be created by removing sections of interior partitions.
- All glass must be removed from the windows and all small, loose objects removed from the room.
- All personnel in the room should be forward of the rear of the TOW.
- All personnel in the room should wear ballistic eye protection and earplugs.
- A clearance of 9 inches (23 centimeters) must exist between the launch tube and the aperture from which it is fired. (See FM 23-34, *TOW Weapon System*, for more detailed safety information.)

Figure B-10. TOW Fired From Inside a Room

(b) To fire a Dragon from inside a room, the following safety precautions must be taken.

- The building must be of sturdy construction.
- The ceiling should be at least 7 feet high.
- The floor size should be at least 15 by 15 feet, or larger if possible.
- At least 20 square feet of ventilation should exist (room openings), preferably to the rear of the weapon. An open 7- by 3-foot door would provide minimum ventilation.
- All glass should be removed from windows, and small, loose objects should be removed from the room.
- The room should be clean or the floors must be wet to prevent dust and dirt (kicked up by the backblast) from obscuring the vision of other soldiers in the room.
- All personnel in the room must be forward of the rear of the weapon.
- All personnel in the room must wear ballistic eye protection and earplugs.
- At least a 6-inch clearance must exist between the launch tube and the aperture from which it is fired.

b. Weapon Penetration. ATGMs can penetrate and destroy heavily armored tanks. They have large warheads that contain shaped charges. Because of their size, these warheads can achieve significant penetration against typical urban targets. Penetration, however, does not mean a concurrent destruction of the structural integrity of a position. The shaped-charge warhead produces relatively little spall. Enemy personnel not standing directly behind or near the point of impact of an ATGM may escape injury.

(1) Standard TOW Missiles. The basic TOW missile can penetrate 8 feet of packed earth, 4 feet of reinforced concrete, or 16 inches of steel plate. The improved TOW (ITOW), the TOW 2, and the TOW 2A have all been modified to improve their penetration. They all penetrate better than the basic TOW. All TOW missiles can defeat triple sandbag walls, double layers of earth-filled 55-gallon drums, and 18-inch log walls.

(2) TOW 2B. The TOW 2B uses a different method of defeating enemy armor. It flies over the target and fires an explosively formed penetrator down onto the top armor, which is thinner. *Because of this design feature, the TOW 2B missile cannot be used to attack nonmetallic structural targets. When using the TOW 2B missile against enemy armor, gunners must avoid firing directly over other friendly vehicles, disabled vehicles, or large metal objects such as water or oil tanks.*

(3) Dragon Missile. The Dragon missile can penetrate 8 feet of packed earth, 4 feet of concrete, or 13 inches of steel plate. It can attain effective short-range fire from upper stories or from the rear or flanks of a vehicle. These engagements are targeted against the most vulnerable parts of tanks and can entrap tanks in situations where they are unable to counterfire. Elevated firing positions increase the first-round hit probability. Firing down at an angle of 20 degrees increases the chance of a hit by 67 percent at 200 meters. A 45-degree downward angle doubles the first-round hit probability, compared to a ground-level shot.

c. Breaching Structural Walls. Firing ATGMs is the least efficient means to defeat structures. Because of their small basic load and high cost, ATGMs are better used against tanks or enemy-fortified firing positions. They can be effective against bunkers or other identified enemy firing positions.

7. Flame Weapons. Flame weapons are characterized by both physical and psychological casualty-producing abilities. Flame does not need to be applied with pinpoint accuracy, but it also must not spread to structures needed by friendly forces. Large fires in built-up areas are catastrophic. If they burn out of control, fires can create an impenetrable barrier for hours. The Marine Corps does not possess flame-generating equipment. However, the M34 WP grenade can be used to ignite and destroy flammable objects, especially wooden structures. The Army's M2A1-7 portable flamethrower is stored in war reserve status as a standard "C" item. Its availability is limited; however, Marines could be issued flame-support weapons for urban operations depending on the status of international agreements on the use of flame weapons that are in effect at the time.

a. Employment. Flame weapons used against fortified positions should be aimed directly at the aperture. Even if the round or burst misses, enough flaming material enters the position to cause casualties and disrupt the enemy occupants. The M34 WP grenade is difficult to throw far or into a small opening such as a bunker aperture. However, its effects are dramatic when thrown into a room or building.

b. Effects. The M34 is used to ignite and destroy flammable objects, especially wooden structures. It is also used to create an immediate smoke cloud to conceal movement across a narrow open space such as a street. Its smoke is not toxic but can cause choking in heavy concentrations.

(1) The grenade's explosion, bright flash, smoke, and burning WP particles all combine to make the M34 one of the most effective psychological weapons available. The M34 hand grenade throws WP fragments up to 35 meters from the point of detonation. These fragments can attach to clothing or skin and continue burning. Because of its weight, most infantrymen can throw this grenade only 30 to 40 meters.

(2) The Marine must avoid injury from friendly use of the M34. The M34 can ignite if the WP inside is exposed to the air. Bullets and shell fragments have been known to strike and rupture M34 grenades; therefore, grenades must be protected from enemy fire.

(3) The M34 WP grenade is an effective weapon against enemy armored vehicles when used in the close confines of combat in built-up areas. It can be thrown or dropped from upper stories onto enemy vehicles. The M34 can be combined with flammable liquids, detonating cords, blasting caps, and fuze igniters to create the eagle fireball, a field-expedient antiarmor device. (See FM 21-75, *Combat Skills of the Soldier*, Appendix H.)

(4) The M34 is also excellent as a screening device. A grenade can be thrown from behind cover into an open street or plaza. When it explodes, the enemy's observation is temporarily obscured. Thus, friendly forces can quickly cross the open area—if the enemy fires, it is unaimed and presents less of a danger. If screening smoke is used to cover a squad's movement across short open areas, it will reduce expected casualties from small-arms fire by about 90 percent.

8. Hand Grenades. Hand grenades are used extensively during combat in built-up areas. Smoke grenades are used for screening and signaling. Riot control grenades are used to drive the enemy out of deep fortifications. Fragmentation and concussion grenades are used to clear the enemy out of rooms and basements. Hand grenades are the most-used explosive munition during intense combat in built-up areas. In World War II, it was common for a battalion fighting in a city to use more than 500 hand grenades each day.

a. Employment. Smoke and riot-control grenades have similar employment techniques. Fragmentation grenades are used to produce enemy casualties.

(1) The AN-M8 HC grenade produces a dense white or grey smoke. It burns intensely and cannot be easily extinguished once it ignites. The smoke can be dangerous in heavy concentrations because it makes breathing difficult and causes choking. The M8 grenade is normally used for screening. It produces a slowly building screen of longer duration than the M34 WP grenade without the problem of collateral damage caused by scattered burning particles.

(2) The M18-series smoke grenades produce several different colors of smoke that are used for signaling. Yellow smoke is sometimes difficult to see in built-up areas.

(3) The M7A3 CS riot control grenade can be used to drive enemy troops out of fortifications when civilian casualties or collateral damage constraints are considerations. Built-up areas often create variable and shifting wind patterns. When using CS grenades, Marines must prevent the irritating smoke from affecting friendly forces. The CS grenade burns intensely and can ignite flammable structures. Enemy troops wearing even rudimentary chemical protective masks can withstand intense concentrations of CS gas.

(4) The MK3A2 offensive hand grenade, commonly referred to as the concussion grenade, produces casualties during close combat while minimizing the danger to friendly personnel. For this reason, it is the preferred hand grenade for use during offensive operations in a MOUT environment. The grenade produces severe concussion effects in enclosed areas. It can be used for light blasting and demolitions and for creating breach holes in interior walls. The concussion produced by the MK3A2 is much greater than that of the fragmentation grenade. It is very effective against enemy soldiers within bunkers, buildings, and underground passages.

(5) The M67 fragmentation grenade is the most commonly available grenade during combat in built-up areas. It provides suppression during room-to-room or house-to-house fighting, and it is used while clearing rooms of enemy personnel. The M67 grenade has a 3- to 5-second delay fuze. When used at close ranges, it can be cooked-off for two seconds to deny the enemy the time to throw it back. The fragmentation grenade can be rolled, bounced, or ricocheted into areas that cannot be reached by 40-mm grenade launchers.

b. Effects. Each type of hand grenade has its own specific effect during combat in built-up areas.

(1) Both HC smoke and CS gas *displace* oxygen from enclosed areas. Gas masks will not prevent this from occurring because they only filter air, not provide oxygen. Smoke grenades produce dense clouds of colored or white smoke that remain stationary in the surrounding area. They can also cause fires if used indiscriminately. If trapped and concentrated within a small space, smoke grenades and CS gas *can suffocate Marines.*

(2) The fragmentation grenade has more varied effects in combat in built-up areas. It produces a large amount of small, high-velocity fragments that can penetrate sheetrock partitions and are lethal at short ranges (15 to 20 meters). Fragments lose their velocity quickly and are less effective beyond 25 meters.

The fragments from a fragmentation grenade cannot penetrate a single layer of sandbags, a cinder block, or a brick building, but they can perforate wood-framed and tin buildings if exploded close to their walls.

(3) Inside rooms, fragmentation barriers consisting of common office furniture, mattresses, doors, or books can be effective against the fragmentation grenade. For this reason, a room should never be considered safe just because one or two grenades have been detonated inside. Fragmentation grenades detonated on the floor not only throw fragments laterally, but also send fragments and spall downward to lower floors. Predicting how much spall will occur is difficult because flooring material varies, but wooden floors are usually affected the most.

(4) Some foreign grenades throw fragments that are much larger than those of the U.S.-made M67. Light barriers and interior walls would probably be less effective against these grenades than against the M67. A major problem with the U.S.-made fragmentation grenade is its tendency to bounce back off of hard targets. Grenades are often directed at window openings on the ground floor or second floor. At ranges as close as 20 meters, a thrower's chances of missing a standard 1-meter by 1-meter window are high. The fragmentation grenade normally breaks through standard window glass and enters a room. If the grenade strikes at a sharp angle or the glass is thick plate, the grenade could be deflected without penetrating.

(5) Hand grenades are difficult weapons to use. They involve a high risk of fratricide. Commanders should conduct precombat training with hand grenades as part of normal preparations. Marines must be very careful when throwing hand grenades up stairs to avoid having the grenade roll back down on them.

(6) Once pulled, the pins of both fragmentation and concussion grenades can be replaced if the thrower decides not to use the weapon. This pin replacement must be done carefully (see FM 23-30, *Grenades and Pyrotechnic Signals*) and prior to the spoon being released or detaching.

(7) An estimate of the situation (METT-T) and existing ROE will dictate what type of grenade will be used to clear each room. Because of the high rate of expenditure of grenades, units should use assault packs to carry additional grenades of all types. Additional grenades can also be carried in empty ammunition or canteen pouches.

9. 25-mm Automatic Gun. The 25-mm automatic gun mounted on the LAV-25 is an effective weapon during combat in built-up areas. The primary roles of LAV-25s during combat in built-up areas are to provide suppressive fire and to breach exterior walls and fortifications.

a. Obliquity. The 25-mm gun produces its best urban target results when fired perpendicular to the hard surface (zero obliquity). In combat in built-up areas, however, finding a covered firing position that permits low-obliquity firing is unlikely unless the streets and gaps between buildings are wide. Most shots impact the target at an angle, which normally reduces penetration. With the armor-piercing, discarding, sabot with tracer (APDS-T) round, an angle of obliquity of up to 20 degrees can actually improve breaching. The rounds tend to dislodge more wall material for each shot but do not penetrate as deeply into the structure.

b. Target Types. The 25-mm gun has different effects when fired against different urban targets.

(1) Reinforced Concrete. Reinforced concrete walls, which are 12 to 20 inches thick, present problems for the 25-mm gun when trying to create breach holes. It is relatively easy to penetrate, fracture, and clear away the concrete, but the reinforcing rods remain in place. These create a "jail window" effect by preventing entry but allowing grenades or rifle fire to be placed behind the wall. Steel reinforcing rods are normally 3/4 of an inch thick and 6 to 8 inches apart—there is no quick way of cutting these rods. They can be cut with demolition charges, cutting torches, or special power saws. Firing with either APDS-T or high-explosive, incendiary with tracer (HEI-T) rounds from the 25-mm gun will not always cut these rods.

(2) Brick Walls. Brick walls are more easily defeated by the 25-mm gun regardless of their thickness, and they produce the most spall.

(3) Bunker Walls. The 25-mm gun is devastating when fired against sandbag bunker walls. Obliquity has the least effect on the penetration of bunker walls. Bunkers with earth walls up to 36 inches thick are easily penetrated. At short ranges typical of combat in built-up areas, defeating a bunker should be easy, especially if the 25-mm gun can fire at an aperture.

c. Burst Fire. The 25-mm gun's impact on typical urban targets seems to be magnified if the firing is in short bursts. At close ranges, the gunner might need to shift his point of aim in a spiral pattern to ensure that the second and third bursts enlarge the hole. Even without burst fire, sustained 25-mm gunfire can defeat almost all urban targets.

d. Weapon Penetration. The penetration achieved by the two combat rounds (HEI-T and APDS-T) differs slightly, but both are effective.

(1) APDS-T. The APDS-T round penetrates urban targets by retaining its kinetic energy and blasting a small hole deep into the target. The APDS-T round gives the best effects behind the wall, and the armor-piercing core often breaks into two or three fragments that can create multiple enemy casualties. The APDS-T needs as few as four rounds to achieve lethal results behind walls. Table B-10 lists the number of APDS-T rounds needed to create different-sized holes in common urban walls.

Target	Loophole	Breach Hole
3-inch brick wall at 0-degree obliquity	22 rounds	75 rounds
3-inch brick wall at 45-degree obliquity	22 rounds	35[1] rounds
5-inch brick wall at 0-degree obliquity	32 rounds	50[1] rounds
8-inch reinforced concrete at 0-degree obliquity	22 rounds	75 rounds (Note: Reinforcing rods still in place)
8-inch reinforced concrete at 45-degree obliquity	22 rounds	40[1] rounds (Note: Reinforcing rods still in place)

[1] Obliquity and depth tend to increase the amount of wall material removed.

Table B-10. Breaching Effects of APDS-T Rounds

(a) When firing single rounds, the APDS-T round provides the greatest capability for behind-the-wall incapacitation. The APDS-T round can penetrate more than 16 inches of reinforced concrete with enough energy left to cause enemy casualties. It penetrates through both sides of a wood-framed or brick veneer building. Field fortifications are easily penetrated by APDS-T rounds. Table B-11 lists the number of APDS-T rounds needed to create different-sized holes in commonly found bunkers.

(b) The APDS-T round creates a hazardous situation for exposed personnel because of the pieces of sabot that are thrown off of the round.

Type Bunker	Obliquity	Penetration	Loophole	Small Breach Hole
36-inch sand/timber	0 degree	1 round	25 rounds	40 rounds
36-inch sand/6-inch concrete	0 degree	6 rounds	6 rounds	20 rounds

Table B-11. Number of APDS-T Rounds Needed To Create Different-Sized Holes in Bunkers

WARNING

Personnel not under cover forward of the 25-mm gun's muzzle and within the danger zone could be injured or killed by these sabots, even if the penetrator passes overhead to hit the target.

The danger zone extends at an angle of about 10 degrees below the muzzle level, out to at least 100 meters, and about 17 degrees left and right of the muzzle. Figure B-11 shows the hazard area of the APDS-T round.

Figure B-11. APDS-T Danger Zone

(2) HEI-T. The HEI-T round penetrates urban targets by blasting away chunks of material.

(a) The HEI-T round does not penetrate an urban target as well as the APDS-T, but it does strip away a greater amount of material for each round. The HEI-T does more damage to an urban target when fired in multiple short bursts because the cumulative impact of multiple rounds is greater than the sum of individual rounds. Table B-12 lists the number of HEI-T rounds needed to create different-sized holes.

Target	Loophole	Breach Hole
3-inch brick wall at 0-degree obliquity	10 rounds	20 rounds
3-inch brick wall at 45-degree obliquity	20 rounds	25 rounds
5-inch brick wall at 0-degree obliquity	30 rounds	60 rounds

8-inch reinforced concrete at 0-degree obliquity	15 rounds	25 rounds
8-inch reinforced concrete at 45-degree obliquity	15 rounds	30 rounds

Table B-12. Number of HEI-T Rounds Needed To Create Different-Sized Holes in Bunkers

(b) The HEI-T round does not provide single-round perforation or incapacitating fragments when used against any external masonry structural wall. It can create first-round fragments behind wood-framed and brick veneer walls. HEI-T rounds cannot penetrate a bunker as quickly as APDS-T rounds, but they can create more damage inside the bunker once the external earth has been stripped away. Against a heavy bunker, about 40 rounds of HEI-T ammunition are needed to strip away the external earth shielding and breach the inner lining of concrete or timber. The HEI-T round is also used for suppression against known or suspected firing ports such as doors, windows, and loopholes.

10. Tank Cannon. The powerful, high-velocity cannon mounted on the M1A1 tank provides Marines with a key requirement for victory in built-up areas—heavy direct-fire support. Although the infantry assumes the lead role during combat in built-up areas, tanks and infantry work as a close team. Tanks move down streets after the infantry has cleared them of any suspected ATGM positions and, in turn, support the infantry with fire. The tank is one of the most effective weapons for heavy fire against structures. The primary role of the tank cannon during combat in built-up areas is to provide heavy direct fire against buildings and strongpoints that are identified as targets by the infantry. The wall and fortification breaching effects of the 120-mm tank cannon are major assets to Marines fighting in built-up areas.

a. Obliquity. The tank cannon produces its best urban target effects when fired perpendicular to the hard surface (zero obliquity). During combat in built-up areas, however, finding a covered firing position that permits low-obliquity firing is unlikely. Most shots strike the target at an angle that would normally reduce penetration. With tank cannon APDS rounds, obliquity angles of up to 25 degrees have little effect, but angles greater than 45 degrees greatly reduce penetration.

b. Ammunition. Armor-piercing, fin-stabilized, discarding sabot (APFSDS) rounds are the most commonly carried tank ammunition. These rounds work best against armored vehicles. The 120-mm cannon also carries an effective high-explosive, antitank multipurpose (HEAT-MP) round.

c. Characteristics. The 120-mm tank cannon has two specific characteristics that affect its employment in built-up areas: limited elevation and depression and short arming ranges. In addition, the M1A1 tank has another characteristic that is not involved with its cannon but that affects Marines working with or around the tank—*extremely hot turbine exhaust.*

(1) The M1A1 tank can elevate its cannon 20 degrees and depress it 10 degrees. The lower depression limit creates a 35-foot (10.8-meter) dead space around a tank. On a street 16 meters wide, this dead space extends to the buildings on each side (Figure B-12). Similarly, there is a zone overhead in which the tank cannot fire (Figure B-13). This dead space offers ideal locations for short-range antiarmor weapons and allows hidden enemy gunners to fire at the tank when the tank cannot fire back. It also exposes the tank's most vulnerable areas: the flanks, rear, and top. Infantry must move ahead, alongside, and to the rear of tanks to provide close protection. The extreme heat produced immediately to the rear of the M1A1 tank prevents dismounted infantry from following closely, but protection from small-arms fire and fragments is still provided by the tank's bulk and armor. The M1A1 tank also has a blind spot caused by the zero-degree depression available over part of the back deck. To engage any target in this area, the tank must pivot to convert the rear target to a flank target.

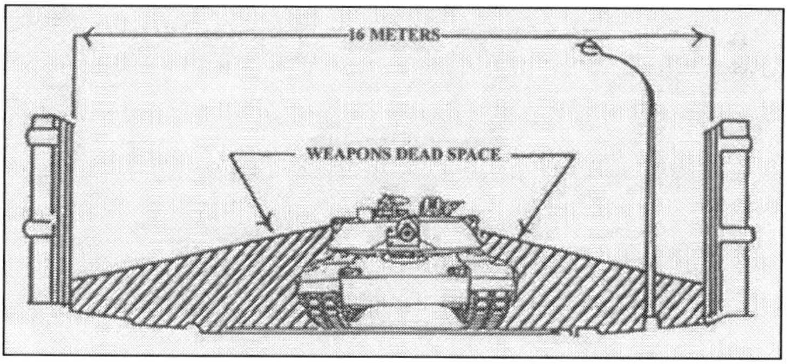

Figure B-12. Tank Cannon Dead Space at Street Level

(2) The 120-mm HEAT-MP round arms at about 36 feet. This arming distance allows the tank to engage targets from short ranges. The armor of the tank protects the crew from both the blowback effects of the round and enemy return fire. The APFSDS round does not need to arm and can, therefore, be fired at almost any range. The discarding portions of the round can be lethal to exposed infantry forward of the tank.

d. Target Effects. HEAT rounds are most effective against masonry walls. The APFSDS round can penetrate deeply into a structure but does not create as large a hole or displace as much spall behind the target. In contrast to lighter HEAT rounds, tank HEAT rounds are large enough to displace enough spall to inflict casualties inside a building. One HEAT round normally creates a breach hole in all but the thickest masonry construction—brick veneer and wood-framed constructions are demolished by a single round. Even the 120-mm HEAT round cannot cut all of the reinforcing rods, which are usually left in place, often hindering entry through the breach hole (Figure B-14). Both HEAT and APFSDS rounds are effective against all field fortifications. Only large earth berms and heavy mass-construction buildings can provide protection against tank fire.

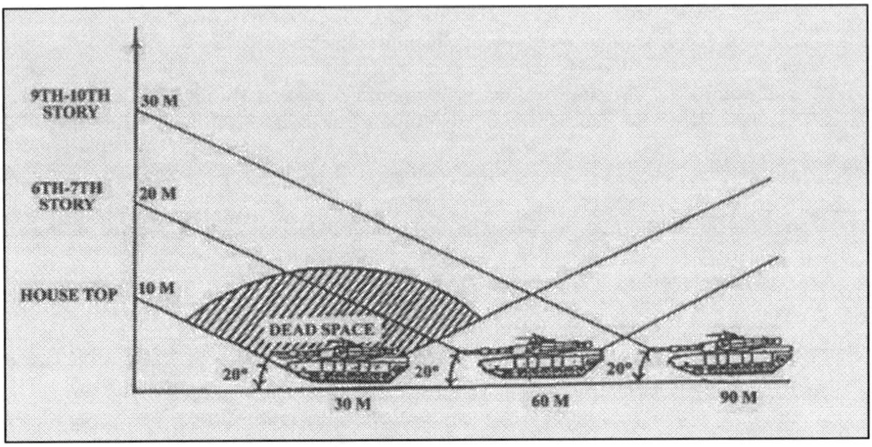

Figure B-13. Tank Cannon Dead Space Above Street Level

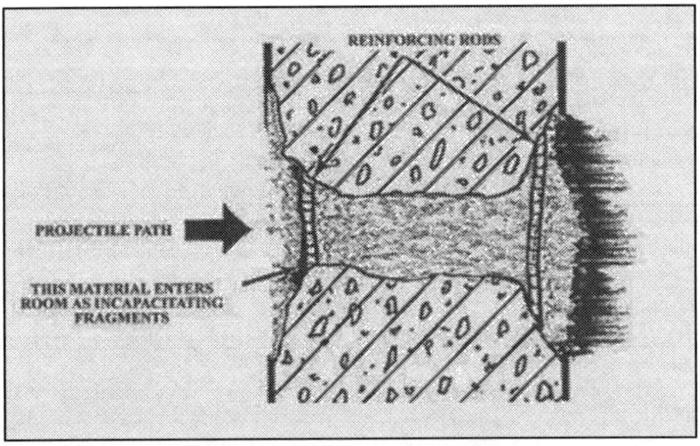

Figure B-14. Tank HEAT Round Effects on Reinforced Concrete Walls

e. Employment. Tank-heavy forces could be at a severe disadvantage during combat in built-up areas, but a few tanks working with the infantry can be most effective, especially if they work well together at the small-unit level. Tank, infantry, and engineer task forces are normally formed to attack a fortified area. The tank platoon or section can work together with rifle squads or platoons.

(1) Tanks need infantry on the ground to provide security in built-up areas and to designate targets. Against targets protected by structures, tanks should be escorted forward to the most covered location that provides a clear shot. On-the-spot instructions by infantry leaders ensure that the tank's fire is accurate and its exposure is limited. The tank commander may have to halt in a covered position, dismount, and reconnoiter his route forward into a firing position.

(2) When the tank main gun fires, it creates a large fireball and smoke cloud. In the confines of a built-up area, dirt and masonry dust are also picked up and added to this cloud. The target is further obscured by the smoke and dust of the explosion. Depending on the local conditions, this obscuration could last as long as two or three minutes. Marines can use this period to reposition or advance unseen by the enemy. Caution must be exercised, however, because the enemy might also move.

(3) A tank cannon creates an overpressure and noise hazard to exposed Marines. All dismounted Marines working near tanks should wear Kevlar helmets and protective vests, as well as ballistic eye protection. If possible, they should also wear earplugs and avoid the tank's frontal 60-degree arc during firing (Figure B-15).

WARNING

- **The overpressure from the tank 120-mm cannon can kill a marine found within a 90-degree arc extending from the muzzle of the gun tube out to 200 meters.**

- **From 200 to 1,000 meters along the line of fire, on a frontage of about 400 meters, dismounted Marines must be aware of the danger from discarding sabot petals, which can kill or seriously injure personnel.**

- **Personnel outside the tank should remain at least 50 meters from the tank in all directions as they may receive damaging effects from firing noise and overpressure.**

- **Personnel must also wear hearing protection when operating within 704 meters of a tank that is firing its main gun.**

TM 9-2350-264-10-2, *Operator's Manual for Operation Under Usual and Unusual Conditions for Tank, Combat, Full-Tracked: 120mm Gun, M1A1 General Abrams*, with Change 4

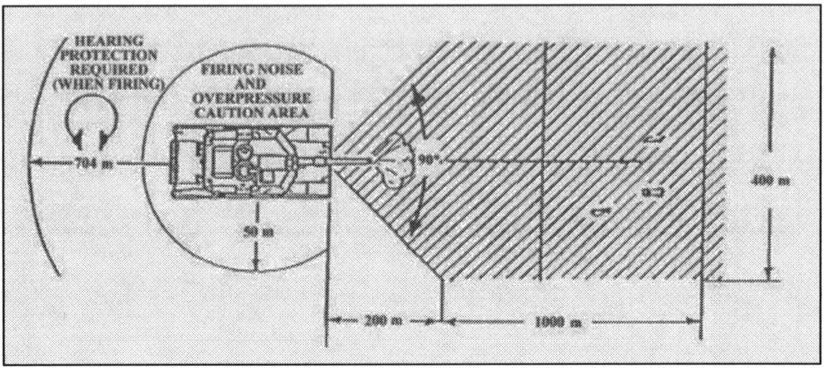

Figure B-15. Danger Areas Around a Tank Firing a 120-mm Gun

(4) Tanks are equipped with powerful thermal sights that can be used to detect enemy personnel and weapons that are hidden in shadows and behind openings. Dust, fires, and thick smoke significantly degrade these sights.

(5) Tanks have turret-mounted grenade launchers that project screening smoke grenades. The grenades use a bursting charge and burning RP particles to create this screen. *Burning particles can easily start uncontrolled fires and are hazardous to dismounted Marines near the tank.* The tank commander and the infantry small-unit leader must coordinate when and under what conditions these launchers can be used. Grenade launchers are a useful feature for protecting the tank but can cause significant problems if unwisely used.

(6) The tank's size and armor can provide dismounted Marines with cover from direct-fire weapons and fragments. With coordination, tanks can provide moving cover for Marines as they advance across small open areas. However, enemy fire that strikes but does not penetrate a tank is a major threat to nearby Marines. Fragmentation generated by antitank rounds and ricochets off of tank armor have historically been a prime cause of casualties while infantry was working with tanks in built-up areas.

11. Artillery. A major source of fire support for infantry forces fighting in built-up areas is field artillery weapons. If the built-up area is near the coast, NGF can be used. Field artillery employment can be in either the indirect- or direct-fire mode.

a. Indirect Fire. Low-angle indirect artillery fire is not effective in attacking targets within walls and masonry structures. It tends to impact on roofs or upper stories rather than on structurally critical wall areas or pillars. High-angle fire can achieve the trajectories necessary to strike targets in urban areas.

Weapons of at least 155 mm are required against thick reinforced concrete, stone, or brick walls. Even with heavy artillery, large expenditures of ammunition are required to knock down buildings of any size. Tall buildings also create areas of indirect-fire dead space, which are areas that cannot be engaged by indirect fire because of a combination of building height and angle of fall of the projectile (Figure B-16). Usually the dead space for low-angle indirect fire is about five times the height of the highest building over which the rounds must pass.

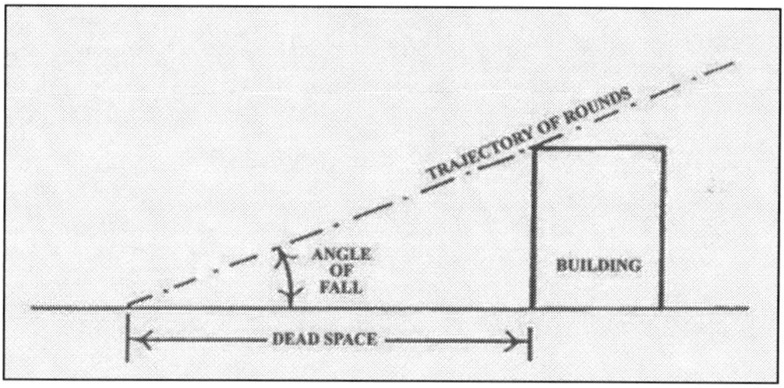

Figure B-16. Indirect-Fire Dead Space (Low Angle)

b. Direct Fire. A likely use of Marine artillery in an urban direct-fire role is to reinforce infantry against tough or important urban targets. Artillery should be used in this role only after an analysis of the need for heavy direct fire and the tradeoff involved in the extreme decentralization of artillery firepower. Artillery has the same need for close security and target designation as tanks.

c. Target Effects. Medium-caliber (155-mm) direct fire has a devastating effect against masonry construction and field fortifications. Smaller artillery (105-mm) pieces in the direct-fire mode are much less destructive than the larger caliber weapons. However, all Marine artillery is towed and, therefore, more difficult to employ in the direct-fire mode.

The 155-mm (M198) towed howitzer is effective because of its rate of fire and penetration. HE rounds can penetrate up to 38 inches of brick and unreinforced concrete. Projectiles can penetrate up to 28 inches of reinforced concrete with considerable damage beyond the wall. HE rounds with concrete-piercing fuzes provide an excellent means of penetrating strong reinforced concrete structures. One round can penetrate up to 46 inches. Five rounds are needed to reliably create a 1.5-meter breach in a wall that is 1 meter thick. About 10 rounds are needed to create the same size breach in a wall that is 1.5 meters thick. Superquick fuzing causes the rubble to be blown into the building, whereas delay fuzing tends to blow the rubble outward into the street. (See Table B-13 for employment considerations for artillery.)

Organization for Combat	Movement/Positioning	Delivery of Fire	Security	Command and Control
Centralized control is required during initial phases; decentralized control is required during later phases to support semi-independent actions of small units.	Movement should occur during night or periods of reduced visibility when possible. There are few displacements, often by platoon or section. Positions should be selected that minimize masking, provide several routes of escape, and afford as much cover and concealment as possible. Use of existing structures (garages, office buildings, highway overpasses) is recommended. Special techniques for emplacing howitzers, such as spades against a curb when the ground is not suitable for emplacement, may be required. Explosives may be required to soften emplacement of howitzers. Reconnaissance, selection, and occupation of position (RSOP) elements should be well armed because they may have to clear areas to be occupied. Extensive route reconnaissance is required. Target acquisition devices are somewhat degraded. Radars should be emplaced to cover likely areas of enemy indirect-fire weapon employment. Radars should not be placed in the midst of an urban area because of masking.	Both direct and indirect fires are delivered for supported units. Destruction of fortifications may require assault fire techniques. High-angle fires may be required. Need for accurate meteorological (MET) and survey data increases because most targets are point targets. Improved conventional munition and variable time (fuze) effects are reduced by structures, although they are effective against personnel on rooftops. HE delay is used for penetration effects. Illumination, chemical incendiary ammunition, and smoke are effective. Ammunition expenditures will be heavy. Lasers and PGMs permit destruction of targets with minimal rubbling of adjacent buildings. Tall buildings may hamper laser use. Batteries must be prepared for hasty survey techniques. Magnetic instruments are impaired.	Positions must be fortified.	Radio communications are impaired by buildings. Wire can usually be run overhead. Make use of civilian communications. A greater use of messengers and prearranged audio and visual signals is required.

Table B-13. Artillery Employment Considerations in Built-Up Areas

12. Mortars. The urban environment greatly restricts low-angle indirect fires because of overhead masking. While all indirect-fire weapons are subject to overhead masking, mortars are less affected than field artillery weapons because of the mortar's higher trajectory. For low-angle artillery fire, dead space is about five times the height of the building behind which the target sits. For mortar fire, dead space is only about one-half the height of the building. Because of these advantages, mortars are even more important than field artillery to the infantry during combat in built-up areas.

a. Employment. Not only can mortars fire into the deep defilade created by tall buildings, but they can also fire out of it. Mortars emplaced behind buildings are difficult for the enemy to accurately locate and even harder to hit with counterfire. Because of their light weight, even heavy mortars can be hand-carried to firing positions that may not be accessible to vehicles.

(1) The 60-mm and 81-mm mortars have limited effects on structural targets. Even with delay fuzes, they seldom penetrate more than the upper stories of light buildings. However, their wide area coverage and multioption fuzes make them useful against an enemy force advancing through streets, through other open areas, or over rubble.

(2) Mortar platoons often operate as separate firing sections during combat in built-up areas. The lack of large open areas can preclude establishing a platoon firing position. Figure B-17 shows how two mortar sections that are separated by only one street can be effective in massing fires and be protected from countermortar fire by employing defilade and dispersion.

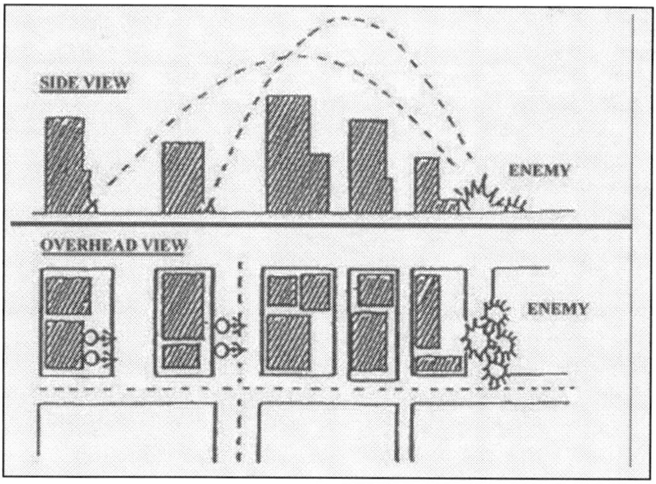

Figure B-17. Split-Section Mortar Operations on Adjacent Streets

(3) All three of the standard mortar projectiles are useful during combat in built-up areas. HE fragmentation rounds are most commonly used. WP is effective for starting fires in buildings and forcing the enemy out of cellars and light-framed buildings, and it is also the most effective mortar round against dug-in enemy tanks. Even near-misses blind and suppress the tank crew, forcing them to button up. Hits are difficult to achieve but are effective.

(4) Because the artificial roughness of urbanized terrain reduces wind speed and increases atmosphere mixing, mortar smoke tends to persist longer and give greater coverage in built-up areas than in open terrain.

(5) Urban masking has an impact on the use of illumination. In built-up areas, it is often necessary to plan illumination behind friendly positions, which places friendly troops in shadows and enemy troops in the light. Illumination rounds are difficult to adjust and are often of limited use because of the deep canyon nature of the urbanized area. Rapidly shifting wind currents in built-up areas also affect mortar illumination, making it less effective.

b. Effects of Mortar Fire. The multioption fuze on newer U.S. mortar rounds makes them effective weapons on urbanized terrain. Delay settings can increase penetration slightly, and proximity bursts can increase the lethal area covered by fragments. Tall buildings can cause proximity-fuzed mortar rounds to detonate prematurely if they pass too closely to the buildings.

(1) 60-mm Mortar. The 60-mm mortar round cannot penetrate most rooftops, even with a delay setting. Small explosive rounds are effective, however, in suppressing snipers on rooftops and preventing roofs from being used by enemy observers. The 60-mm WP round is not normally a good screening round because of its small area of coverage. In combat in built-up areas, however, the tendency of smoke to linger and the small areas to be screened make it more effective. During the battle for Hue in South Vietnam, 60-mm WP rounds were used to create small, short-term smoke screens to conceal movement across open areas such as parks, plazas, and bridges. Fragments from 60-mm HE rounds landing as close as 10 feet cannot penetrate a single sandbag layer or a single-layer brick wall. The effect of a 60-mm mortar HE round that achieves a direct hit on a bunker or fighting position is equivalent to 1 - 2 pounds of TNT. Normally, the blast will not collapse a properly constructed bunker, but it can cause structural damage. The 60-mm mortar will not normally create craters in a hard-surfaced road.

(2) 81-mm Mortar. The 81-mm mortar has much the same effect against urban targets as the 60-mm mortar. It has a slightly greater lethal area, and its smoke rounds (WP and RP) are more effective. A direct hit is equivalent to about 2 pounds of TNT. The 81-mm round cannot cause significant craters in a hard-surfaced road. With a delay setting, the 81-mm round can penetrate the roofs of light buildings.

13. Naval Gunfire. The primary NGF weapon, the 5-inch/54, has a high rate of fire and is roughly equivalent to the 155-mm howitzer in target effect. Target engagement considerations are the same as for artillery; however, the 5-inch/54's flat trajectory is even more affected by terrain masking and a constantly changing gun-target line. NGF can achieve greater target engagement by firing reduced charges at high angles; however, range will be reduced.

14. Aerial Weapons. Both rotary- and fixed-wing aircraft can quickly deliver large volumes of firepower over large built-up areas. Specific targets are hard to distinguish from the air. Good ground-to-air communications are vital in successfully employing aerial firepower. Planners have historically tended to overestimate the effects of HE on defenders. Modern, large buildings are remarkably resistant to damage from bombs and rocket fire.

a. Rotary-Wing Aircraft. Attack helicopters can be used to engage targets in built-up areas. Enemy armored vehicles in small parks, boulevards, or other open areas are good targets for attack helicopters.

(1) The HELLFIRE missile has a larger warhead and greater range than the TOW, but it too is a shaped-charge warhead and is not specifically designed for use against masonry targets. Laser target designation for the HELLFIRE may be difficult as a result of laser reflections off of glass and shiny metal surfaces. The use of attack helicopters to deliver ATGMs against targets in the upper stories of high buildings is sometimes desirable.

(2) The 2.75-inch folding-fin aerial rocket and the 20-mm cannon on Marine attack helicopters are good area weapons to use against enemy forces in the open or under light cover. They are usually ineffective against a large masonry target. The 20-mm cannon produces many ricochets, especially if armor-piercing ammunition is fired into built-up areas.

b. Fixed-Wing Aircraft. Close air support to ground forces fighting in built-up areas is a difficult mission for fixed-wing aircraft. Targets are hard to locate and identify, enemy and friendly forces could be intermingled, and enemy short-range air defense weapons are hard to suppress.

(1) Because enemy and friendly forces may be separated by only one building, accurate delivery of ordnance is critical. Marking panels, lights, electronic beacons, smoke, or some other positive identification of friendly forces is needed.

(2) General-purpose bombs that weigh from 500 to 2,000 pounds are moderately effective in creating casualties among enemy troops located in large buildings. High-dive-angle bomb runs increase accuracy and penetration but also increase the aircraft's exposure to antiaircraft weapons. Low-dive-angle bomb runs using high-drag (retarded) bombs can be used to get bombs into upper stories. Penetration is not good with high-drag bombs. Sometimes aerial bombs pass completely through light-clad buildings and explode on the outside.

(3) Aerial rockets and 20-mm cannons are only moderately effective against enemy soldiers in built-up areas because rockets lack the accuracy to concentrate their effects. The 20-mm cannon rounds penetrate only slightly better than the .50-caliber round, 20-mm armor-piercing rounds can ricochet badly, and tracers can start fires.

(4) The AC-130 aircraft has weapons that can be most effective during combat in built-up areas. This aircraft can deliver accurate fire from a 20-mm Vulcan cannon, 40-mm rapid-fire cannon, and 105-mm howitzer. The 105-mm howitzer round is effective against the roofs and upper floors of buildings. The AC-130 is accurate enough to concentrate its 40-mm cannon and 105-mm howitzer fire onto a single spot to create a rooftop breach that allows fires to be directed deep into the building.

(5) Laser and optically guided munitions can be effective against high-value targets. Specially developed, heavy, laser-guided bombs can penetrate hardened targets. However, problems associated with dense smoke and dust clouds hanging over the built-up area and laser scatter can restrict their use. If the launching aircraft can achieve a successful laser designation and lock-on, these weapons have devastating effects, penetrating deep into reinforced concrete before exploding with great force. If launched without a lock-on, or if the laser spot is lost, these weapons are unpredictable and can travel long distances before they impact.

15. Demolitions. Combat in built-up areas requires the extensive use of demolitions. All Marines, not just engineers, should be trained to employ demolitions. See FM 5-250, *Explosives and Demolitions*, for specific information on the safe use of demolitions.

a. Demolitions. Demolitions come in two types: TNT and C4. Exposed Marines must take cover or move at least 300 meters away from bulk explosives that are being used to breach walls.

(1) TNT comes in 1/4-, 1/2-, and 1-pound blocks. About 5 pounds of TNT are needed to breach a nonreinforced concrete wall that is 12 inches thick if the explosives are laid next to the wall and are not tamped. If the explosives are tamped, about 2 pounds are sufficient.

(2) C4 comes in many different-sized blocks. Normally, it is found in 1 1/4- and 2 1/2-pound blocks within prepackaged satchel charges. About 10 pounds of C4 placed between waist and chest high will blow a hole in the average masonry wall large enough for a man to walk through.

b. Shaped Charges. There are two sizes of shaped charges: a 15-pound M2A3 and a 40-pound M3A3. The M3A3 is the shaped charge that is most likely to be used in built-up areas. It can penetrate 5 feet of reinforced concrete. The hole tapers from 5 inches down to 2 inches. The amount of spall thrown behind the target wall is considerable. There is also a large safety hazard area for friendly forces.

c. Satchel Charges. There are two standard satchel charges: the M183 and the M37. Both come in their own carrying satchel with detonators and blasting cords. Each weighs 20 pounds. The M183 has 16 individual 1 1/4-pound blocks that can be used separately. When used untamped, a satchel breaches a 3-foot-thick concrete wall. Satchel charges are very powerful. Debris is thrown great distances. Friendly forces must move away and take cover before detonation.

d. Cratering Charges. The standard cratering charge is a 43-pound cylinder of ammonium nitrate. This explosive does not have the shattering effect of bulk TNT or C4. It is more useful in deliberate demolitions than in hasty ones.

Appendix C

Attacking and Clearing Buildings

Offensive operations on urbanized terrain will require Marines to attack and clear buildings.

1. **Introduction.** At the rifle platoon and squad level, the major offensive tasks for combat in built-up areas are attacking and clearing buildings. This involves:

- Suppressing defensive fires
- Advancing Marine assault forces
- Assaulting buildings
- Clearing buildings
- Reorganizing assault forces.

To accomplish these tasks, the assaulting unit task organizes into assault, security, and support forces or elements. (See Appendix A)

2. **Requirements.** There are five interrelated requirements for attacking a defended building: *fire support, movement, assault, clearing,* and *reorganization.* Proper application and integration of these requirements reduce casualties and hasten accomplishment of the mission. The proper integration of these requirements are determined by METT-T. For example, larger cities may have high-rise apartment buildings and industrial complexes that are separated by parking areas or parks. Increased fire support may be required to suppress and obscure enemy gunners covering the open terrain and spaces between buildings. Conversely, a built-up area with few open areas can provide numerous covered routes that may decrease fire support requirements.

3. **Fire Support.** Fire support necessary to advance the assault force may include:

- Suppressing and obscuring enemy gunners within the objective building(s) and adjacent structures
- Isolating the objective building(s) with fires to prevent enemy withdrawal, reinforcement, or counterattack
- Breaching walls en route to and in the objective structure
- Destroying enemy positions with direct-fire weapons

- Securing cleared portions of the objective
- Providing replacements for the assault force
- Providing resupply of ammunition and explosives
- Evacuating casualties and prisoners.

a. The size of the support force is determined by the type and size of the objective building(s), whether the adjacent terrain provides open or covered approaches, and the organization and strength of enemy defenses.

b. The support force for a Marine rifle squad may consist of one fire team that could be reinforced with heavy machine guns, antitank weapons, and/or armored vehicles. In situations involving a larger assault force, a rifle platoon or company reinforced with armored vehicles, engineers, and/or artillery may be required to support the movement and assault of an adjacent platoon or company.

c. After seizing objective buildings, the assault force reorganizes and may be required to provide supporting fires for a subsequent assault. Each weapon is assigned a target or area to cover. Individual small-arms weapons place fires on likely enemy weapon positions—loopholes, windows, and roof areas. Snipers are best employed in placing accurate fire through loopholes or engaging long-range targets. M203 fires are directed through windows or loopholes.

d. AT4s and demolitions are employed to breach walls, doors, barricades, and window barriers on the ground level of structures. The tank cannon or LAV-mounted 25-mm gun engages first-floor targets and breaches walls for attacking Marines. Armored vehicle- or wheeled vehicle-mounted machine guns can engage suspected enemy positions on upper floors and in adjacent structures. In addition to destroying or weakening structures, the tank's main gun rounds cause casualties by their explosive effects and by hurling debris throughout the interior of structures.

e. NGF, artillery, and mortars should initially use time fuzes to initially clear exposed enemy personnel, weapons, OPs, and radio sites from rooftops. They should then use delayed fuze action to cause casualties among the defenders inside the structure as a result of shrapnel and falling debris. They may also provide smoke to cover the movement and assault phases of the attack. Artillery can also be used in the direct-fire mode.

f. In support of the movement and assault phases of the attack, buildings can be attacked by fixed-wing aircraft or by attack helicopters.

4. Movement. The assault force minimizes the effects of enemy defensive fires during movement by:

- Using covered routes
- Moving after defensive fires have been suppressed or obscured
- Moving at night or during other periods of reduced visibility
- Selecting routes that will not mask friendly suppressive fires
- Crossing open areas (e.g., streets and spaces between buildings) quickly under the concealment of smoke and suppressive fires
- Moving on rooftops that are not covered by enemy direct fires.

a. The requirement for speed may dictate movement through streets and alleys without clearing all buildings. When moving in this manner, the maneuver force should employ infantry to lead the column, and supported by AAVs, LAVs, wheeled vehicle-mounted machine guns, or tanks following.

b. Marines should move along each side of a street with the leading squad staying abreast of armored vehicles. When not accompanied by armored vehicles, Marines move in a single file along one side of the street under cover of fires from supporting weapons. Individual Marines should be detailed to observe and cover a specific area, such as second-floor windows on the opposite side of the street.

5. Assault. Assault forces quickly and violently execute their assault and subsequent clearing operations. Momentum is maintained to prevent the enemy from organizing a more determined resistance on other floors or in other rooms. Small-unit leaders should keep the assault force moving without allowing the operation to become disorganized.

a. An assault in a built-up area involves the elementary skills of close combat. Leaders must:

- Be trained in the required techniques to defeat the enemy in a face-to-face encounter
- Keep themselves in excellent physical condition
- Have confidence in their abilities and those of their unit.

b. The composition of the assault force will vary depending on METT-T; however, its basic organization at the squad level consists of two assault elements carrying a fighting load of equipment and as much ammunition as possible, especially grenades.

6. Clearing. Entering at the top and fighting downward is the preferred method of clearing a building. Clearing a building is easier from an upper story because gravity and building construction become assets to the assault elements when throwing hand grenades and moving from floor to floor. However, this method is only feasible when access to an upper floor or rooftop can be gained from the windows or roofs of adjoining buildings and assault elements can be transported to the rooftops by helicopter (Figure C-1). Helicopters should land only on those buildings that have special heliports or parking garages on the roofs. When these are unavailable, Marines can rappel or fastrope onto the roof or dismount as the helicopter hovers a few feet above the roof. Assault elements then breach the roof or common walls with explosives and use ropes to enter the lower floors. Stairs are guarded by a covering team or security force when not in use to protect the assault force from a flank attack.

Figure C-1. Helicopters Used To Clear Buildings

a. Although the top-to-bottom method is preferred for clearing a building, assaulting the bottom floor and clearing upward is a common method. When using this technique, the assault force closes on the flank(s) or rear of the building. It then clears each room on the ground floor and then, moving upward, begin a systematic clearance of the remaining floors.

b. Entry can be gained through walls breached by explosives or gunfire. Assault elements should avoid windows and doors because they are usually covered by fire or boobytrapped. If tanks are attached to the company, they can breach the wall by firing their main gun (Figure C-2).

Figure C-2. Main Gun Used To Breach Exterior

c. Shortly before the assault, suppressive fires on the objective should be increased until masked by the advancing assault force. Once masked, fires are shifted to upper windows and continued until the assault force has entered the building. At that time, fires are shifted to adjacent buildings to prevent enemy withdrawal or reinforcement.

d. Hand grenades should be used to clear the space before assault element members enter through a breached wall. Immediately after the explosion, the assault element clearing team enters and uses three round bursts to fire into the interior of the room, concentrating on areas that are possible enemy positions.

e. Once inside the building, the assault force must cover staircases and seize rooms that overlook approaches to the building. These actions are required to isolate enemy forces within the building and to prevent reinforcement. Assault elements should clear each ground floor room and then the basement. Sequence for the assault is that:

(1) The assault force leader determines which room(s) to clear first.

(2) The support force provides suppressive fire while the assault force systematically clears the building. It also provides suppressive fire on adjacent buildings to prevent enemy reinforcements or withdrawal. The support force destroys any enemy trying to exit the building.

(3) The security force isolates the immediate area.

(4) After assault elements establish a foothold in the building, the assault force leader proceeds with the systematic clearing of the building.

f. When clearing a building from the top floor down, assault forces should gain entrance through a common wall or the roof of an adjoining building (Figure C-3). Demolition charges are used to breach the wall and gain entrance to the top floor. Access to lower floors and rooms may be gained by breaching holes in the floor and having the Marines jump or slide down ropes to the lower floors. Stairs can be used if they are first cleared.

g. When using the top-to-bottom method of clearing, requirements remain the same as for other methods. After the floor is breached to gain access to a lower floor, a grenade is allowed to cook-off and is dropped to the lower room. A Marine then sprays the lower room with gunfire using three-round bursts and drops through the mousehole.

Figure C-3. Upper Floors Secured

h. Marines should avoid predictability when entering and clearing rooms. They should vary entry through doors and mouseholes to keep the enemy off guard (Figure C-4). As rooms are cleared, doors should be left open, and a predetermined mark (cloth, tape, spray paint) should be placed on the doorjamb or over the door, as established by unit SOPs.

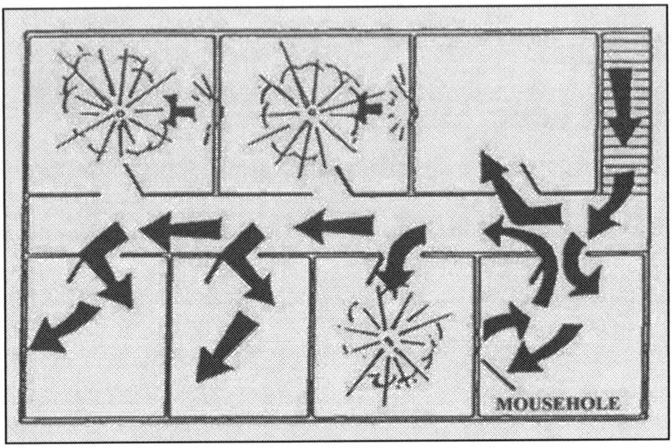

Figure C-4. Varying Techniques for Clearing Rooms

i. If there is a basement, it should be cleared as soon as possible, preferably at the same time as the ground floor. The procedures for clearing a basement are the same as for any room or floor. However, basements often contain entrances to tunnels such as sewers and communications cable tunnels. These should be cleared and secured to prevent the enemy from infiltrating back into cleared areas.

j. Types of buildings that may be cleared include brick buildings and houses, box-wall buildings, heavy-clad framed buildings, and light-clad framed buildings.

(1) Brick Buildings and Houses. The best way to enter a brick building is to blow a breaching hole in the side by using a tank firing HEAT ammunition (Figure C-5). If tanks are not available, a door or window in the rear of the building usually provides better cover and concealment for entry than one in the front. If there is enough cover and concealment, the assault force should enter the rear of the building at an upper level by using a fire escape or grappling hook.

Brick houses usually have similar floor plans on each floor (Figure C-6); therefore, ground floors are cleared the same way as upper floors.

Converely, the floor plans in larger brick buildings typically are different on ground-floor levels than on upper levels (Figure C-7).

Figure C-5. Entering a Brick Building

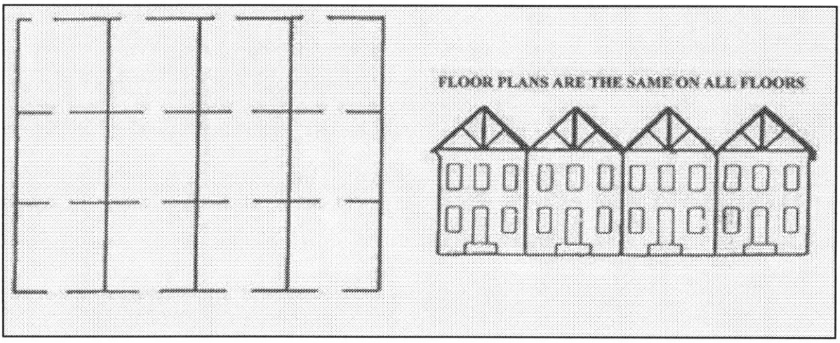

Figure C-6. Brick Houses Have Similar Floor Plans

To clear from building to building, the best method is to move from rooftop to rooftop because the roofs of brick buildings are usually easy to breach. The walls between buildings are at least three bricks thick (total of six bricks between buildings) and require large quantities of demolitions to breach. Walls are normally easier to breach on an upper floor than on a lower floor because the walls are thinner on upper floors. If rooftops are covered by fire and there are not enough demolitions to breach walls between buildings, clearing from rear to rear of buildings is safer than clearing from front to front.

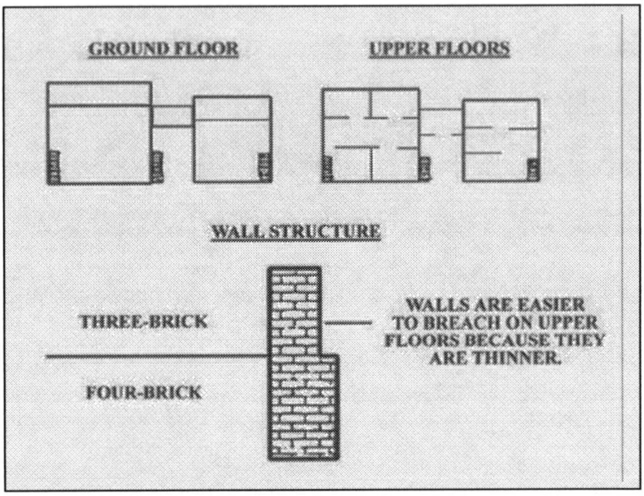

Figure C-7. Floor Plans of Brick Buildings

(2) **Box-Wall Buildings.** Box-wall buildings often have reinforced concrete walls (Figure C-8), which are difficult to breach because of reinforcing bars. Therefore, the best way to enter is to blow down the door or to blow in one of the side windows. The floor plans of these buildings are predictable; clearing rooms is usually done from one main hallway. Interior walls are also constructed of reinforced concrete and are difficult to breach. The stairways at the ends of the building must also be secured during clearing.

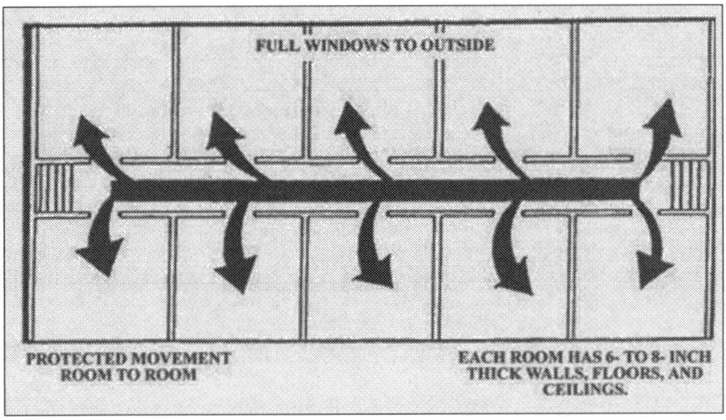

Figure C-8. Box-Wall Buildings

(3) Heavy-Clad Framed Buildings. Tanks can be used to breach a hole in heavy-clad framed buildings. Their floor plans are normally oriented around a stairway or elevator. this must be secured during clearing (Figure C-9). The interior walls of these buildings can be breached, although they may require use of demolitions.

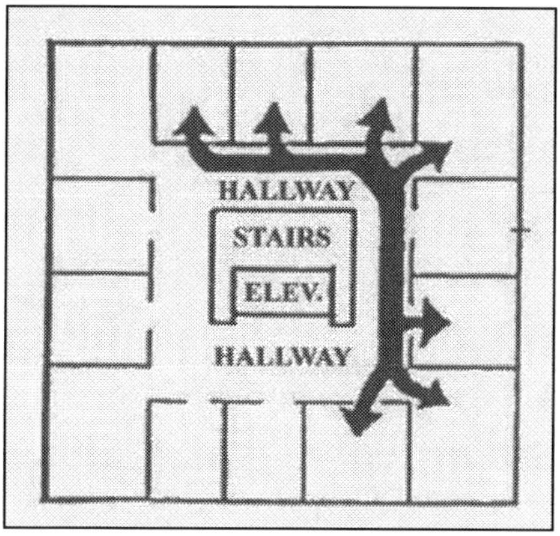

Figure C-9. Heavy-Clad Framed Buildings

(4) Light-Clad Framed Buildings. Light-clad framed buildings (Figure C-10), can be cleared in the same manner as heavy-clad buildings. Secure the central stairway and clear in a circular pattern. Walls are usually thin enough to be breached with an ax.

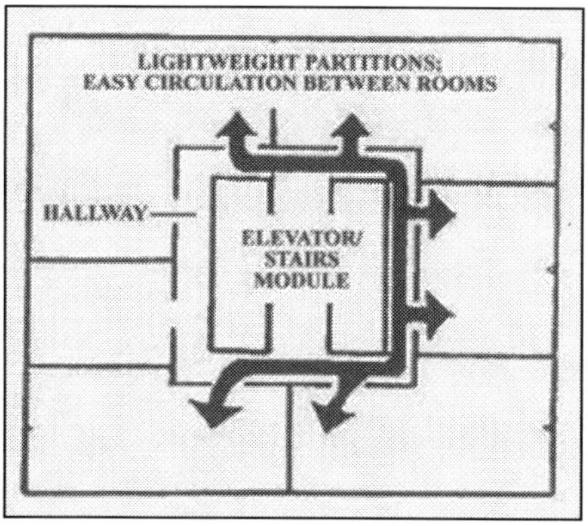

Figure C-10. Light-Clad Framed Buildings

7. Consolidation and Reorganization. Consolidation and reorganization of the assault force in a cleared building must be accomplished quickly to repel enemy counterattacks and to prevent the enemy from infiltrating back into the cleared building. After securing a floor (bottom, middle, or top), selected members of the assault force are assigned to cover potential enemy counterattack routes to the building. Those sentinels alert the assault force and place a heavy volume of fire on enemy forces approaching the building. They should cover:

- Enemy mouseholes between adjacent buildings
- Covered routes to the building
- Underground routes into the basement
- Approaches over adjoining roofs.

As the remainder of the assault force completes search requirements, defensive positions are assigned and tasks are accomplished as established by the priorities of work in the defense. The following actions may then be taken:

- Resupplying and redistributing ammunition
- Marking the building to indicate to friendly forces that the building has been cleared
- Assuming an overwatch mission and supporting an assault on another building
- Treating and evacuating wounded personnel
- Further developing a defensive position if the building is to be occupied for any period.

Appendix D

Fighting Positions

A critical defensive task in combat in built-up areas is the selection and preparation of fighting positions.

1. Considerations. Leaders should consider the following factors when establishing fighting positions:

a. Protection. Leaders should select buildings that provide protection from direct and indirect fires. Reinforced concrete buildings with three or more floors provide suitable protection, while buildings constructed of wood, paneling, or other light materials require reinforcement to gain sufficient protection. One- to two-story buildings without strongly constructed cellars are vulnerable to indirect fires and require construction of overhead protection for each firing position.

b. Dispersion. Normally, a position should not be established in a single building when it is possible to occupy two or more buildings that permit mutually supporting fires. A position in one building without mutual support is vulnerable to bypass, isolation, and subsequent destruction from any direction.

c. Concealment. The fighting position should be concealed. Buildings that are obvious defensive positions should not be selected. The occupation of exposed buildings may be necessary because of requirements for security and fields of fire. Therefore, reinforcement of the structure must be accomplished to provide suitable protection within the building.

d. Fields of Fire. Positions should be mutually supporting and have interlocking fields of fire in all directions. Clearing fields of fire may require the destruction of adjacent buildings by using explosives, engineer equipment, and field expedients. Care should be taken to avoid highlighting the building as a defensive position.

e. Covered Routes. Defensive positions should have at least one covered route that permits resupply, MEDEVAC, reinforcement, or withdrawal from the building. The route can be established by one of the following means:

- Through walls to adjacent buildings
- Through underground systems
- Through communications trenches
- Behind protective buildings.

f. Observation. The building should permit observation of enemy avenues of approach and adjacent defensive sectors.

g. Fire Hazard. Avoid selecting positions in buildings that are a fire hazard. If flammable structures must be occupied, the danger of fire can be reduced by wetting the structure, laying an inch of sand on the floors, and providing fire extinguishers and firefighting equipment. Routes of escape should also be prepared in case of fire.

h. Time. The time available to prepare the defense could be the most critical factor. If enough time is insufficient, buildings that require extensive preparation should not be used. Conversely, buildings located in less desirable areas that require little improvement may be a better choice for a defensive position.

2. Preparation. Preparation of fighting positions depends on proper selection and construction.

a. Selecting Positions. Each weapon should be assigned a primary sector of fire to cover enemy approaches. Alternate positions that overwatch the primary sector should also be selected. These positions are usually located in an adjacent room on the same floor (Figures D-1A, D-1B, and D-1C). Each weapon must be assigned a supplementary position (to engage attacks from other directions) and an FPL.

b. Building Positions. There are many ways to establish a fighting position in a building.

(1) Window Position. Marines should kneel or stand on either side of a window. To fire downward from upper floors, tables or similar objects can be placed against the wall to provide additional elevation, but they must be positioned to prevent the weapon from protruding through the window (Figure D-2). Leaders should inspect positions to determine the width of sector that each position can engage.

(2) Loopholes. To avoid establishing a pattern of always firing from windows, loopholes should be prepared in walls. Marines should avoid firing directly through loopholes to enhance individual protection.

(a) Several loopholes are usually required for each weapon (primary, alternate, and supplementary positions). The number of loopholes should be carefully considered because they can weaken walls and reduce protection. Engineers should be consulted before numerous loopholes are made. Loopholes should be made by punching or drilling holes in walls and should be placed where they are concealed. Blasting loopholes can result in a large hole that can easily be seen by the enemy.

(b) Loopholes should be cone-shaped to obtain a wide arc of fire, to facilitate engagement of high and low targets, and to reduce the size of the exterior aperture (Figure D-3). The edges of a loophole splinter when hit by bullets; therefore, protective linings such as an empty sandbag held in place by wire mesh will reduce spalling effects. When not in use, loopholes should be covered with sandbags to prevent the enemy from firing into or observing through them.

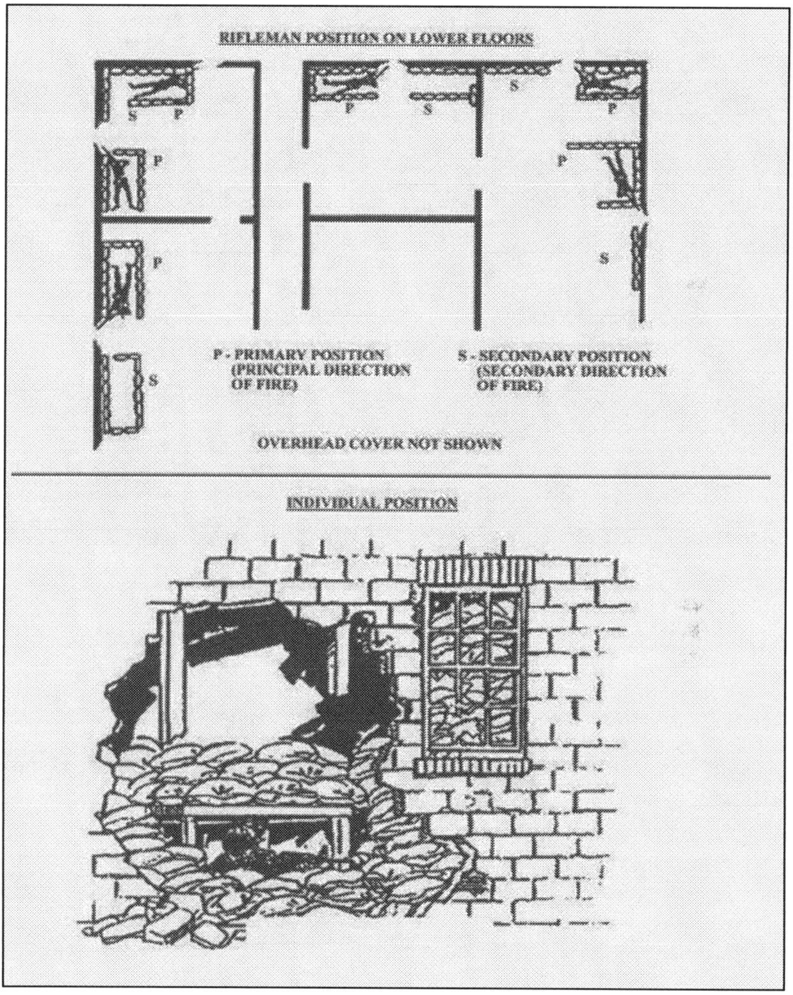

Figure D-1A. Weapon Positions

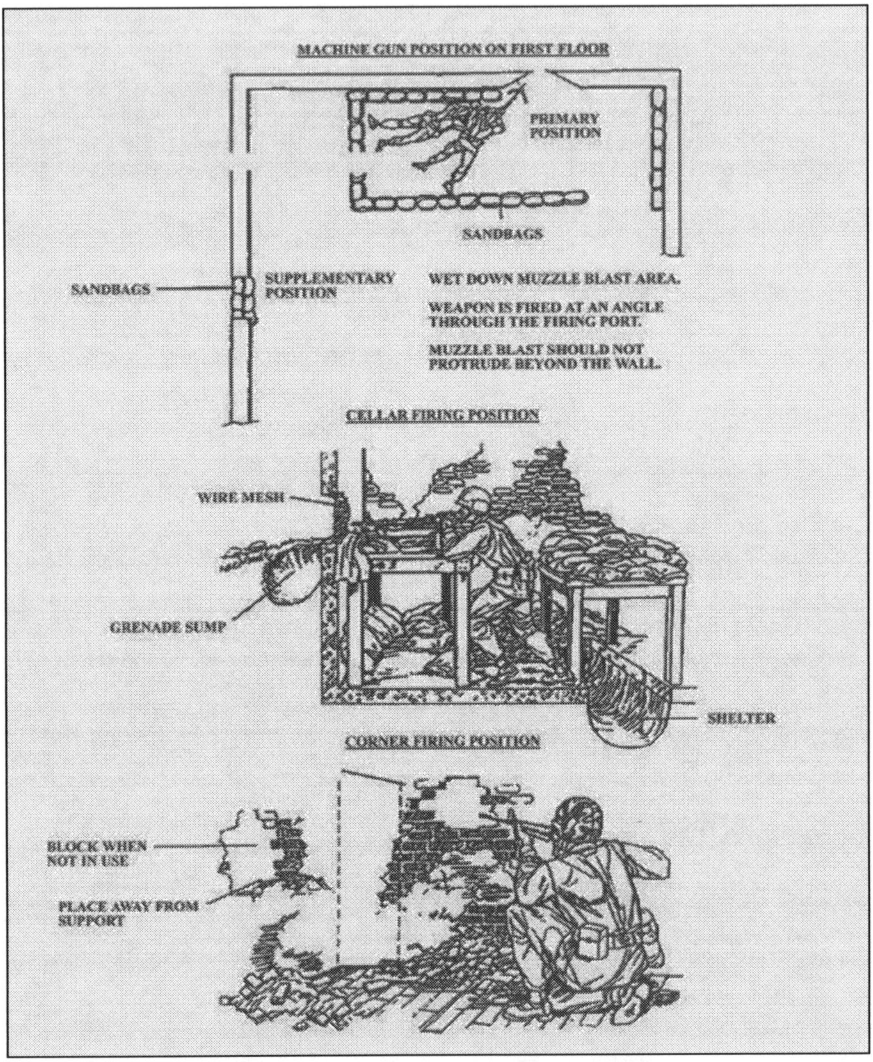

Figure D-1B. Weapon Positions (Continued)

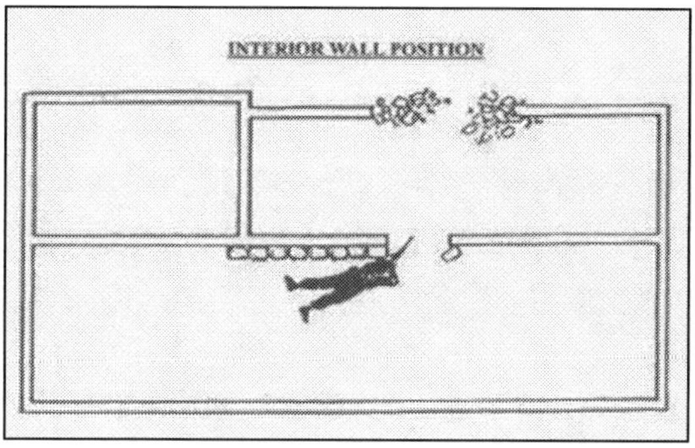

Figure D-1C. Interior Wall Position

Figure D-2. Window Position

(c) Loopholes should also be prepared in interior walls and ceilings of buildings to permit fighting within the position. Interior loopholes should enable overwatch of stairs, halls, and unoccupied rooms and be concealed by pictures, drapes, or furniture. Loopholes in floors permit the defender to engage enemy personnel on lower floors with small-arms fire and grenades.

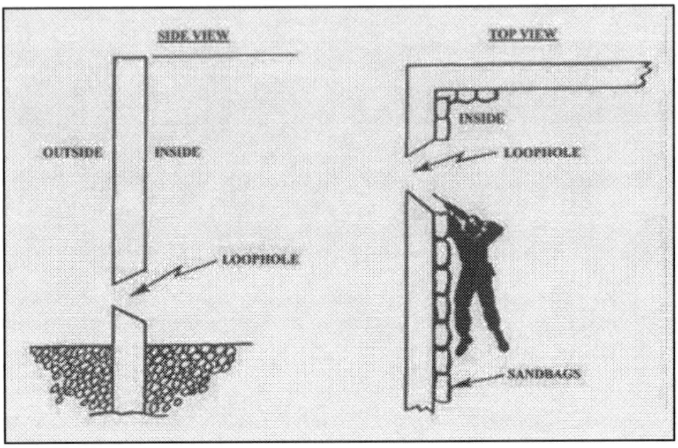

Figure D-3. Cone-Shaped Loopholes

(d) Although walls provide some frontal protection, they should be reinforced with sandbags, furniture filled with dirt, or other expedients. Each position should have overhead and all-around protection (Figure D-4).

Figure D-4. Position With Overhead and All-Around Protection

c. Other Construction Tasks. Other construction tasks may need to be performed. These include work on:

(1) Basements and Ground Floors. Basements require preparation similar to that of the ground floor. Any underground system that is not used by the defender but that could provide enemy access to the position must be blocked.

(a) Doors. Unused doors should be locked, nailed shut, blocked, and reinforced with furniture, sandbags, or other field expedients. Outside doors can be boobytrapped by engineers or other personnel.

(b) Hallways. If not required for the defender's movement, hallways should be blocked with furniture and tactical wire (Figure D-5). If authorized, boobytraps should be employed to impede enemy movement.

(c) Stairs. Defenders should block stairs not used by the defense with furniture and tactical wire (Figure D-5) or remove them. If possible, all stairs should be blocked, and ladders should be used to move from floor to floor and then removed when not being used. Boobytraps should also be employed on stairs.

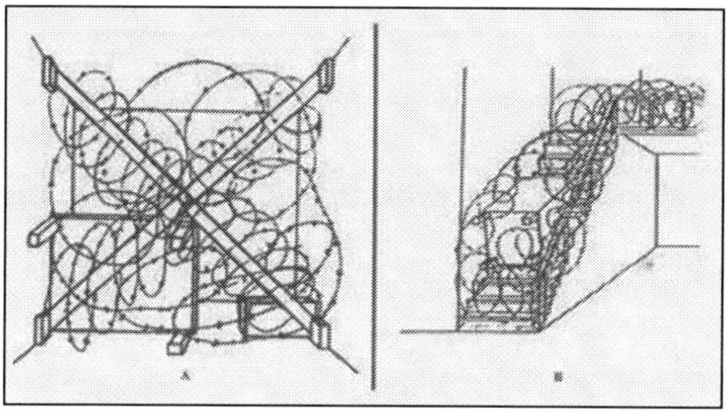

Figure D-5. Blocking Hallways and Stairs

(d) Windows. All glass should be removed. Windows not being used should be blocked with boards or sandbags.

(e) Floors. Fighting positions should be constructed on the ground or basement floors. If there is no basement, fighting positions on the first floor must be constructed to provide additional protection from heavy direct-fire weapons.

(f) Ceilings. Supports that can withstand the weight of rubble from upper floors should be placed under ceilings to provide overhead cover (Figure D-6).

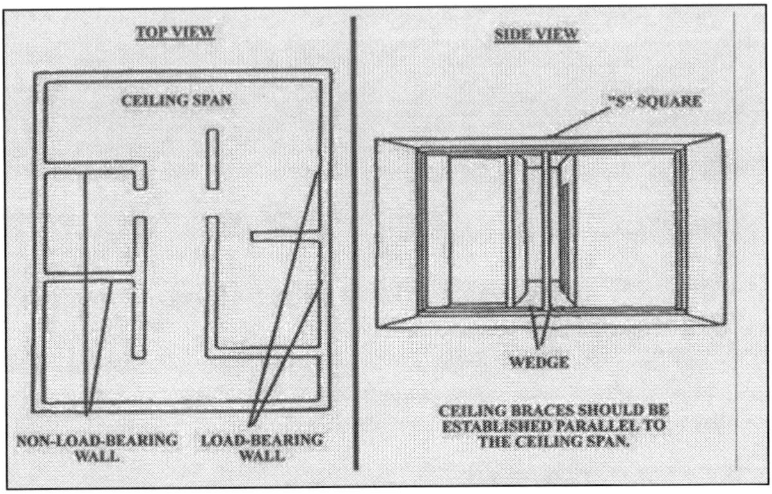

Figure D-6. Ceiling Reinforcement

(g) Unoccupied Rooms. Rooms not required for defense should be blocked with tactical wire, boobytrapped, or both.

(2) Upper Floors. Upper floors require the same preparation as ground floors. Windows need not be blocked, but they should be covered with wire mesh to block grenades thrown from the outside. The wire should be loose at the bottom to permit the defender to drop grenades on enemy forces below.

(3) Interior Routes. Routes are required that permit defending forces to move within the building to engage enemy forces from any direction or reinforce other forces (Figure D-7). Escape routes should also be planned and constructed to permit rapid evacuation of a room or the building. Small holes (called mouseholes) should be made through interior walls to permit movement between rooms. Once the defender has withdrawn to another level, such holes should be clearly marked for both day and night identification. All personnel must be briefed as to where the various routes are located. Rehearsals should be conducted so that everyone becomes familiar with the routes.

(4) Fire Prevention. Buildings that have wooden floors and raftered ceilings require extensive fire prevention measures. The attic and other wooden floors should be covered with about 1 inch of sand or dirt, and buckets of water should be positioned for immediate use. Firefighting materials (dirt, sand, fire extinguishers, and blankets) should be placed on each floor for immediate use. Water basins and bathtubs should be filled as a reserve for firefighting. All electricity and gas should be turned off. Fire breaks can be created by destroying buildings adjacent to the defensive position.

Figure D-7. Movement Between Floors

(5) Communications. Telephone lines should be laid through adjacent buildings or underground systems or buried in shallow trenches. Radio antennas can be concealed by placing them among civilian television antennas, along the sides of chimneys and steeples, or out windows that direct FM communications away from enemy electronic warfare sources and ground observation. Telephone lines within the building should be laid through walls and floors.

(6) Rubbling. Rubbling parts of the building may provide useful cover and concealment for weapons emplacements. Rubbling should be performed only by trained engineers.

(7) Rooftops. Positions in flat-roofed buildings require obstacles that restrict helicopter landings (See Figure F-2). Rooftops that are accessible from adjacent structures should be covered with tactical wire or other expedients and must be guarded. Entrances to buildings from rooftops can be blocked if compatible with the overall defensive plan. Any structure on the outside of a building that could assist in scaling the buildings to gain access to upper floors or to the rooftop should be removed or blocked.

(8) Obstacles. Obstacles should be positioned adjacent to buildings to stop or delay tanks and infantry.

(9) Fields of Fire. Fields of fire should be improved around the defensive position. Selected buildings may be destroyed to enlarge fields of fire. Obstacles to ATGMs, such as telephone or electrical wires, should be cleared. Dead space should be covered with mines and obstacles.

3. Tank and Other Armored Vehicle Positions. Fighting positions for tanks and other armored vehicles are essential to a complete and effective defensive plan in built-up areas.

a. Armored Vehicle Positions. Armored vehicle positions are selected and developed to obtain the best cover, concealment, observation, and fields of fire while retaining the vehicle's ability to move.

(1) If fields of fire are restricted to streets, hull-down positions should be used to gain cover and to fire directly down streets (Figure D-8). From those positions, armored vehicles are protected and can rapidly move to alternate positions. Buildings collapsing from enemy fires are a minimal hazard to armored crews.

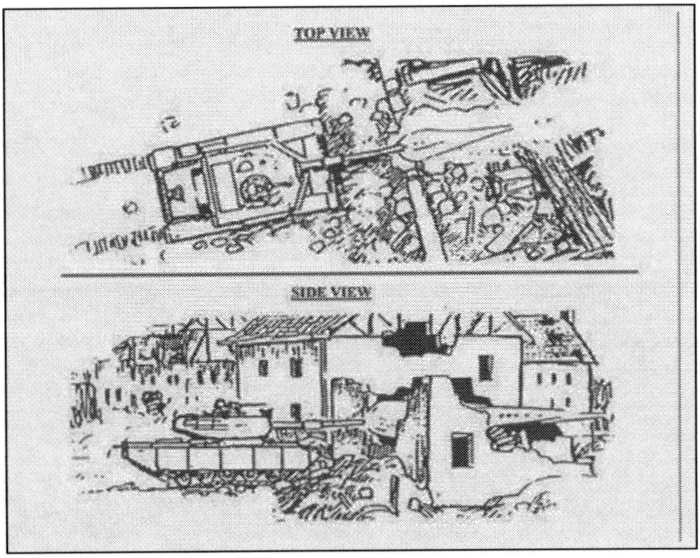

Figure D-8. Hull-Down Position

(2) The hide position (Figure D-9) covers and conceals the vehicle until time to move into position for engagement of targets. Because the crew will not be able to see advancing enemy forces, an observer from the vehicle or a nearby infantry unit must be concealed in an adjacent building to alert the crew. The observer acquires the target and signals the armored vehicle to move to the firing position and fire. After firing, the vehicle moves to an alternate position to avoid compromising one location.

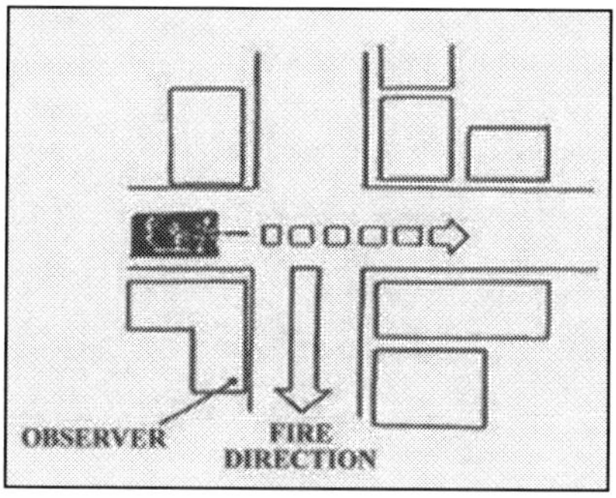

Figure D-9. Hide Position

(3) The building hide position (Figure D-10) conceals the vehicle inside a building. If basement hide positions are inaccessible, engineers can evaluate the building's floor strength and prepare for the vehicle. Once the position is detected, it should be evacuated to avoid enemy fires.

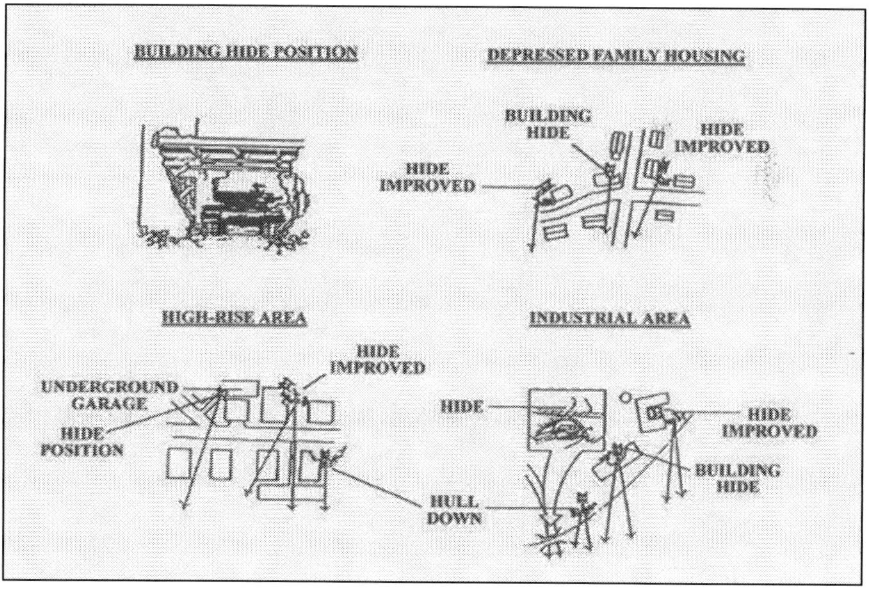

Figure D-10. Building Hide Position

4. Antitank Guided Missile Positions. HMMWVs with ATGMs are employed in areas that maximize their capabilities in the built-up area. Figure D-11 shows mounted and dismounted firing positions.

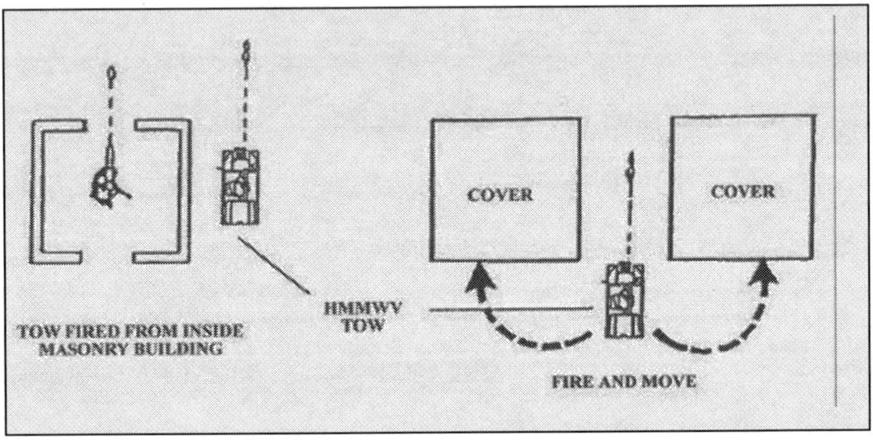

Figure D-11. ATGM Positions

a. Rubble may interfere with missile flight when ATGMs are fired from the HMMWV or from street-level or bottom-floor fighting positions. When firing down streets, missiles must have at least 1 meter of clearance over rubble. Other obstacles to missile flight include trees and brush, vehicles, television antennas, buildings, power lines and wires, walls, and fences (See Figure B-7).

b. An AT4 is best suited for built-up areas because its short arming distance allows employment at close range. AT4s and other light and medium antitank weapons are not effective against the front of modern battle tanks. Because tanks have the least armor protection on the top and rear deck and the tank presents a larger target when engaged from above, AT4s should fire down onto tanks.

5. Sniper Positions. Snipers contribute to combat in built-up areas by firing on selected enemy targets. Effective sniper employment can trouble the enemy far more than its relative personnel numbers reflect.

a. General areas (a building or group of buildings) are designated as sniper positions (Figure D-12), but the sniper selects the best position for engagement. Masonry buildings that offer the best protection, long-range fields of fire, and all-around observation are preferred. The sniper also selects several secondary and supplementary positions to cover his areas of responsibility.

b. Engagement priorities for snipers are determined by the relative importance of the targets to the effective operations of the enemy. Sniper targets may include tank commanders, crewmen of direct fire support weapons, crewmen of crew-served weapons, officers, FOs, and radiotelephone operators.

c. The urban environment often limits snipers to firing down or across streets. Snipers can be employed to cover rooftops, obstacles, dead space, and gaps in final protective fires.

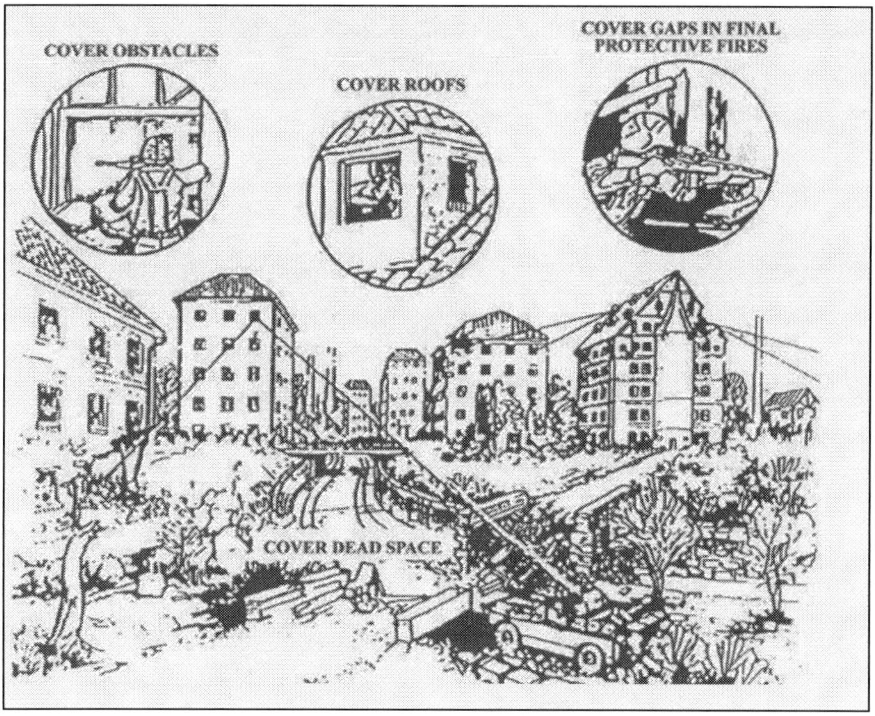

Figure D-12. Sniper Positions

Appendix E

Subterranean Operations

Knowledge of the nature and location of underground facilities is valuable to both the urban attacker and defender.

1. Tactical Value. Fighting in MOUT is multidimensional. Larger cities may have subterranean features that include sunken garages, underground passages, subway lines, utility tunnels, sewers, and storm drains (Figure E-1). Most of these features allow for the movement of units. Even in smaller European towns, sewers and storm drains permit Marines to move beneath the fighting and surface behind the enemy.

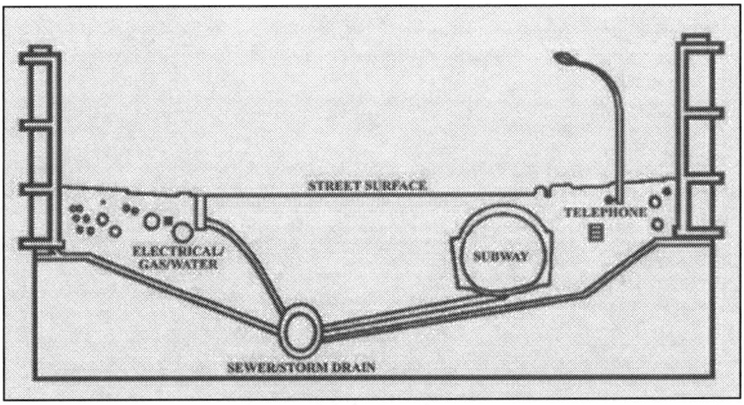

Figure E-1. Potential Underground Passages

a. Subterranean passages may provide the attacker with covered and concealed routes into and through built-up areas. This may allow an enemy to launch his attack along roads that lead into the city while infiltrating a smaller force into the defender's rear. An enemy unit in the defender's rear area may disrupt his defense and obstruct the avenues of withdrawal for his forward defense.

b. An attack along a subterranean avenue of approach could become the main effort. Even if the subterranean effort is not immediately successful, it forces the defender to fight on atleast two levels and to extend resources to more than just street-level fighting.

c. The existence of subterranean passages forces the defender to defend the built-up area above and below ground. While these passages are usually more of a disadvantage to the defender than to the attacker they may offer some advantages in the defense. When thoroughly reconnoitered and controlled by the defender, subterranean passages provide excellent covered and concealed routes to move reinforcements or to launch counterattacks. They also provide ready-made covered lines of communication for the movement of supplies and evacuation of casualties and provide places to cache supplies for forward companies. Subterranean passages also offer the defender a ready-made conduit for communications wire, with protection from armored vehicles and indirect fires.

2. Denial to the Enemy. Subterranean passages are useful to the defender only to the extent that the attacker can be denied their use. The defender has an advantage in that, given the confining, dark environment of these passages, a small group of determined Marines in a prepared position can defeat a numerically superior force.

a. Tunnels afford the attacker little cover and concealment except for the darkness and any manmade barriers. The passageways provide tight fields of fire and amplify the effects of grenades. Obstacles at intersections in tunnels support excellent ambush sites and may turn the subterranean passages into a maze. These obstacles can be created quickly by using chunks of rubble, furniture, and parts of abandoned vehicles interspersed with M18 claymore mines.

b. A thorough reconnaissance of the subterranean or sewer system is required. To be effective, obstacles must be located at critical intersections in the passage network so that they trap attackers in a kill zone but allow defenders freedom of movement (Figure E-2).

3. Subterranean Reconnaissance Techniques. The mission of local reconnaissance for a platoon- or company-sized area of responsibility should be given to a squad-sized element.

a. The squad leader tasks one fire team with providing forward security. One fire team follows in support, and one fire team provides rear security (Figure E-3). The squad leader moves directly behind the pointman and records data collected by the patrol. If available, an engineer should be with the lead fire team. The rear security fire team protects the ground- and subterranean-level point of entry. This fire team is responsible for detecting enemy movement on the patrol's rear and serves as the communications link between the squad leader and higher headquarters. Organization of the patrol may vary based on METT-T.

b. The squad leader should carry a map, compass, street plan, and notebook in which he has detailed the information that he must gather. One Marine is assigned to carry the tools needed to open manhole covers. If the patrol is to move more than 200 to 300 meters, or if the platoon leader directs, the squad may carry a sound-powered telephone (TA-1) and wire dispenser (XM-306A) for communications. Radios are often unreliable in this environment. Marines (especially the pointman, squad leader, engineer, and fire team leaders) should be equipped with night vision devices to maintain surveillance within the sewer. Thermal night devices are also useful as there may not be enough ambient light for accurate visibility with night vision devices.

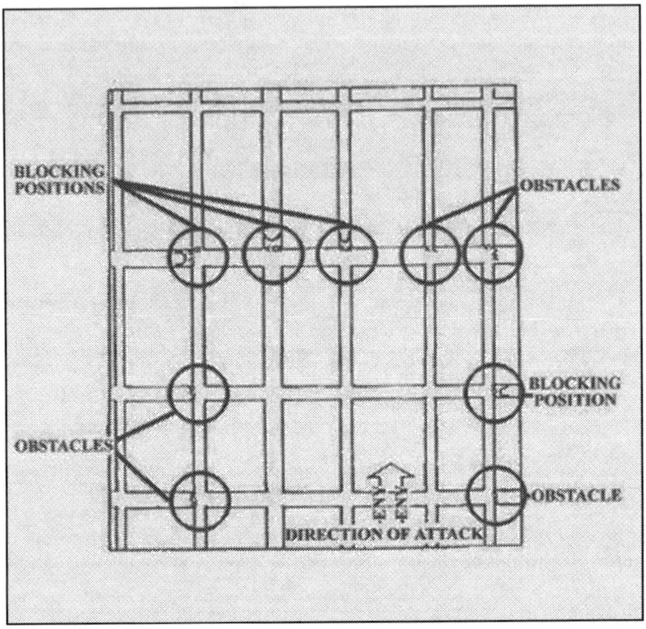

Figure E-2. Defense of a Sewer System

Figure E-3. Organization of the Patrol Unit

c. Marines entering the sewer should carry a sketch of the sewer system that includes magnetic north, azimuths, distances, and manholes. They should also carry protective masks, flashlights, gloves, and chalk (or chemiluminescent lights) for marking features along the route. The patrol should also be equipped with a safety rope to which each man is tied. To improve their footing in slippery sewers and storm drains, the squad members may wrap chicken wire or screen wire around their boots.

d. A constant concern to units conducting a subterranean patrol is chemical defense. Enemy chemical agents used in tunnels may be suddenly encountered in dense concentrations with little dispersion. The pointman should carry an automatic chemical agent alarm system to provide instant warning of the presence of chemical agents. M8 detector paper can also be used to detect chemical agents. At the first indication that harmful gases are present, the squad should mask.

e. In addition to enemy chemical agents, noxious gases from decomposing sewage can also pose a threat. These gases are not detected by chemical agent alarm systems, nor are they completely filtered by the protective mask. Physical signs that indicate their presence in harmful quantities are nausea and dizziness. The squad leader should be constantly alert to these signs and know the shortest route to the surface for fresh air. Some sewer gases deplete oxygen, a situation against which the field protective mask provides no protection.

f. Once the squad is organized and equipped, it moves to the entrance of the tunnel. This may be a manhole. With the manhole cover removed, the squad waits 15 minutes before entry to allow any gases that may be present to dissipate. The pointman then descends into the tunnel to determine whether the air is safe to breathe and whether movement is restricted. The pointman should remain in the tunnel for 10 minutes before the rest of the squad follows. If he becomes ill or is exposed to danger, he can be pulled out by the safety rope.

g. When the squad is moving through the tunnel, the pointman moves about 10 meters in front of the squad leader. Other squad members maintain 5-meter intervals. If water in the tunnel is flowing faster than 2.5 meters per second or the sewer contains slippery obstacles, those intervals should be increased to prevent all patrol members from falling if one man slips. All squad members should stay tied into the safety rope so that they can easily be retrieved from danger. The last Marine in the patrol marks the route with chalk or chemiluminescent lights so other units can find the squad.

h. The squad leader should record the azimuth and pace count of each turn he takes in the tunnel. When he encounters a manhole to the surface, the pointman should open it and determine the location, which the patrol leader then records. The use of recognition signals (Figure E-4) prevents friendly forces from accidentally shooting the pointman as he appears at a manhole.

Figure E-4. Recognition Signals

i. Once the patrol has returned and submitted its report, the platoon leader must decide how, or whether, to use the tunnel. In the offense, a tunnel might provide a covered route to move behind the enemy's defenses. In the defense, a tunnel might provide a covered passage between positions. In any case, the squad members that provided the reconnaissance should act as guides along the route.

j. If the tunnel is to be blocked, the platoon may emplace concertina wire, early warning devices, and antipersonnel mines. A security force should be established at the entrance of the sewer (Figure E-5) to provide warning against any enemy trying to approach the platoon's defense. It should be equipped with command-detonated illumination. The overpressure from grenades, mines, and boobytraps exploding in a sewer or tunnel can have adverse effects on friendly forces, including ruptured eardrums and wounds from flying debris. Also, gases found in sewers can be ignited by the blast effects of these munitions. For these reasons, small-arms weapons should be employed in tunnels and sewers. Friendly personnel should be outside of tunnels or out of range of the effects when mines or demolitions are detonated.

4. Psychological Considerations. Combat operations in subterranean passages are similar to night combat operations. The psychological factors of night operations reduce confidence, cause fear, and increase a feeling of isolation. This feeling of isolation is further magnified by the tight confines of the tunnels. The layout of tunnels could require greater dispersion between positions.

Figure E-5. Security Position Established at the Entrance to a Sewer

a. Leaders must enforce measures to help dispel the feelings of fear and isolation experienced when operating in tunnels. These measures include leadership training, physical and mental fitness, sleep discipline, and stress management.

b. Leaders maintain communication with Marines manning positions in tunnels either through personal visits, messengers, or field telephone. Communication increases situational awareness and provides insight into the mental state of Marines. Marines manning positions below ground should be given as much information as possible on the organization of the tunnels and the importance of the mission. They should be briefed on contingency plans and alternate positions should their primary positions become untenable.

c. Physical and mental fitness can be maintained by periodically rotating Marines out of tunnels so they can stand and walk in fresh air and sunlight. Stress management is also a factor of operations in tunnels. Historically, combat in built-up areas has been one of the most stressful forms of combat. Fighting in continuous darkness and restricted maneuver space causes even more stress to Marines than street fighting.

Appendix F

Obstacles, Mines, and Demolitions

Obstacles are used extensively in combat in built-up areas to allow the defender to canalize the enemy, impede his movement, and disrupt his attack.

Section I

Obstacles

1. Introduction. Obstacles are designed to prevent movement by personnel, to separate infantry from tanks, and to slow or stop vehicles. (See MCRP 3-17A, *MAGTF Breaching Operations.*)

2. Types. Antipersonnel mines, barbed wire, boobytraps, and exploding flame devices are used to construct antipersonnel obstacles (Figure F-1). (See FMFRP 13-5/FM 5-34, *Engineer Field Data*; FM 5-250, *Explosives and Demolitions*; and FM 3-11, *Flame, Riot Control Agent, and Herbicide Operations,* for more detailed information.) These obstacles are used to block the following infantry approaches:

- Streets
- Buildings
- Roofs
- Open spaces
- Dead space
- Underground systems.

a. The approval authority to boobytrap buildings is normally the GCE commander. This authority may be delegated to subordinate commanders.

b. The three types of wire obstacles used in defensive operations are protective, tactical, and supplementary. Wire obstacles should be integrated with other obstacles and covered by fire and observation.

(1) Protective wire is usually located beyond hand-grenade range (40 to 100 meters) from the defensive position.

(2) Tactical wire is positioned to increase the effectiveness of friendly weapons fire. Tactical wire is usually positioned on the friendly side of machine gun FPLs.

(3) Supplementary wire is used to break up the pattern of tactical and protective wire to prevent the enemy from easily locating friendly weapons.

c. Obstacles may be placed in dead spaces to restrict infantry movement in areas that cannot be observed and that are protected from direct fires.

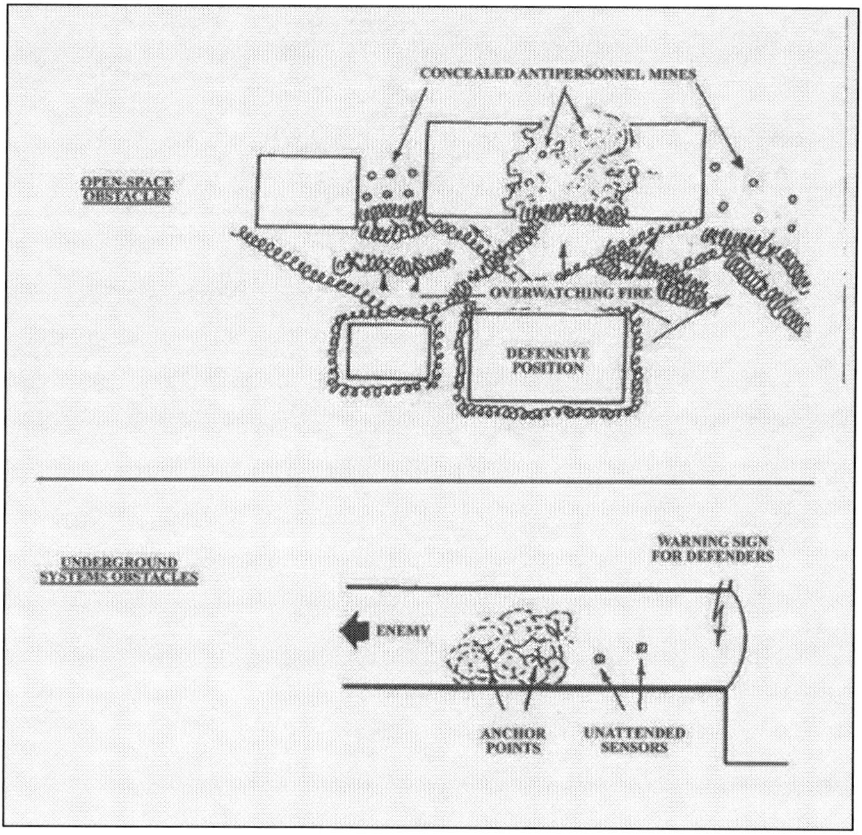

Figure F-1. Antipersonnel Obstacles

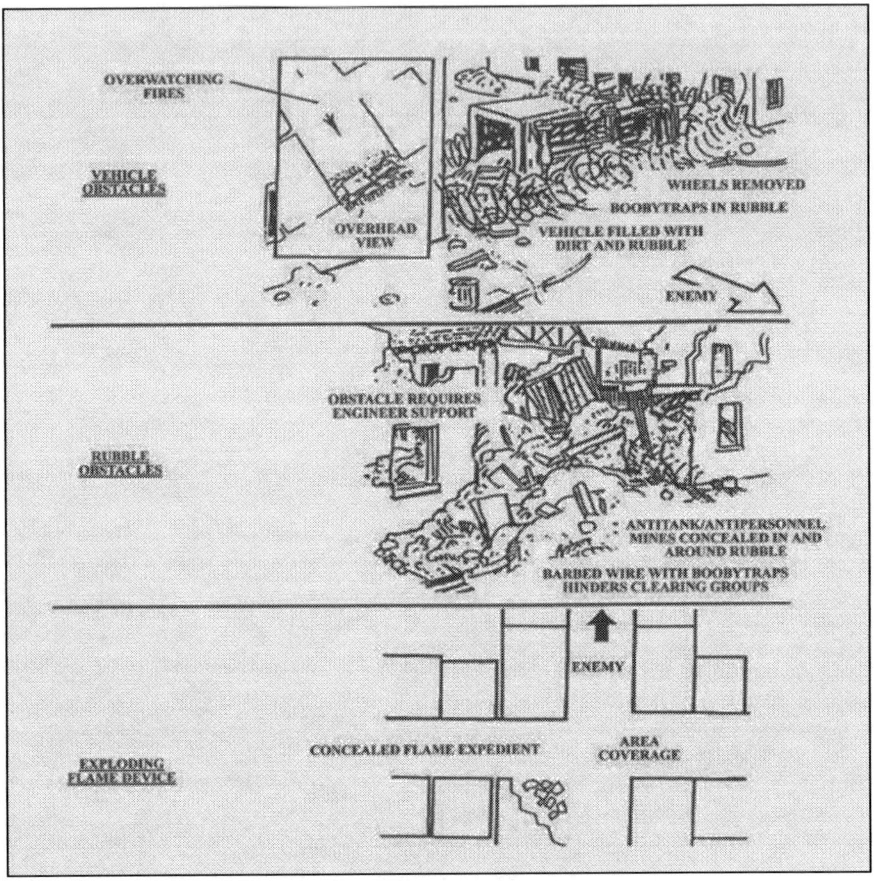

Figure F-1. Antipersonnel Obstacles (Continued)

d. Antiarmor obstacles are mainly restricted to streets and other armored avenues of approach (Figure F-2).

e. Some antitank mines may be sited and employed as antihelicopter mines to cover probable HLZs (Figure F-2).

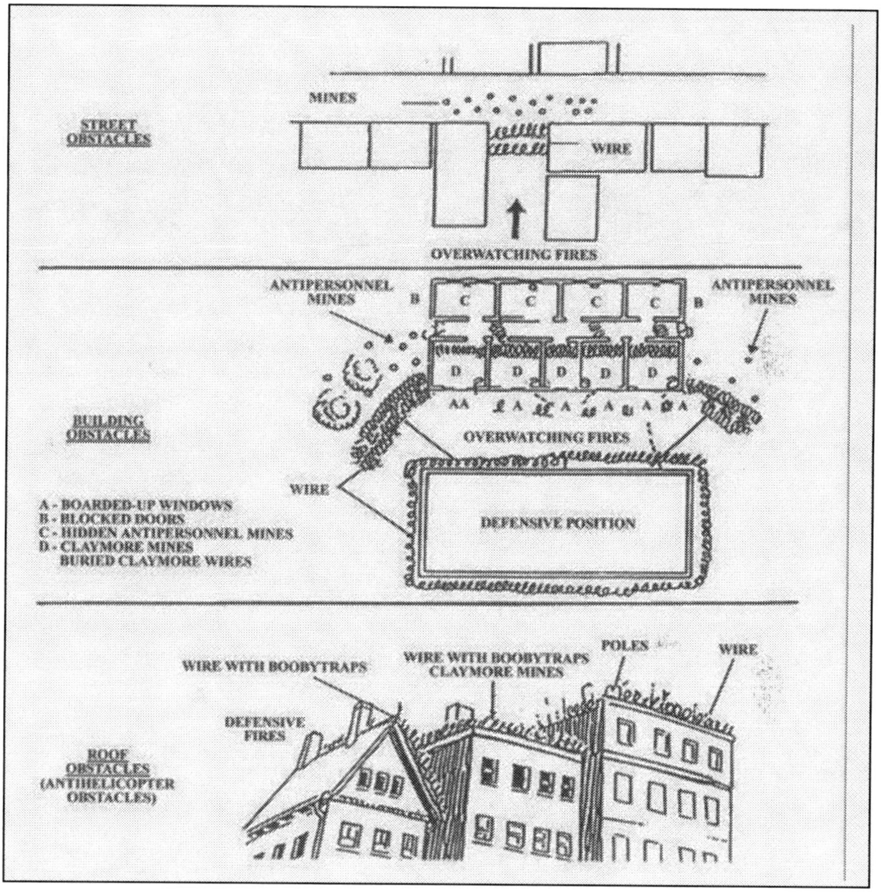

Figure F-2. Antiarmor Obstacles

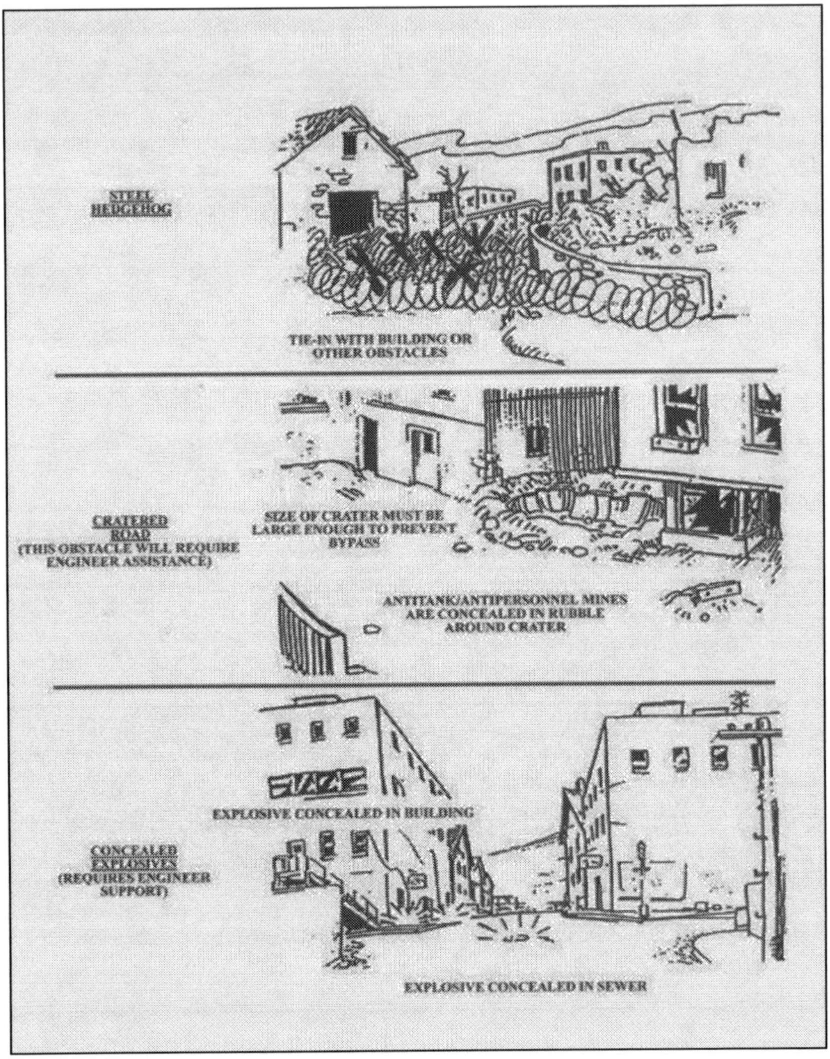

Figure F-2. Antiarmor Obstacles (Continued)

3. Construction of Obstacles. Obstacles are constructed in buildings to deny enemy infantry covered routes and weapons positions that are close to friendly defensive positions. They can be constructed of explosives, by using wire, or by using boobytraps within buildings. The building may be prepared as an explosive or flame trap for command detonation after enemy occupation.

Section II

Mines

4. Introduction. The employment or discovery of mines in built-up areas should be recorded on a building sketch (Figure F-3). (See FM 20-32.) The sketch should include the number of the building (taken from a city map, if available) and all floor plans. It should also include the type of mine and firing device. When possible, mined buildings should be marked on the friendly side (Figure F-4). Clearing areas or buildings that have been mined is extremely difficult and dangerous. Therefore, such areas should be considered "no go" areas for infantry. This factor must be carefully considered when planning and authorizing the placement of mines. (See Table F-1 for notional approving authority for minefields, but with the understanding that information in that table is subject to change with international agreements and ROE.)

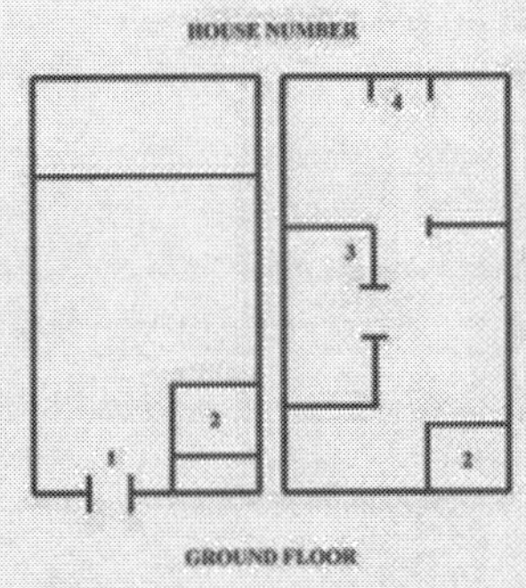

MINE	LOCATION	TYPE	FIRING DEVICE
1	FRONT DOOR	CLAYMORE	COMMAND
2	STAIRS	CLAYMORE	COMMAND
3	CLOSET	CLAYMORE	COMMAND
4	DRESSER	CLAYMORE	COMMAND

Figure F-3. Building Sketch Showing Mines

Figure F-4. Marking Mined Buildings

	Minefields (Type)		FASCAM (Duration)		Other Obstacles	
Level	Tactical	Protective	Long	Short	Tactical	Protective
MEF	X	X	X	X	X	X
Division	X	X	X	X	X	X
Regiment	X	X	X	X	X	X
Battalion	X	X		X	X	X
Company		X				X

Table F-1. Notional Minefield Employment Authority Matrix

5. **Types.** Several types of mines may be encountered or employed in built-up areas.

a. The M14 mine (a plastic type of antipersonnel landmine) cannot currently be used by American forces due to the Leahy Amendment of 1995 and the Certain Conventional Weapons Convention, of which the United States is a signatory nation. It may, however, be encountered in MOUT or be employed by coalition or enemy forces. For this reason, the following information is given as a guide for what U.S. forces may encounter. This type of antipersonnel landmine is normally used in conjunction with metallic antipersonnel, antitank, or chemical mines to confuse and hinder enemy breaching attempts. When encountered, it must be carefully handled because its light weight makes it easy to displace (Figure F-5). However, its size makes it ideal for covering obscure places such as stairs and cellars.

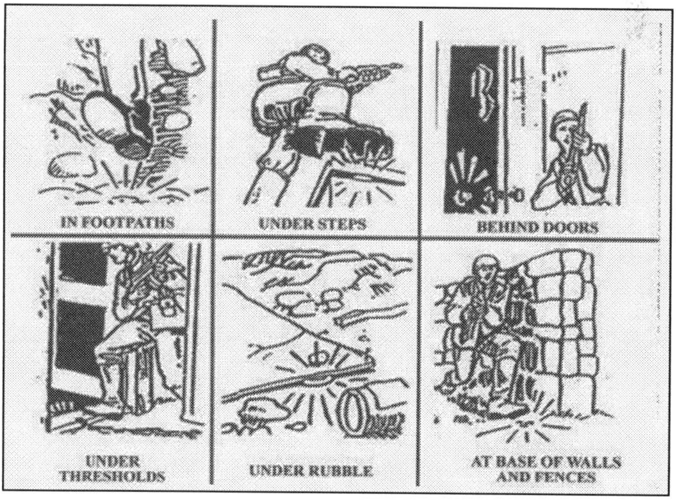

Figure F-5. Typical Employment of the M14 Antipersonnel Mine

b. The M16A2 antipersonnel landmine is subject to the same legal limitations as the M14. It is a boundary type of antipersonnel landmine, which makes it ideal for covering large areas such as rooftops, backyards, parks, and cellars. It can be expediently rigged for command detonation by attaching a rope or piece of communications wire to the pull ring (Figure F-6).

Figure F-6. Emplacements of M16A2 Antipersonnel Mine

c. The M18A1 claymore mine can be employed during the reorganization and consolidation phase on likely enemy avenues of approach. It does not have to be installed in the street, but can be employed on the sides of buildings or any other sturdy structure. The claymore mine, in command-detonated mode, is currently legal for employment by U.S. forces.

(1) Claymore mines can also be used for demolition against thin-skinned buildings and walls.

(2) Claymore mines may be configured for detonation by tripwire. They can help fill the dead space in the final protective fires of automatic weapons (Figure F-7).

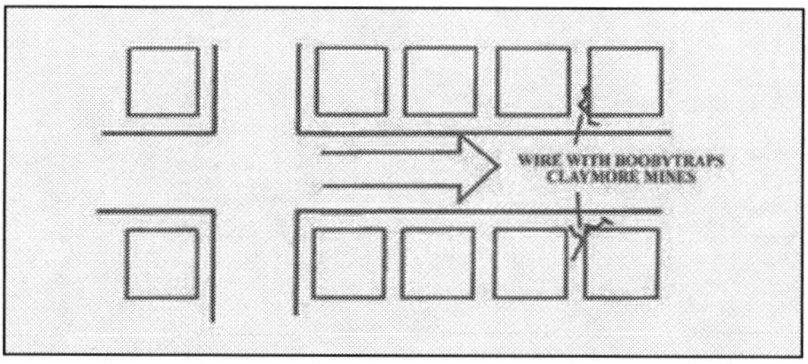

Figure F-7. Claymore Mine Used To Cover Dead Space of Automatic Weapons

(3) Claymore mines can be used in several ways in the offense. For example, if friendly forces are advancing on a city, claymore mines can be used in conjunction with blocking positions to cut off enemy avenues of escape (Figure F-8).

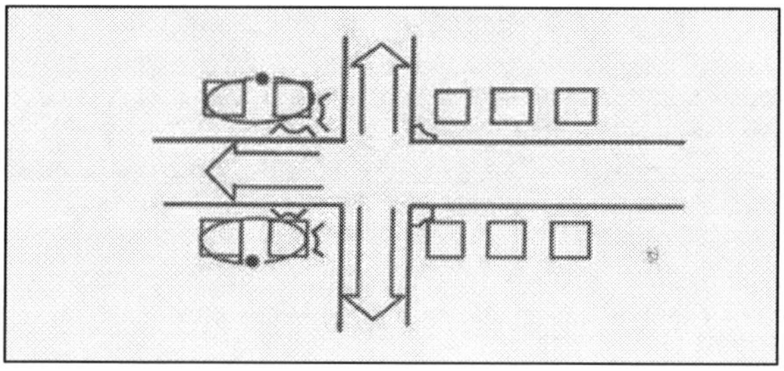

Figure F-8. Claymore Mines Used To Block Enemy Avenues of Escape

d. The M15, M19, and M21 antitank mines are used to stop or turn enemy armor and should be employed (Figure F-9):

- In conjunction with other manmade obstacles and covered by observation and fire
- In streets or alleys to block routes of advance in narrow defiles
- As command-detonated mines with other demolitions.

6. Enemy Mines and Boobytraps. Buildings contain many areas and items that are potential hiding places for boobytraps, for example, doors, windows, telephones, stairs, books, canteens, and so on.

Figure F-9. Emplacement of Antitank Mines

When moving through a building, Marines must not pick up anything—food, valuables, enemy weapons, or other items. Such items could be rigged with pressure devices that explode when they are moved. Marines must be well-dispersed so that if a boobytrap explodes, the number of casualties will be few. Many different types of mines and boobytraps could be encountered during combat in built-up areas (Figure F-10).

a. The equipment commonly used to detect mines in clearing operations includes:

- Mine detectors
- Probes
- Grappling hooks
- Ropes
- Explosives
- Flak vests
- Eye protection
- Engineer tape.

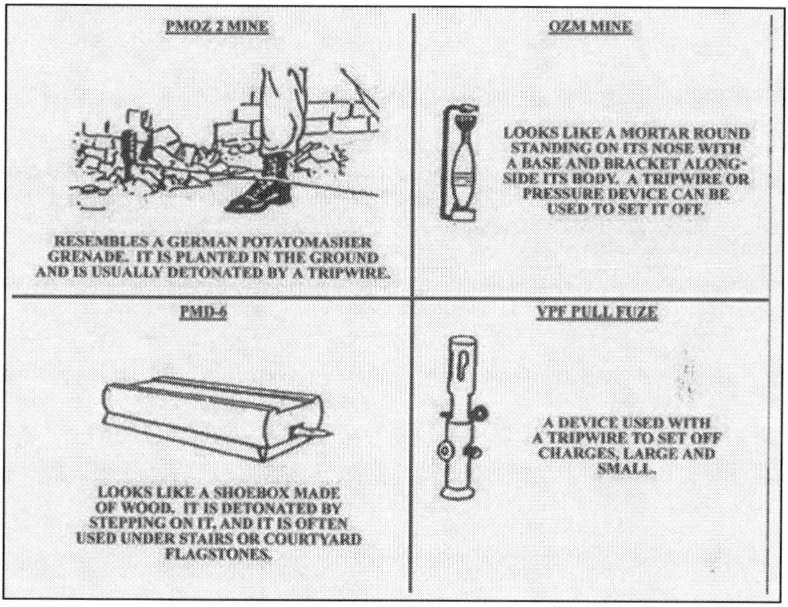

Figure F-10. Common Threat Antipersonnel Mines and Boobytraps

b. To detect tripwires, a Marine can use a 10-foot pole with 5 feet of string tied on one end. A small weight attached to the loose end of the string will snag on the tripwire. This allows the lead Marine to easily detect a tripwire (Figure F-11). Marines can also use a flashlight to cause reflections and shadows to find tripwires. A wand with a dangling thread can be swept over the ground ahead with the thread being watched for any disturbance by a tripwire.

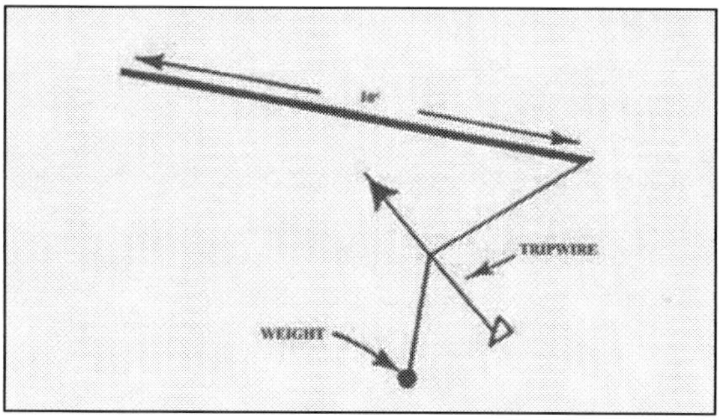

Figure F-11. Tripwire Detection

c. Many standard antipersonnel mines are packed in boxes and crates. If a Marine discovers explosive storage boxes, he should sketch them and turn the sketch over to the platoon commander for forwarding up the chain of command.

d. Most boobytraps should be neutralized by explosive ordnance disposal (EOD) personnel. If EOD teams are not available, boobytraps may be blown in place. Personnel should be protected by adequate cover. If the boobytrap is in a building, all personnel should go outside before the boobytrap is destroyed. Engineer tape placed around the danger area is one method of marking boobytraps. If tape is not available, strips ripped from bedsheets can be used. If possible, a guide should lead personnel through boobytrapped areas. EPWs and civilians may be a good source of information on where and how boobytraps are employed.

Section III

Demolitions

7. Introduction. Demolitions are used more often during combat in built-up areas than during operations in open natural terrain. Demolition operations should be directed by the engineers that support the GCE. However, if engineers are involved in the preparation and execution of the barrier plan, other trained Marines can prepare mouseholes, breach walls, and rubble buildings themselves if assisted and advised by attached engineers.

8. Offensive Use. When assaulting or clearing a built-up area, demolitions may enable the GCE commander to create an avenue of approach through buildings.

a. Every other man in an assault force should carry demolitions, and other selected personnel should carry blasting caps. The same man should *not* carry both the explosives and the blasting caps because of the danger of sympathetic detonation. As the demolitions are expended by the assault force, they should be replaced by explosives carried by the support force.

b. One of the most difficult breaching operations faced by the assault force is the breaching of masonry and reinforced concrete walls. When demolitions must be used, composition C4 is the ideal explosive to use. Normally, building walls are 15 inches thick or less. Assuming that all outer walls are constructed of reinforced concrete, a rule of thumb for breaching is to place 10 pounds of C4 against the target between waist and chest height. When detonated, this normally blows a hole large enough for a man to go through. The amounts of TNT and C4 required to breach concrete are shown in Table F-2.

c. Mouseholes provide the safest method of moving between rooms and floors. They can be created with C4. Because C4 comes packaged with an adhesive backing or can be emplaced using pressure-sensitive tape, it is ideal for this purpose. When using C4 to blow a mousehole in a lath and plaster wall, one block or a strip of blocks should be placed on the wall from neck to knee height. Charges should be primed with detonating cord or electrical blasting caps to obtain simultaneous detonation, which will blow a hole large enough for a man to fit through.

9. Defensive Use. The use of demolitions in defensive operations is similar to their use in offensive operations. When defending a built-up area, demolitions are used to create covered and concealed routes through walls and buildings that can be used for withdrawals, reinforcements, or counterattacks. Demolitions are also used to create obstacles and clear fields of fire.

a. Marines can use demolitions for creating mouseholes and constructing command-detonated mines. Expedient C4 satchel charges can be concealed in firing positions and along movement routes for detonation in the vicinity of the enemy.

	Reinforced Concrete		
Thickness of Material	**TNT**	**C4**	**Size of Opening**
Up to 10 cm (4 inches)	.792 lbs	.591 lbs	10 to 15 cm (4 to 6 inches)
10 to 20 cm (4 to 8 inches)	3.4 lbs	2.54 lbs	15 to 30 cm (6 to 12 inches)
	Nonreinforced Concrete Masonry		
Thickness of Material	**TNT**	**C4**	**Size of Opening**
Up to 35 cm (14 inches)	3.66 lbs	2.73 lbs	35 cm (14 inches)
35 to 45 cm (14 to 18 inches)	7.5 lbs	5.6 lbs	45 cm (18 inches)
45 to 50 cm (18 to 20 inches)	10.2 lbs	7.61 lbs	50 cm (20 inches)

Table F-2. Amounts of TNT and C4 Required To Breach Concrete

b. The engineers should furnish technical assistance for selective rubbling. Normally, buildings can be rubbled by using shaped charges or C4 on the supports and major beams.

c. Charges should be placed directly against the surface to be breached unless a shaped charge is used. Whenever possible, demolitions should be tamped to increase their effectiveness. Tamping materials could be sandbags, rubble, or desks and chairs (Figure F-12).

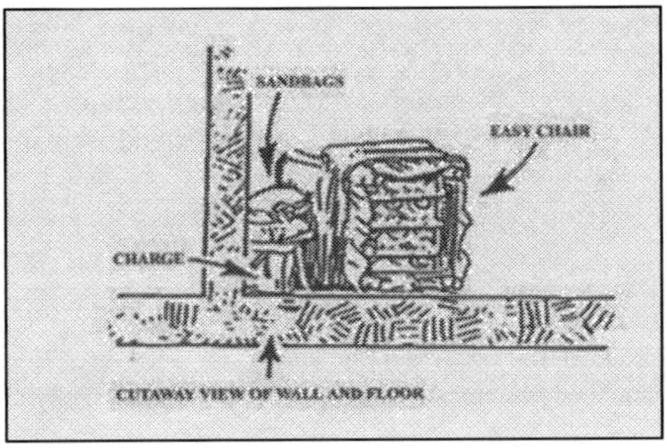

Figure F-12. Chair Used To Tamp Breaching Charge

d. For exterior walls, tamping of breaching charges may be impossible because of enemy fire. Thus, the untamped charge normally requires twice the explosive charge to produce the same effect as a tamped charge (Figure F-13).

METHODS OF PLACEMENT				
THICKNESS OF CONCRETE				
FEET	POUNDS OF TNT	POUNDS OF C4	POUNDS OF TNT	POUNDS OF C4
2	14	11	28	21
2 1/2	27	21	54	41
3	39	30	78	59
3 1/2	62	47	124	93
4	93	70	185	138
4 1/2	132	99	263	196
5	147	106	284	211
5 1/2	189	141	376	282
6	245	186	490	366

Figure F-13. Breaching Reinforced Concrete

e. When enemy fire prevents an approach to the wall, the breaching charge can be attached to a pole and slid into position for untamped detonation at the base of the wall (Figure F-14).

Figure F-14. Charge Placement When Small-Arms Fire Cannot Be Suppressed

f. The internal walls of most buildings function as partitions rather than as load-bearing walls. Therefore, smaller explosive charges can be used to breach them. In the absence of C4 or other military explosives, internal walls can be breached by using a claymore mine (Figure F-15). The claymore should be tamped to increase its effectiveness and to reduce the amount of explosive force directed to the rear.

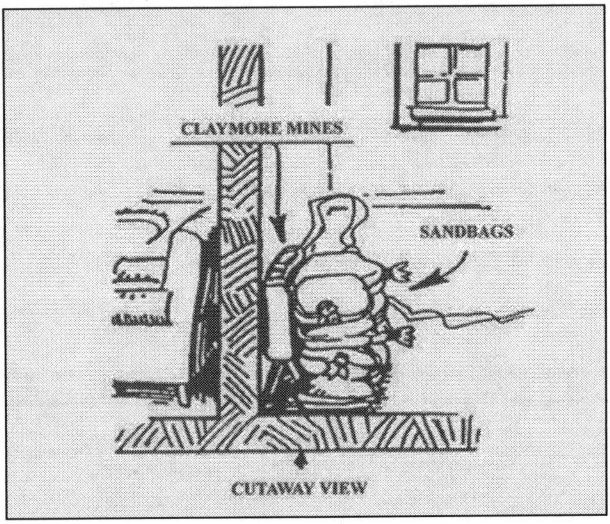

Figure F-15. Tamping a Claymore Mine To Breach Internal Walls

g. The Molotov cocktail (Figure F-16) is an expedient device for disabling both wheeled and tracked vehicles. It is easy to make because the materials are readily available. The results are most effective because of the close engagement in built-up areas. The objective is to ignite a flammable portion of the vehicle such as the fuel or ammunition that it is transporting. The following materials are needed to make a Molotov cocktail:

- Bottle or other glass container
- Gas (60 percent)
- Oil (40 percent)
- Rag for use as a wick.

The gas and oil are mixed thoroughly (60 percent gas to 40 percent oil). The rag is soaked with the mixture, then the mixture is placed into the bottle. The rag is then inserted in the opening of the bottle as a wick. When a target is sighted, the wick is lit and the bottle is thrown hard enough to break.

Figure F-16. Molotov Cocktail

WARNING

Ensure that a safe distance is maintained when throwing a Molotov cocktail.

h. The bunker bomb (Figure F-17) is an expedient, explosive flame weapon that is best used against fortified positions or rooms. This expedient munition should be used with a nonelectrical firing system. The following materials are required to make a bunker bomb:

- 1 metal small-arms ammunition container
- 1 gallon of gasoline
- 50 feet of detonating cord
- 1 nonelectrical blasting cap
- 1 M60 fuze igniter
- 7 1/2 feet of M700 time fuze
- 1 M49 trip flare or M34 WP grenade.

(1) Step 1. Fill the ammunition can 3/4 full with fuel and secure the lid.

(2) Step 2. "Hasty whip" the device with 15 turns around the center of the container using 44 feet of detonating cord. Leave 2-foot "pigtails" for attaching the igniter and fuze igniter.

(3) Step 3. Tape the igniter (M49 trip flare or M34 WP grenade) to the container handle.

(4) Step 4. Place one detonating cord pigtail end under the igniter spoon handle. Tape it in place.

(5) Step 5. Attach the M60 fuze igniter and the nonelectrical blasting cap to the M700 time fuze.

(6) Step 6. Attach the nonelectrical firing system to the other pigtail by making a loop in the detonating cord and attaching the blasting cap to it.

(7) Step 7. Remove the safety pin from the igniter (M49 trip flare or M34 WP grenade). The device is ready to be fired.

WARNING

Never carry the device by the handle or igniter. Remove the igniter safety pin only when it is time to use the device. Use extreme care when handling or carrying nonelectrical firing systems. Protect blasting caps from shock and extreme heat. Do not allow the time fuze to kink or become knotted. Doing so may disrupt the powder train and may cause a misfire. Prime detonating cord and remove the time fuze igniter safety pin only when it is time to use the device.

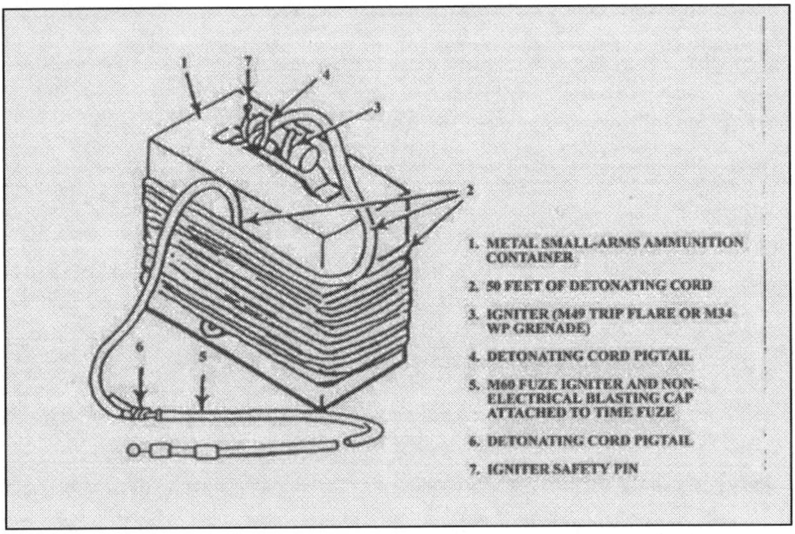

Figure F-17. Bunker Bomb Made From an Ammunition Can

10. Safety. The greatest danger to friendly personnel from demolitions is the debris thrown by the explosion. Leaders must ensure that protective measures are enforced. The safe distance listed in Table F-3 indicates the danger of demolition effects.

Pounds of Explosive	Safe Distance in Meters[1]	Pounds of Explosive	Safe Distance in Meters[1]
1 to 29	300	150	534
30	311	175	560
35	327	200	585
40	342	225	609
45	356	250	630
50	369	275	651
60	392	300	670
70	413	325	688
80	431	350	705
90	449	375	722
100	465	400	737
125	500	425	750

[1]These distances will be modified in combat when troops are in other buildings, around corners, or behind intervening walls.

Table F-3. Minimum Safe Distances for Personnel in the Open

a. The following are the guidelines for using demolitions:

- Keep the blasting machine under the control of a noncommissioned officer.
- Wear helmets and flack jackets at all times while firing explosives.
- Handle misfires with extreme care.
- Clear the room and protect personnel when blowing interior walls.

b. Some charges should be prepared beforehand, without detonators, to save time. Examples include 10- or 20-pound breaching charges of C4 and expedient-shaped charges in No. 10 cans.

- Use C4 to breach hard targets (masonry construction).

- Do not take chances.
- Do not divide responsibility for demolition work.
- Do not mix demolitions and detonators until ready to use.
- Do not carry demolitions and caps together.

Appendix G

Nuclear, Biological, and Chemical Considerations

Current U.S. policy regarding lethal or incapacitating agents is that their use against an armed enemy requires approval at the NCA level. Potential enemies may not operate under the same restrictions. Field commanders must be prepared to assume an adequate NBC defensive posture when engaged in urban fighting.

1. Protection From Nuclear, Biological, and Chemical Weapons. Generally, the lowest floor or basement of a reinforced concrete or steel-formed building offers good protection from nuclear hazards and initial liquid chemical contamination although some chemical agents tend to collect in lower areas. Tanks and LAVs/AAVs also provide protection. (See FMFM 11/MCWP 3-37 series manuals for greater detail.)

a. Biological attacks are difficult to detect and recognize. Biological agents can be disseminated by using aerosols, vectors, and covert methods. Because biological agents can be sprayed or dropped in bomblets, personnel who observe such indicators should promptly report them. Prompt reporting and treatment of the sick speed the employment of medical countermeasures. Although buildings and shelters provide some protection against spraying, they provide little protection against biological agents.

b. Chemical agents cause casualties by being inhaled or absorbed through the skin. They may afford Marines a few seconds to mask. Buildings have a channeling effect and tend to contain the effects of an agent, causing great variations in chemical concentration from room to room or from building to building. Chemical agents usually settle in low places, making sewers and subways hazardous hiding places. A prepared defender should include some collective protective measures in the defensive network. Personnel using fans may be able to put enough overpressure into tunnels to keep some chemical agents from entering. The protective mask and battle dress overgarment provide the best protection against chemical agents.

c. Personal hygiene is a critical defensive measure against infection and disease. Unfortunately, built-up areas are characterized by sophisticated sanitation systems. When those systems are destroyed, the resulting sanitary conditions become much worse than those in areas where sanitation facilities do not exist. Commanders must ensure that personnel employ sanitation measures and that their immunizations are current.

d. Commanders should plan their mission-oriented protective posture (MOPP) with the realization that increased logistics demands in built-up areas also apply to NBC equipment. Protective clothing, detection and decontamination equipment, and sealed containers of food and water must be stockpiled the same as other critical supplies. When operating in protective clothing, commanders must make allowances for the strenuous activities normally associated with combat in built-up areas.

(1) Detection. After an NBC attack, battalions should dispatch their detection and survey teams. Detection in built-up areas is complicated by the numerous rooms within buildings. Teams should conduct tests and surveys of major streets, intersections, and buildings in their area for inclusion in initial NBC reports. A systematic survey of all buildings, rooms, and underground facilities must be accomplished before occupation by unmasked personnel. All data should be forwarded using the appropriate NBC report.

(2) Decontamination. Personnel begin decontamination operations as soon after an NBC attack as the mission allows. Personnel should conduct individual decontamination of themselves and their personal equipment. Unit commanders determine the need for MOPP gear exchange and the requirements for a hasty or deliberate decontamination operation.

(a) Nuclear. Personnel should wear wet-weather gear for certain decontamination operations (hosing down buildings) to prevent radioactive material from touching the skin.

(b) Chemical and Biological. Roads, sidewalks, and other hard surfaces are best decontaminated by weathering, if time permits. Agents can also be covered with several inches of dirt or sand to provide protection. Fragment testing should be conducted periodically to ensure that the agent has not seeped through the covering. For critical sections of roads, a truck-mounted M12A1 (power-driven decontamination apparatus (PDDA)) can be used to spray supertropical bleach (STB) slurry; this aids rapid decontamination. Buildings are difficult to decontaminate, especially wooden ones. Some techniques for their decontamination are scrubbing with STB slurry; washing with hot, soapy water; washing or spraying with a soda solution; and airing.

2. Smoke Operations. The use of smoke is an integral part of either offensive or defensive operations in built-up areas. In the offense, smoke can support the maneuver of combat units and deception operations. Smoke employed in the defense obscures enemy air and ground observation, limiting the accuracy of enemy fires and intelligence collection.

a. Smoke should not be used when it degrades the effectiveness of friendly forces. Likewise, an extremely dense concentration of smoke in a closed area displaces the oxygen, suffocating Marines even when they are wearing protective masks.

b. Smoke pots, generators, or artillery smoke munitions should be used to cover the withdrawal of defending forces or the movement of attacking forces. Artillery-delivered WP can be effective against enemy forces by screening, in addition to causing casualties and fires. Consideration should be given to the incendiary effects of WP, smoke (HC and M825), and illumination munitions on the litter and debris of built-up areas.

c. Smoke grenades can be massed to provide a hasty screen for concealing personnel moving across streets and alleys. Smoke grenades can also be used for signaling; those launched by an M203 can be used to mark targets for fixed and rotary-wing air support.

d. The use of smoke in built-up areas is affected by the complex wind patterns caused by buildings. When covering a built-up area with a smoke haze or blanket, personnel must include all buildings. Failure to obscure tall buildings, towers, and steeples provides enemy observers with reference points for fire placement within the built-up area.

3. Riot Control Agents. When authorized, riot control agents such as CS and CN can be used to drive enemy forces from positions or to deny them areas for occupation. Riot control agents are incapacitating but generally have no lasting effects. They may be appropriate when preventing civilian casualties is a planning consideration. However, riot control agents may not be effective against an enemy that is well trained in chemical defense.

Appendix H

MOUT Under Limited-Visibility Conditions

To be successful, leaders must use limited-visibility conditions to their advantage.

1. Advantages. When fighting in built-up areas during night or periods of limited visibility, attacking or defending forces have several advantages.

a. In many cases, Marines may have a technological advantage in thermal imagery and light intensification over their opponents. This enables Marines to identify, engage, and destroy enemy targets before being detected by the enemy.

b. Possessing the capability to conduct operations as well at night or during periods of reduced visibility as during daylight gives Marines a tremendous advantage over an enemy with reduced capability. It also aids the generation of tempo and allows Marines to gain or retain the initiative.

c. Generally, ranges of direct-fire engagements are greatly reduced in the MOUT environment. During periods of limited visibility, effective unaided target acquisition ranges are further reduced. This enables attacking forces to close to shorter ranges, thus increasing the lethality and accuracy of weapons. Attacking forces can also take advantage of the enemy's reduced ability to see and can more effectively engage the enemy before being detected with thermal imagery or light intensification devices.

d. Helicopterborne assaults are best conducted during periods of limited visibility because the enemy's air defenses are degraded.

e. Attacking during periods of limited visibility gives the attacker a greater chance of achieving surprise.

2. Disadvantages. When fighting in built-up areas during periods of limited visibility, attacking and defending forces also face some disadvantages.

a. Command and control is difficult in any operation in a built-up area; periods of limited visibility increase this difficulty.

b. Marines have an instinctive tendency to move close together during periods of limited visibility. Constant attention must be given to prevent Marines from "bunching up."

c. Marines may become disoriented easily because of the combined effects of low visibility and the characteristics of built-up areas.

d. Target identification becomes more difficult in limited-visibility conditions. Depending on the training of the individual, the Marine may fire at anything seen or may hesitate too long before firing. Improper target identification is one of the leading causes of fratricide, so leaders must pay close attention to individual target engagement.

3. Fratricide Avoidance. The risk of fratricide is much greater during periods of limited visibility. The key to avoiding fratricide is increased situational awareness by leaders and individuals, coupled with realistic training in target identification. (See *Fratricide: Reducing Self-Inflicted Losses*, Newsletter, No. 92-4, April 92, Center for Army Lessons Learned, U.S. Army Combined Arms Command, Fort Leavenworth, Kansas.) Other considerations include:

a. Graphic control measures should be clearly defined and obvious. Examples include distinctive buildings, large boulevards, rivers, and so forth.

b. Leaders must exercise firm control when engaging targets. Movements should also be coordinated and controlled.

c. Cleared rooms and buildings should be distinctly marked to identify cleared areas and friendly forces to base-of-fire elements supporting the maneuver.

d. Visible markers (for example, glint tape or thermal strips) should be attached to individual Marines for rapid identification.

e. Far and near recognition symbols should be coordinated and used properly.

f. Units employing close air support must exercise firm control. Failure to do so may lead to the pilot becoming disoriented and engaging friend and foe alike.

4. Urban Environmental Effects on Night Vision Devices. Built-up areas affect standard night vision devices and sights differently than do open areas. This may cause some confusion because the images Marines receive through their night vision devices are unusual compared to those they may be used to.

a. Most built-up areas have electric power; therefore, street lights and/or building lights may "white out" some light intensification devices unless the power is disrupted or lights are turned off.

b. Fires may be burning in the urban area. Open flames cause problems not only for light intensification devices, but also for thermal devices.

c. Subterranean areas and the interiors of buildings will not have ambient light if the power is off. Passive night vision devices must then use an artificial light source, such as infrared radiation, to provide enough ambient light for the devices to work effectively.

d. The many reflective surfaces found in built-up areas may cause false images, particularly for laser range finders and laser target designators.

e. Thermal imaging devices may not be able to see through large amounts of dust particles suspended in the air.

f. Smoke and dust degrade the effectiveness of night vision devices.

g. Fog degrades long-range target acquisition from thermal sights.

h. Weapons flashes within enclosed areas appear to be much brighter. This causes Marines to lose their night vision and washes out light intensification devices.

5. Considerations. The environment of built-up areas presents special challenges and considerations during periods of limited visibility.

a. The use of glint tape, thermal tape, or chemiluminescent lights is an important consideration. These can be used to mark the FLOT, casualties, cleared buildings and rooms, weapons positions, and individual Marines. Their use must be clearly addressed in the unit's SOP. When markers are used for extended periods, their meanings should change because the enemy may be able to capture or manufacture and use these marking devices to their advantage.

b. The use of tracer and incendiary ammunition may be restricted to prevent fires. The light of fires whites out some night vision devices and may interfere with or confuse thermal devices.

c. The control of power stations may be essential to operations during periods of limited visibility. Control of power stations enables friendly forces to control, to a degree, background illumination. Shutting off the power to the street lights is much easier than shooting out the lights. Commanders must balance the tradeoff between force protection and maintaining law and order after the battle is over. During cold weather, the control of power stations may be critical to the welfare of the civilian population.

d. Identifying friendly units, noncombatant civilians, and enemy troops becomes more difficult during limited-visibility operations.

e. Locating the source of sounds becomes more difficult because of the natural echoing in built-up areas and the tendency of sounds to carry farther at night.

f. Locating boobytraps and obstacles also becomes more difficult at night. Movement rates are normally slower than during periods of normal visibility.

6. Special Equipment. Fighting during periods of limited visibility requires some specialized equipment to maximize advantages.

a. As a rule, thermal imaging devices such as the AN/PAS-7 infrared viewer and the AN/TAS-5 Dragon infrared sight are better for limited-visibility operations than are light intensification devices such as the AN/PVS-7B. Light intensification devices are more easily washed out by background light, weapons flashes in enclosed areas, and fires. Thermal devices, while also affected by fires, are not as easily washed out.

b. The AN/PAQ-4A infrared aiming device is similar to its off-the-shelf laser aiming sight counterparts, except it is not visible to the naked eye. Infrared pen lights can also be attached to weapons and, when used in conjunction with night vision devices, can provide a quick sight picture, illuminate rooms and hallways, identify obstacles and boobytraps, and identify friendly forces.

c. Other night sights for weapons include the AN/TVS-5 crew-served weapon night vision sight, the AN/PVS-4 individual weapons night vision sight, and the AN/UAS-12 night vision sight, which mounts to the modular universal laser equipment (MULE) and the TOW missile system.

d. Trip flares, flares, illumination from mortars, and artillery and spotlights (visible light or infrared radiation) can be used to blind enemy night vision devices or to artificially illuminate the battlefield (Figure H-1). (See FMFM 6-8, *Supporting Arms Observer, Spotter, and Controller*, for more information on illumination from mortars and artillery.)

e. Spare batteries for the night vision devices should be carried to keep the devices operational. Soft, clean rags should be used to clean the lenses.

7. Combat Support. Coordinating employment of forces and fire support in limited-visibility conditions is a major concern to leaders during MOUT. The concentration of forces and fires at the point of decision is facilitated by the technological edge that night vision devices provide.

a. Any degradation of accuracy in artillery fire will likely be a result of the limitations of target acquisition assets. While FOs and FACs may have thermal sights and laser range finders, most Marines on the battlefield do not yet have devices that will enable them to acquire targets accurately. The following are some devices and techniques to improve target acquisition for indirect fires:

(1) If the target is within LOS, tanks and LAVs can rapidly identify the target and provide an accurate range.

(2) Preregistered targets are effective if the target reference point can be observed and the observer has clear communications with the firing unit.

(3) Fixed-wing and rotary-wing aircraft can be used to identify targets and adjust supporting arms.

Figure H-1. Use of Indirect-Fire Illumination During MOUT

8. Combat Service Support. Maneuver unit commanders and their Marines are not the only individuals who must adjust to combat under limited-visibility conditions in MOUT. Logisticians at every level must anticipate requirements for these conditions.

a. Units conducting resupply operations during periods of limited visibility should consider:

- Issuing drivers and vehicle commanders night vision devices so that the vehicles going to and from logistic release points do not require illumination (This also prevents the enemy from acquiring resupply locations by following vehicles with blackout lights on.)

- Maintaining strict noise and light discipline

- Following a clearly marked route to avoid obstacles and prevent the resupply vehicle(s) from becoming disoriented

- Providing radios to resupply vehicles

- Issuing each vehicle a map of the AO (preferably a city map with the street names).

b. Combat units operating for an extended time under conditions of limited visibility should have enough batteries to keep the night vision devices functioning at optimum power and sensitivity.

c. A large operational readiness float of night vision devices should be maintained.

d. Casualty collection during periods of limited visibility is more difficult. Clear methods for marking any casualties must be established before the operation begins.

e. CSS operations located in existing structures should not be visible from a distance. This includes limiting vehicle traffic to an absolute minimum, sealing doors and windows to prevent light leakage, and dispersing assets as much as possible.

9. Operational Considerations. Marine units conduct attacks during periods of limited visibility to gain or sustain the momentum of the attack. Before conducting a limited-visibility attack, the commander must balance the risks and ensure that every Marine understands the mission, intent, and control measures. Rehearsals and strict command and control reduce casualties and greatly enhance the chances for mission accomplishment.

a. To reduce confusion, Marines should clear buildings and rooms using the same techniques they use during the day.

b. Movement rates are slower in the dark. Each Marine must remain alert for mines, boobytraps, and enemy positions.

c. Rifle squads and fire teams should be equipped with a mixture of both thermal imaging and light intensification devices whenever possible. This enables the squads and fire teams to obtain a better picture of the night environment and enables Marines to balance the strengths and weaknesses of each type of night vision device for maximum results.

d. When moving through buildings, assault forces mark cleared rooms and buildings and communicate with the support forces. Marking cleared rooms is especially critical if more than one assault force or element is in the same building. Communications with supporting forces are imperative to avoid fratricide.

e. If flashlights or chemiluminescent lights are used, they should be held away from the head or chest area. This will make it harder for enemy soldiers firing at the light to kill the Marine holding the flashlight or chemiluminescent light.

f. Units must know where everyone is during offensive operations. This not only reduces the risk of fratricide, but also decreases the time spent identifying, locating, and treating casualties. Also, it greatly reduces the chance of Marines becoming disoriented and separated from the unit.

g. Assault units should be aware of adjacent fires that may diminish the effectiveness of night vision devices. Weapons flashes within small rooms cause Marines to lose their night vision and can wash out light intensification devices. Also, enemy soldiers may use flares inside and outside of buildings to deliberately render night vision devices ineffective.

h. Leaders must ensure that all Marines follow the ROE and the laws of land warfare. This is critical if the enemy is intermixed with the local civilian population. Also, leaders must follow all control measures.

i. Enemy forces can be expected to use periods of limited visibility to the same advantages as Marine forces do.

j. Enemy forces may have access to sophisticated night vision devices manufactured in Europe, the United States, Japan, Korea, and the former Soviet Union.

Appendix I

Urban Building Analysis

As in other types of operations, success in urban combat depends largely on the ability to analyze the military aspects of terrain. Marines must be able to recognize certain terrain features when evaluating urbanized terrain.

1. Types of Mass-Construction Buildings. Mass-construction buildings are those in which the outside walls support the weight of the building and its contents. Additional support, especially in wide buildings, comes from using load-bearing interior walls, strongpoints (called pilasters) on the exterior walls, cast-iron interior columns, and arches or braces over the windows and doors (Figure I-1). Modern types of mass-construction buildings are wall and slab structures, such as many modern apartments and hotels, and "tilt-up" structures commonly used for industry or storage. Mass-construction buildings are built in many ways:

- The walls can be built in place by using brick, block, or poured-in-place concrete.
- The walls can be prefabricated and then tilted up, or they can be reinforced-concrete panels.
- The walls can be prefabricated and assembled like boxes.

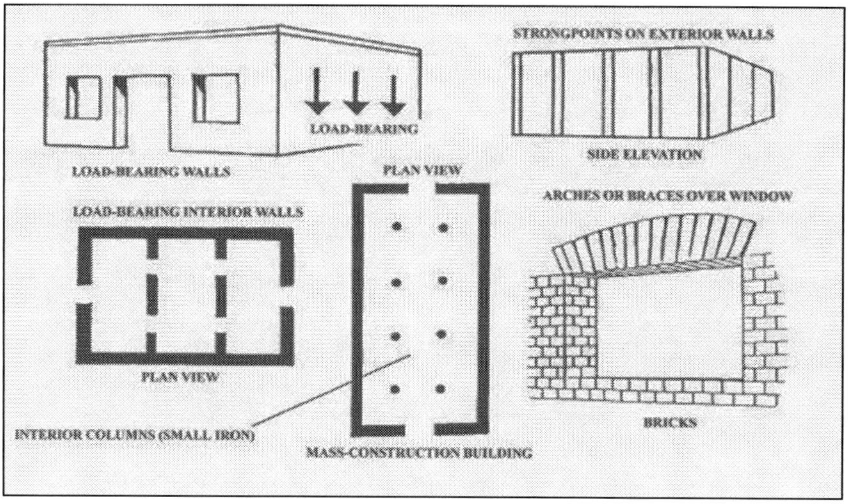

Figure I-1. Mass-Construction Buildings

a. Brick buildings are, in some regions, the most common and most important of the mass-construction buildings. In Europe, brick buildings are commonly covered with a stucco veneer so that bricks do not show (Figure I-2). One of the most common uses of brick buildings is the small store. These buildings are found in all built-up areas but are most common in the core periphery (Figure I-3).

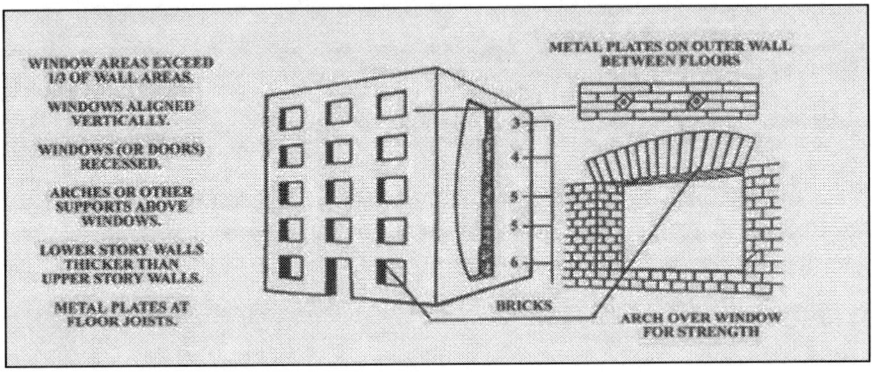

Figure I-2. Brick Buildings

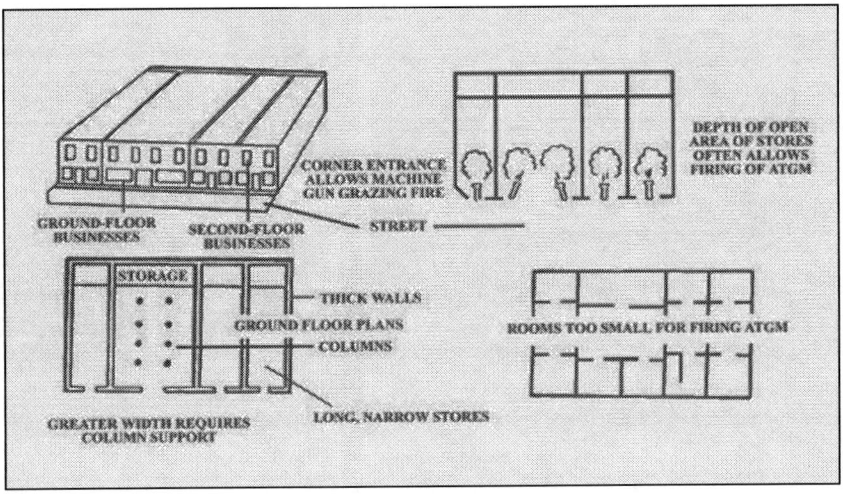

Figure I-3. Brick Store

b. Another common mass-construction building found mainly in industrial areas and along commercial ribbons is the warehouse. It is built of poured-in-place concrete reinforced with steel bars or of prefabricated tilt-up walls. The walls of warehouses provide good cover, although the roofs are generally light and vulnerable. The warehouses' large open bays permit firing of ATGMs or surface-to-air missiles and, because they are normally found in outlying areas, often afford adequate fields of fire for ATGMs. These buildings are built on slabs that can normally support the weight of vehicles and can provide excellent cover and concealment for tanks (Figure I-4).

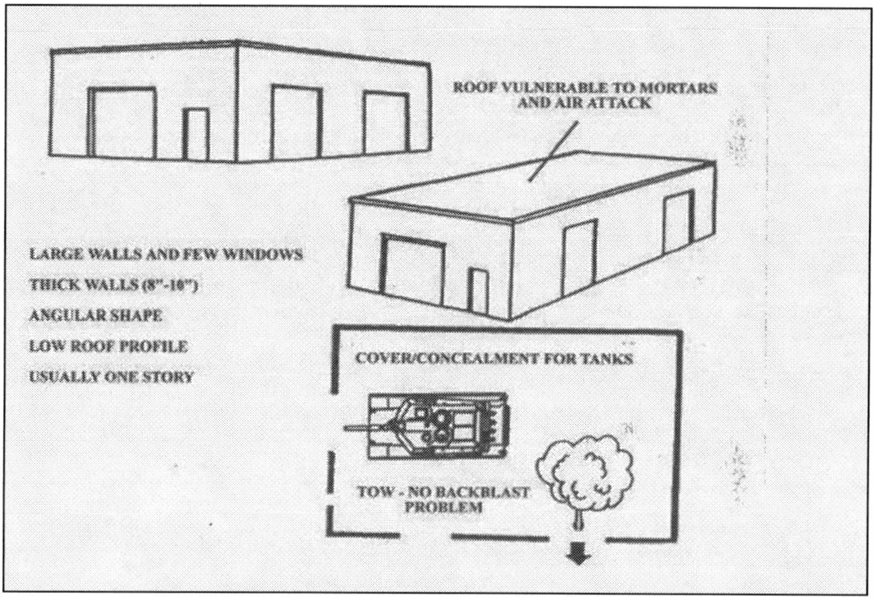

Figure I-4. Warehouse

c. Another type of mass-construction building is the box-wall-principle type. It is made from prefabricated panels of 6- to 8-inch-thick reinforced concrete. The outside wall is often glass. The box-wall-principle building provides good cover, except at the glass wall. The rooms are normally too small for ATGMs to be fired. A good circulation pattern exists from room to room and from floor to floor. These buildings are commonly used as hotels or apartments and are located mainly in residential and outlying areas (Figure I-5).

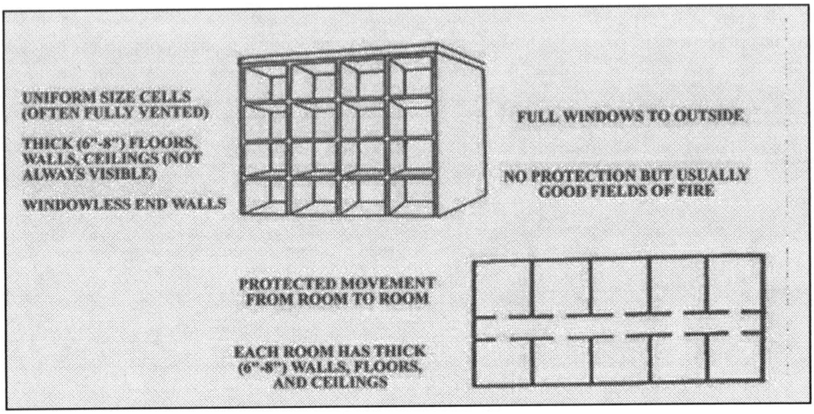

Figure I-5. Box-Wall-Principle Building

d. Public gathering places (churches, theaters) are mass-construction buildings with large, open interiors. The walls provide good cover, but the roof does not. The interior walls are not load-bearing and are normally easy to breach or remove. These buildings have adequate interior space for firing ATGMs or surface-to-air missiles. They are often located next to parks or other open areas and, therefore, have long fields of fire. Public gathering places are most common in core, core periphery, residential, and outlying high-rise areas (Figure I-6).

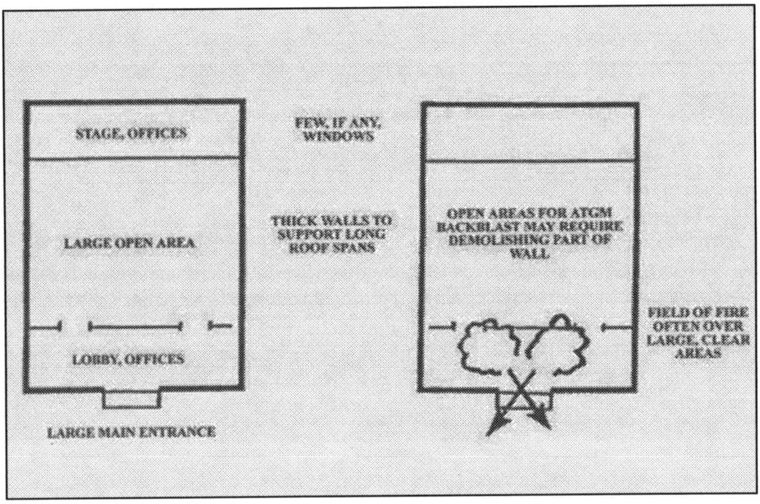

Figure I-6. Public Gathering Places

2. Types of Framed Buildings. Framed buildings are supported by a skeleton of columns and beams and are usually taller than frameless buildings (Figure I-7). The exterior walls are not load-bearing and are referred to as either heavy clad or light clad. Another type of framed building often found in cities is the garage, which generally has no cladding.

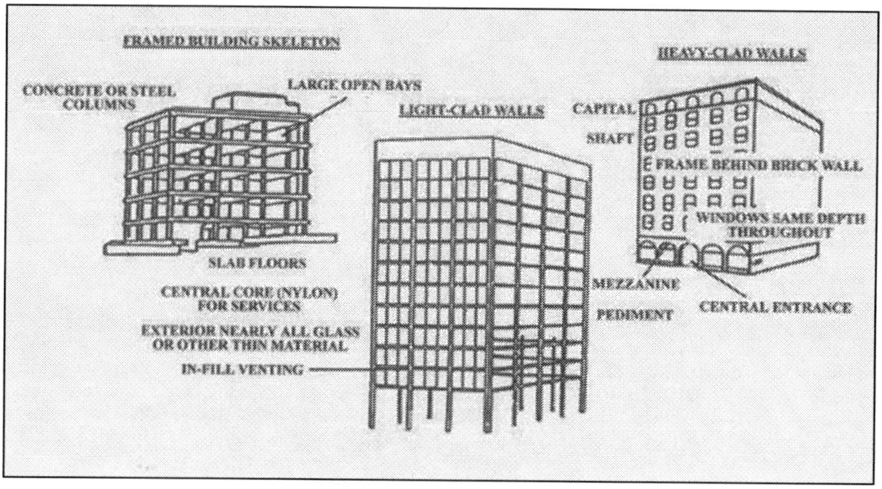

Figure I-7. Framed Buildings

a. Heavy-clad buildings were common when framed buildings were first introduced. Their walls are made of brick and block that are sometimes almost as thick as frameless brick walls, although not as protective. Heavy-clad framed buildings are found in core and core periphery areas. They can be recognized by a classic style or architecture in which each building is designed with three sections: the pediment, shaft, and capital. Unlike the brick building, the walls are the same thickness on all floors, and the windows are set at the same depth throughout. Often the frame members (the columns) can be seen, especially at the ground floor. The cladding, consisting of layers of terra cotta blocks, brick, and stone veneer, does not provide as good a cover as the walls of brick buildings. It protects against small-arms fire and light shrapnel but does not provide much cover against heavy weapons (Figure I-8).

(1) The floor plans of these buildings depend on their functions. Office buildings normally have small offices surrounding an interior hall. These offices have the same dimensions as the distance between columns (some large offices are as large as two times the distance between columns). These rooms are too small to permit firing of ATGMs but do provide some cover for snipers or machine gunners (Figures I-9 and I-10).

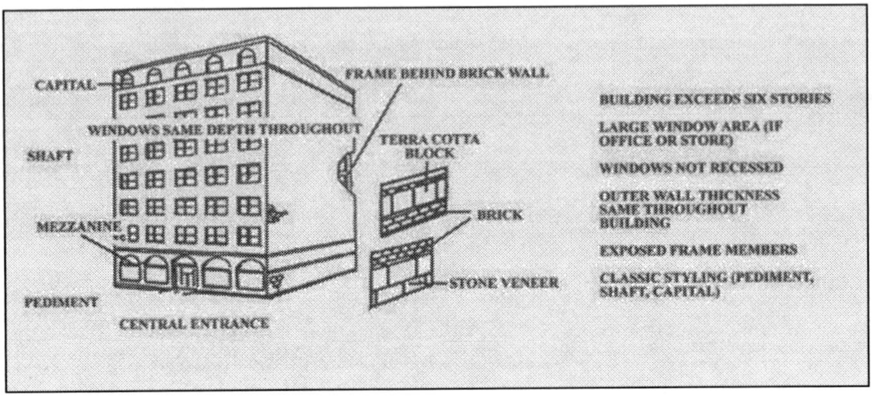

Figure I-8. Heavy-Clad Framed Building

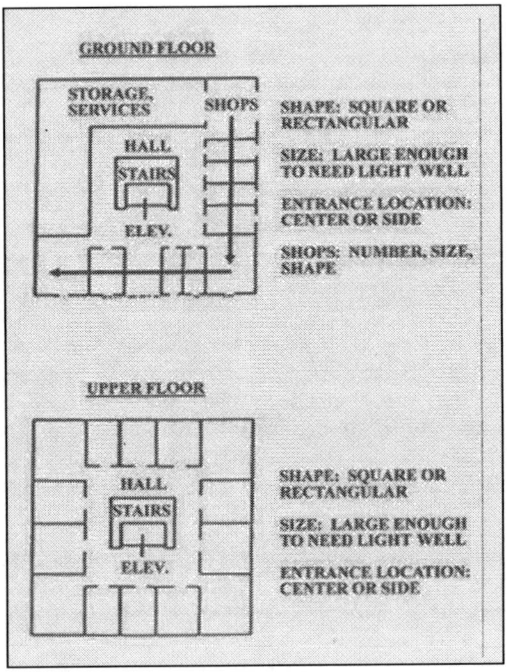

Figure I-9. Floor Plan of Heavy-Clad Framed Office Building

(2) Department stores normally have large, open interiors (Figure I-11). Such areas permit firing of ATGMs or man-portable surface-to-air missiles (if there are adequate fields of fire). Often a mezzanine level with a large backblast area permits firing down onto tanks. Steel fire doors often exist between sections of the store. The steel fire doors are activated by heat. Once closed, they are difficult to breach or force open, but they effectively divide the store into sections (Figure I-12).

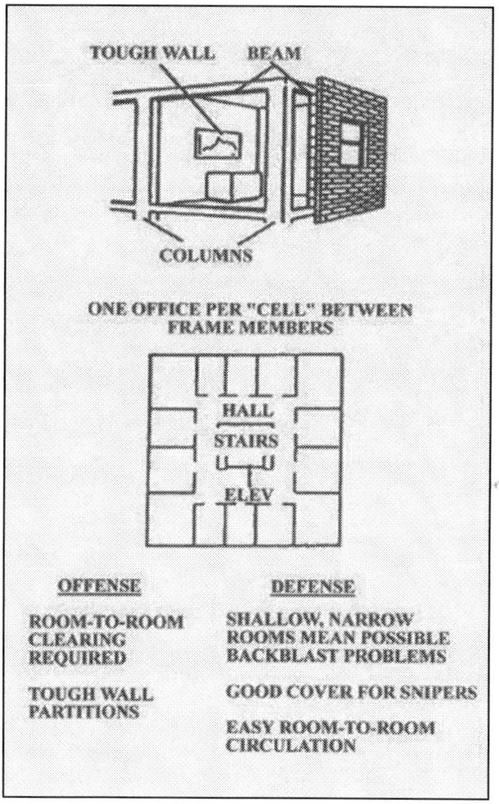

Figure I-10. Heavy-Clad Framed Office

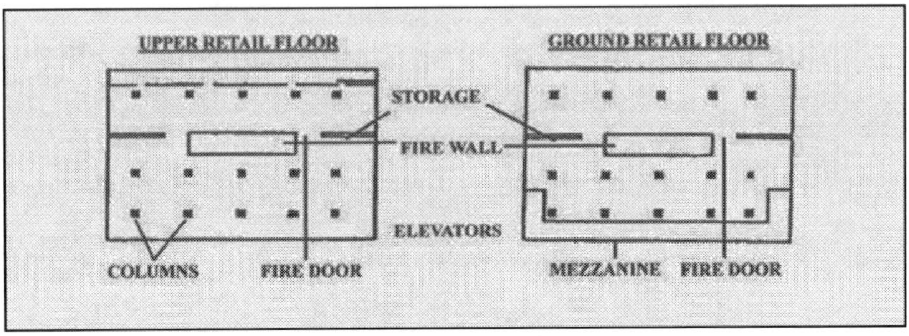

Figure I-11. Heavy-Clad Framed Department Store

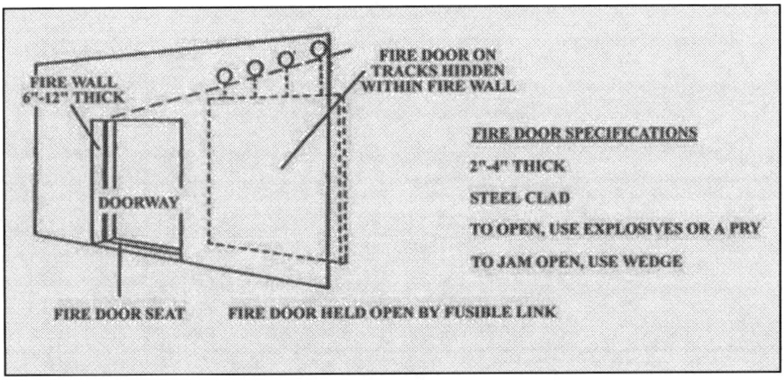

Figure I-12. Fire Wall and Fire Door

(3) Another type of heavy-clad framed building is used as a high-rise factory (Figure I-13). Such buildings are normally easily recognized because the concrete beams and columns are visible from the outside. They are usually located in older industrial areas. The large windows and open interior favor the use of ATGMs. Because the floors are often made to support heavy machinery, these buildings provide good overhead cover.

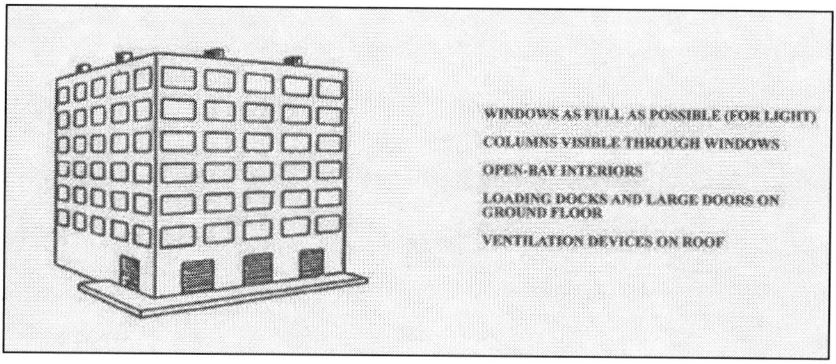

Figure I-13. High-Rise Factory

b. Light-clad buildings are more modern and may be constructed mostly of glass (Figure I-14). Most framed buildings built since World War II are light-clad buildings. They are found in both core and outlying high-rise regions. Their walls consist of a thin layer of brick, lightweight concrete, or glass. Such materials provide minimal protection against any weapon. However, the floors of the buildings are much heavier and provide moderate overhead cover. The rooms in light-clad framed buildings are much bigger than those in heavy-clad buildings. This feature, along with the fact that the buildings usually stand detached from other buildings, favors the employment of ATGMs. The interior partitions are thin, light, and easy to breach (Figure I-15).

Figure I-14. Light-Clad Framed Building

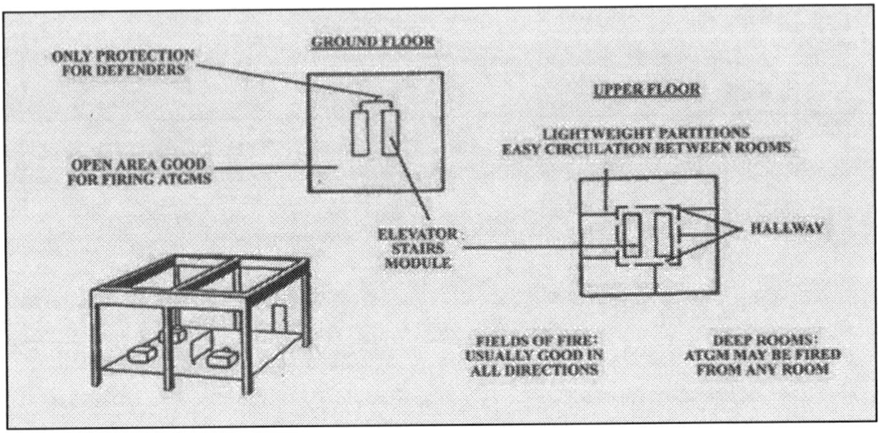

Figure I-15. Light-Clad Framed Room

c. The garage is one of the few buildings in an urban area in which all floors support vehicles. It provides a means to elevate vehicle-mounted TOWs, and the open interiors permit firing of ATGMs. Garages are normally high enough to provide a 360-degree field of fire for antiaircraft weapons. For example, a Stinger could hide under the top floor of the garage, come out to engage an aircraft, and then take cover again (Figure I-16).

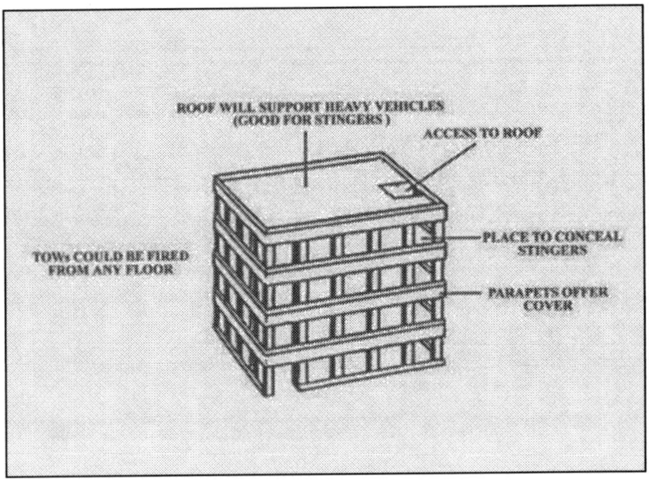

Figure I-16. Garage

3. Floor Plans. Floor plans in buildings follow predictable patterns. One of the factors that determines floor plans is building shape (Figure I-17). The basic principle governing building shape is that rooms normally have access to outside light. This principle helps Marines to analyze and determine the floor plans of large buildings.

4. Residential Areas. The two basic types of houses in the western world are located in and around cities and in rural areas. City-type houses are normally mass-construction brick buildings. Rural-type buildings in the continental United States, South America, and Southeast Asia are commonly made of wood, while in continental Europe, Southwest Asia, and sub-Saharan Africa, where wood is extremely scarce, rural buildings are normally constructed of concrete blocks. Rural-type buildings can also be found within cities (Figure I-18).

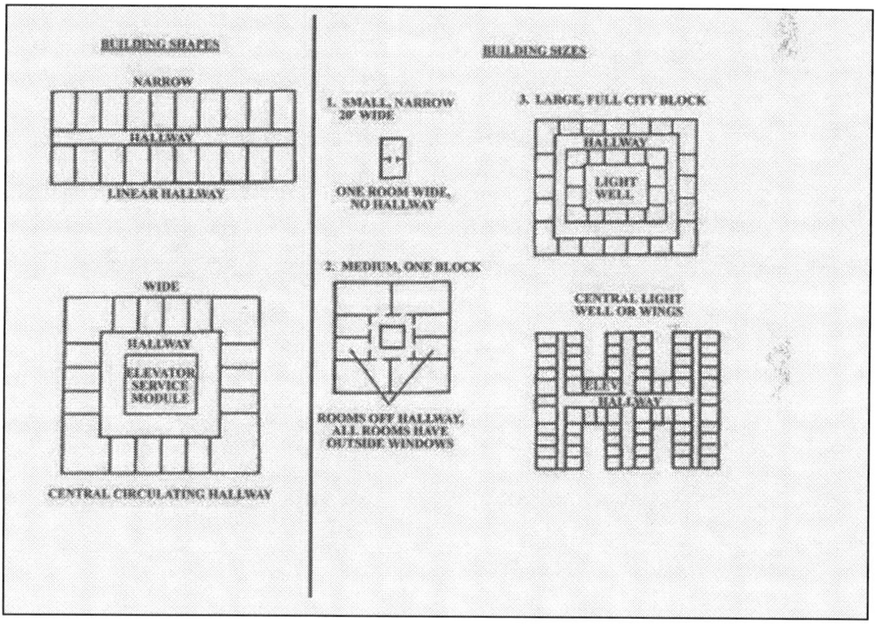

Figure I-17. Building Shapes and Sizes

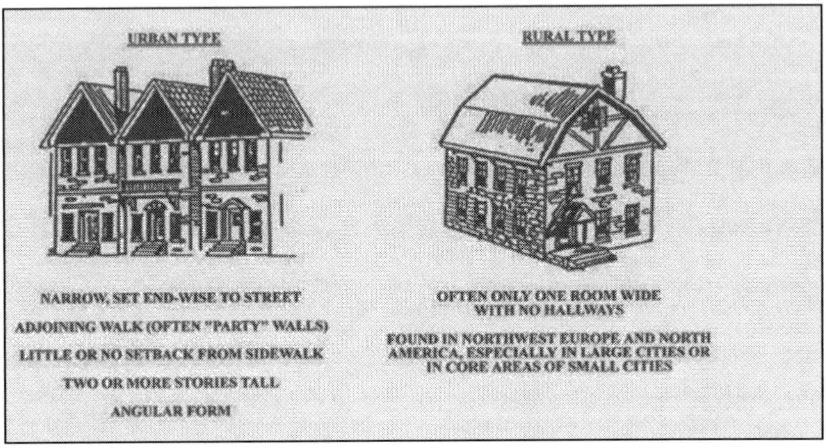

Figure I-18. Types of Housing

a. Another common type of building structure in cities with European influence is called the Hof-style apartment building (Figure I-19).

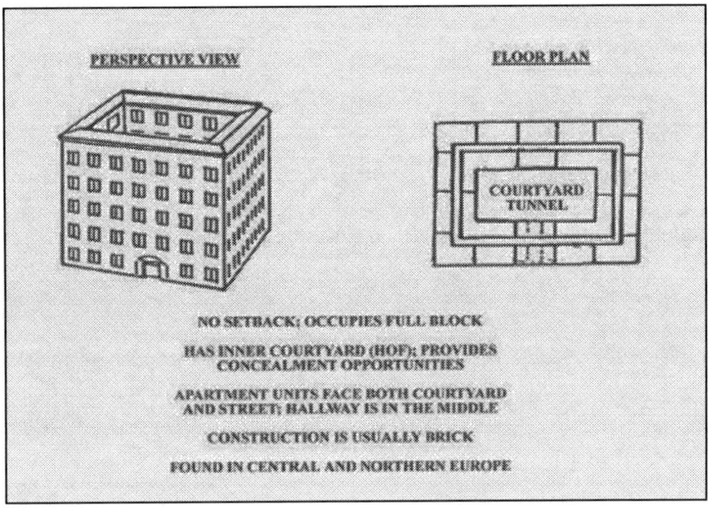

Figure I-19. Hof-Style Apartment Building

b. In the Middle East and tropical regions, the most common type of housing is the enclosed courtyard. Houses are added to one another with little regard to the street pattern. The result is a crooked, narrow maze that is harder to move through or fire in than dense European areas. (Figure I-20).

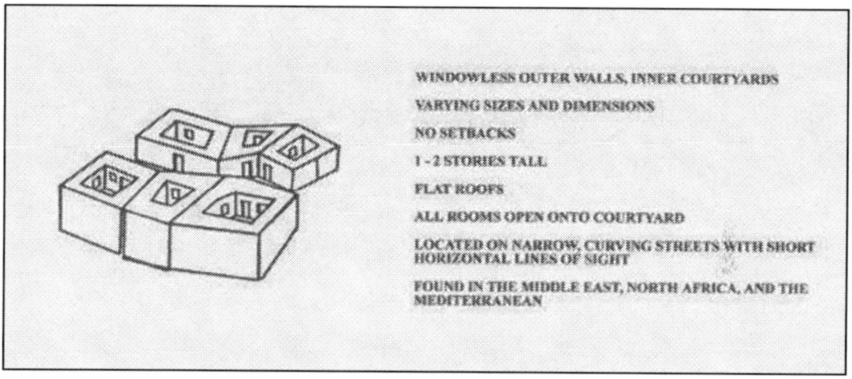

Figure I-20. Enclosed Courtyard

5. Characteristics of Buildings. Certain characteristics of both mass-construction and framed buildings can be helpful in analyzing a built-up area. Leaders can use Table I-1 to determine how to defend or attack a certain building given the unit's available weapons systems.

Type of Construction	Building Material	Height (Stories)	Average Wall Thickness (cm)
Mass	Stone	1 to 10	75
Mass	Brick	1 to 3	22
Mass	Brick	3 to 6	38
Mass	Concrete block	1 to 5	20
Mass	Concrete wall and slab	1 to 10	22 to 38
Mass	Concrete tilt-ups	1 to 3	18
Framed	Wood	1 to 5	3
Framed	Steel (heavy cladding)	3 to 50	30
Framed	Concrete/steel (light cladding)	3 to 100	2 to 8

Table I-1. Characteristics of Buildings

6. Distribution of Building Types. Certain types of buildings dominate certain parts of a city; this establishes patterns within a city. Analysis of the distribution and nature of these patterns has a direct bearing on military planning and weapon selection (Figure I-21).

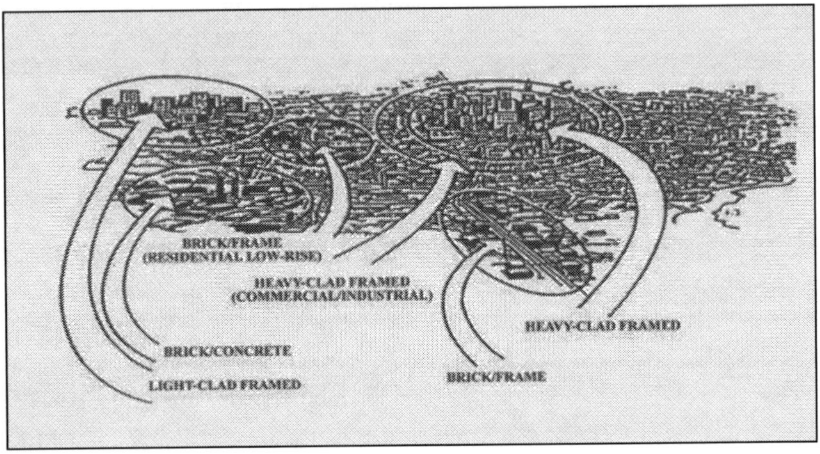

Figure I-21. Distribution of Building Types

a. Mass-construction buildings are the most common structures in built-up areas, forming about two-thirds of all building types. Brick structures account for nearly 60 percent of all buildings, especially in Europe.

b. Steel- and concrete-framed multistory buildings have an importance far beyond their one-third contribution to total ground floor area. They occupy core areas—a city's most valuable land—and as centers of economic and political power, they have a high potential military significance.

c. Open space accounts for about 15 percent of an average city's area. Many open spaces are grass-covered and are used for parks, athletic fields, and golf courses; some are broad, paved areas. The largest open spaces are associated with suburban housing developments where large tracts of land are recreation areas.

d. Streets serving areas consisting of mostly one type of building normally have a common pattern. In downtown areas, for example, high land values result in narrow streets. Street widths are grouped into three major classes: 7 to 15 meters, located in medieval sections of European cities; 15 to 25 meters, located in newer, planned sections of most cities; and 25 to 50 meters, located along broad boulevards or set far apart on large parcels of land. When a street is narrow, observing or firing into windows of a building across the street can be difficult because an observer is forced to look along the building rather than into windows.

When the street is wider, the observer has a better chance to look and fire into the window openings (Figure I-22).

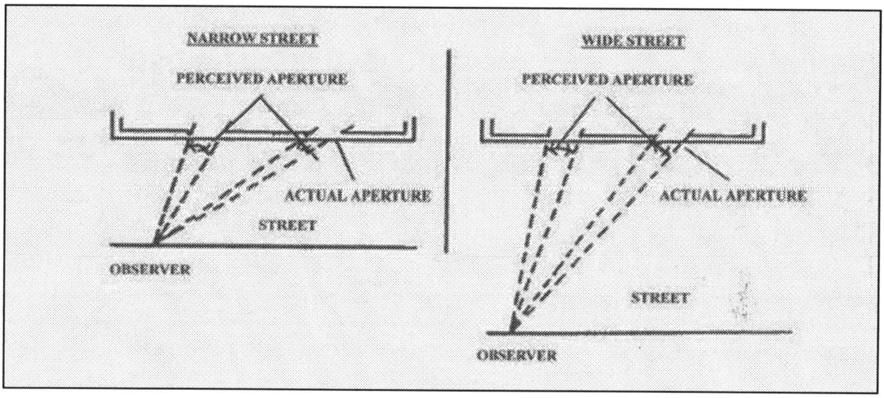

Figure I-22. LOS Distances and Angles of Obliquity

Appendix J

Lessons Learned from Russian Military Operations in Chechnya 1994-1996

1. Introduction. Many historical studies of urban combat have become dated. This is primarily the result of the impact that technology has had on military capabilities and evolving tactics. However, certain tenets of urban combat remain constant. The battles for the city of Grozny during the Russian intervention in the Republic of Chechnya represent a recent and critically important example of large scale operations in urban combat. Combat in Grozny was characterized by a large, technologically sophisticated military force (Russian) engaging and ultimately being defeated by a small, relatively primitive irregular force (Chechen). Grozny provides a number of fresh insights, and reinforcement of time honored tenets of urban warfare, across the scope of activities germane to modern urban combat.

2. Operations in Chechnya. Russia's war in Chechnya essentially began with the collapse of the Soviet Union, immediately after which pro-independence factions in Chechnya began agitating to free the Republic from Russian domination. This eventually led to civil war with pro-Russian factions within the Republic. Russia committed troops to Chechnya on 11 December 1994, for what many in the Russian government and military thought would be a quick campaign to restore Russian primacy. What followed was a bitter, costly, and protracted struggle that ended with the complete withdrawal of Russian military forces from Chechnya.

A primary focus of Russian operations in Chechnya was the capture of the capital city, Grozny. The initial attack on Grozny was conducted by a mechanized force, consisting of 6,000 troops mounted in tanks, BMP infantry fighting vehicles, and BTR armored personnel carriers. The Russians anticipated light opposition, but instead met a determined and heavy resistance from Chechens armed with a large number of antitank weapons. The attack was repulsed with significant Russian losses, both in personnel and armored vehicles (105 of 120 vehicles were lost). A second mechanized attack conducted on New Years Eve, 1994, was also repulsed with the loss of 140 of 200 tanks employed. The Russians captured Grozny some two months later, primarily through the use of excessive, overwhelming firepower, and at great cost to themselves and the local Chechen population.

The capture of Grozny marked the end of the first phase of Russian combat operations in Chechnya. Beginning in March 1995, the Russians became heavily engaged in anti-partisan operations as they tried to gain control of the country. Though the Russians controlled several major Chechen cities and a portion of the countryside by May 1995, they were never able to fully isolate the Republic. Chechen separatist fighters continued to receive a steady flow of arms and supplies from neighboring countries. Consequently, the Chechen fighters were able to maintain the initiative throughout this period, engaging Russian forces whenever and wherever they chose.

Even late in the campaign, Russian intelligence failed to detect Chechen fighters infiltrating Grozny, which enabled the Chechens to launch a major attack against the Russian-controlled city in early August 1996. The Chechens succeeded in capturing the city less than two weeks later.

The recapture of Grozny was a significant loss for the Russians, precipitating a general cease-fire and the withdrawal of Russian forces from the Republic.

Besides the military aspects of combat in Chechnya, other issues such as ethnic, cultural, and religious divides impacted upon the nature of the conflict. The Russians failed to culturally orient their troops, which resulted in many serious cultural mistakes when dealing with the Chechen civilians. Once insulted or mistreated, the Chechen civilians became active participants in the struggle against Russia. The war eventually degenerated into one of abject ferocity and brutality on both sides. The Chechens are reported to have executed Russian prisoners, decapitated bodies, and boobytrapped the dead. Russian wounded and bodies were hung upside down outside of Chechen defensive positions in buildings, causing the Russians to have to shoot through their own people to engage the Chechens. The Russians, initially hesitant to cause civilian casualties, increasingly engaged Chechen units with a growing disregard for the safety of the local populace. As Russian discipline degraded, the incidences of Russian troops engaging in looting, arson, torture, even the summary execution of civilians, rose significantly. The result was the deaths of tens of thousands of civilians.

3. Lessons Learned. The intervention in Chechnya and the battles for Grozny generated many important lessons across the warfighting functions of combat in urbanized terrain. The following information is primarily from the Russian point of view, though some Chechen capabilities are included. The lessons learned are arrayed as warfighting functional considerations.

a. Command and Control.

(1) Russian military commanders lacked clear policy guidance. The Russian military's ability to attain and maintain steady, logical operational progress was hampered, if not negated, by a lack of clear governmental policy. The Russian government was split at the highest levels over the wisdom of committing forces to Chechnya, and the conduct of the intervention.

(2) Poor Russian command and control structure throughout all levels. There was no direct, unified chain of command for the operations in Chechnya. Operational directives often flowed from a number of governmental agencies directly to local unit commanders, often bypassing intermediate military commanders. This resulted in a lack of overall situational awareness, and exceptionally poor coordination among tactical forces.

(3) Russians were unable to contain/prohibit Chechen operations against Russia. When the Russian forces finally began to achieve progress in late May 1995, the Chechens initiated attacks against targets inside Russia. The result was a major propaganda victory for the Chechens.

(4) Excessive senior officer (flag level) interference contributed to operational confusion. There were eight major changes at the senior command level during the course of the intervention. Additionally, there were approximately one hundred general officers on the operational scene, about one general officer per 3,000 troops, which resulted in confusion as they got in each others way.

(5) Poor small unit leadership, particularly at the NCO level, was a primary cause of Russian tactical failures in Grozny. While the Russians recognize that urban combat is small unit intensive, they do not possess a professional NCO corps. Additionally, Russian forces lacked some 12,000 platoon leaders when the intervention began. This had a crippling effect on small unit operations.

(6) Russian inability to isolate Chechnya/Grozny allowed a constant flow of reinforcements and supplies into the country/city. Russian control of any given area was at best, porous. The Chechens continued to receive outside support throughout the campaign, as well as, a large supply of arms and equipment from poorly disciplined Russian troops inside Chechnya.

(7) Both sides used commercial off-the-shelf communication assets. The information war was as vitally important as any other aspect of the campaign. The Chechens had to make do with commercially available communication equipment. They fully exploited the use of cellular telephones, Motorola radios, improvised television stations, video cameras, and the internet. The Russians lacked sufficient military communication equipment, and had to supplement with commercial assets.

(8) Russian tactical communication was very difficult in Grozny. Urban structures in Grozny interfered with Russian military radios, severely hindering tactical communications. The Russians had to establish ground-based and aircraft-based relay stations to overcome the interference.

(9) Boundaries between units were tactical weak points. Units had to think of boundaries in both a horizontal and vertical context. There were many instances when the Chechens held the third floor and above, while the Russians held the first two floors and maybe the roof. This caused a lack of coordination, and mutual support. Additionally, due to poor command and control and coordination between units, boundaries became seams through which the Chechens could pass unopposed.

b. Intelligence.

(1) Lack of high quality intelligence made operations much more difficult and dangerous. Russian intelligence collection and dissemination capabilities were poor throughout the intervention. In March 1996, the Russian Minister of the Interior complained that poor reconnaissance and intelligence allowed Chechen fighters to move in and out of Grozny without opposition, which contributed to the heavy Russian losses.

(2) Directives to minimize civilian casualties ultimately put the Russians at a disadvantage. An advantage accrues to the side with less concern for the safety of the civilian population. Initially, the Russians attempted to limit civilian casualties. The Chechen

.3

fighters took advantage of this reticence, often using civilians, schools, and apartment buildings as a shield. While this may seem to be cold-blooded of the Chechen fighters, what must be understood is that the majority of the populace in Grozny were actually ethnic Russians, and that the ethnic Chechens had no remorse in using them in this manner.

(3) Russian concerns for the safety of the Chechen populace and property declined. As the Russian troops became increasingly frustrated in their attempts to separate Chechen fighters from the local populace, and their casualties began to mount, concerns for civilian safety waned, resulting in greater civilian casualties and further alienating the Chechen people.

(4) Rigorous communications security is essential, even against less sophisticated enemies.Much of the Russian tactical radio traffic was broadcast in the clear, which allowed the Chechens to intercept it. The Chechens were able to redirect Russian airstrikes onto Russian targets, and engage Russian artillery forward observers who broadcast their own position coordinates.

(5) The Russians made extensive use of RPVs for reconnaissance. The Chechen campaign saw the first Russian use of unmanned aerial vehicles for reconnaissance and intelligence gathering. There were primarily used by Russian airborne forces.

c. Maneuver.

(1) Inadequate training in the most basic maneuver and combat skills inhibited Russian operations. Overall, poor Russian combat performance could be traced to a lack of training in fundamental military skills, a situation that was then exacerbated by an additional lack of training in specific urban combat skills. Individual skills training was so inadequate that some units were entirely incapable of undertaking combat operations. The Russian's ability to conduct large scale operations had been seriously eroded by a complete cessation of exercises at the division level in the 2 years prior to the Chechen campaign. Even regimental and battalion level exercises had been curtailed by 75 percent or more. Additionally, The air forces were not any better trained or prepared for their operational role in Chechnya.

(2) Lack of cohesiveness among the Russian forces led to confusion and antagonism. The Russian ground forces consisted primarily of regular army units and Ministry of Interior Internal Troops, which were essentially a police force. With no real unified command and control, the two groups were usually conducting uncoordinated missions, which led to confusion and fratricide.

(3) The nature of cities channels combat operations along narrow lanes of activity. Combat conditions in Grozny were characterized by narrow fields of view, limited fields of fire, and constricted avenues of approach. For the Russians, this meant movement along approaches well known and heavily defended by the Chechens.

(4) Night fighting was the most difficult activity for Russian infantry. The Russian infantry, at all levels, was inadequately trained for night operations, and lacked night vision equipment.

(5) Armored vehicles cannot operate in cities without extensive dismounted infantry support. The Chechens formed antitank hunter-killer teams equipped with RPG-7 and RPG-18 antitank missiles. They engaged Russian tanks with volley fire from above, behind, and the sides. Armor columns not supported by dismounted infantry suffered great losses.

(6) Armored engineer vehicles are indispensable for removing obstacles. The Russians found that armored engineer vehicles were critical to restoring or improving mobility in Grozny. Eventually, the Russians formed special assault groups that included armored engineer vehicles in each group.

(7) The Chechens preferred small unit, hit-and-run ambush tactics. The Chechens relied on small units (15 to 20 troops) to conduct quick, sharp ambushes, from which they would disengage before the Russians could bring significant firepower to bear. Ambushes were often conducted in three tiers, with Chechens in the underground, on the ground floor, and on the roof, with each element having a different task in the ambush. They also used ambush tactics against Russian helicopters.

(8) Urban combat is extremely manpower intensive, and results in significant attrition of personnel and materiel for the attacker. The Russians discovered that a 5 to 1 manpower ratio was often not enough, due to the high casualty rates consistent with urban combat and the requirement to guard virtually every building taken.

(10) Tracked vehicles are preferable to wheeled vehicles in the urban environment. Tracked vehicles provided better mobility than wheeled vehicles due to their significantly greater ability to negotiate obstacles and rubble.

(11) Armored vehicles required additional protection (i.e., reactive armor, sandbags, wire mesh, etc.) to negate attacks from the sides, rear, and top. Russian tank design emphasized greater armor protection on the frontal arc, and less on other areas of the vehicle. This proved inadequate for urban combat, where attacks can come from any direction. The majority of lethal hits came from above, easily penetrating turrets and engine decks, and from the rear.

(12) The spatial qualities and perspectives of urban combat are more "vertical" than non-urban combat. The Chechens regularly fought from basements, rooftops, and upper-floor rooms, making them difficult to locate and engage. Russians soldiers trained in open areas, had to reorient to a confined, highly vertical environment.

.3

(13) Standard Russian military unit configurations were inappropriate for urban combat. Typical Russian army unit organizations, oriented to open spaces and mechanized operations, proved to be inadequate for urban combat. They eventually had to adopt specially configured assault teams that included, among other assets, significantly greater firepower and engineer assets. Essentially, they moved towards a better combined arms team.

(14) The Russians found their helicopters to be far too vulnerable to rooftop snipers and ambushes in the urban setting. While Russian doctrine had long specified the capture of a building from the top-down, it became apparent in Grozny that this tactic was too expensive in terms of helicopter losses. The Chechens made effective use of snipers and RPG-equipped ambush teams to damage or destroy helicopters supporting Russian ground forces.

d. Fires.

(1) Artillery was invaluable in the direct fire mode. The Russians found artillery, to include rocket launchers and mortars, to be very effective for direct fire against point targets , usually at ranges of 150 to 200 meters.

(2) Obscurants (smoke, white phosphorous, tear gas) are especially useful in urban combat. The Russians used significant amounts of smoke and white phosphorus to mask the movement of forces. Every fourth or fifth artillery or mortar round was either smoke or WP. WP also had a toxic effect, easily penetrating Chechen protective masks.

(3) The Chechens effectively used RPGs against helicopters. The Chechens made effect use of ambush teams equipped with RPGs to engage Russian helicopters, either by damaging or destroying the helicopter or simply prohibiting helicopterborne operations.

(4) Air defense guns (i.e., ZSU-23-4) were valuable in suppressing ground targets. The Russians found air defense guns effective against multi-story buildings because they had sufficient elevation to hit targets in the upper stories. Air defense guns became a primary asset in the Russian special assault groups.

(5) The Russians determined that overwhelming firepower can make-up for organizational and tactical deficiencies in the short run if collateral damage is disregarded. Whenever Russian military units proved incapable of effective urban operations, the Russians simply fell back on massive, overwhelming firepower to reduce a strong point. It was through the use of excessive, unrestrained firepower, and a complete lack of regard for collateral damage, that finally enabled the Russians to gain control of Grozny.

(6) The Chechens negated Russian supporting fires by "hugging" Russian units. Whenever possible, Chechen fighters adopted the tactic of staying as physically close to Russian units as possible, so that the Russians could not employ supporting arms without the risk of injuring their own troops.

(7) The Russians found "bunker busting" weapons to be invaluable. The Russians came to rely heavily on "bunker busting" weapons for engaging Chechens, primarily snipers, entrenched in buildings.

(8) Russian helicopters needed stand-off weapons, that would allow them to engage targets while outside the range of Chechen weapons. The Chechens employed 23mm cannons and heavy machine guns against Russian helicopters, causing the Russian's helicopter gunships to fire from ranges of 3,000 meters or more. This required the use of wire-guided munitions and large caliber (240mm) rockets.

(9) The Russians made extensive use of precision-guided weapons. When the weather allowed, the Russians were able to use precision guided weapons, such as laser-guided bombs and missiles. They had great effect against priority targets such as bridges, major road intersections, and buildings.

e. Logistics.

(1) Deficiencies in the Russian logistics system degraded combat effectiveness. Significant inadequacies in the Russian logistics apparatus effected virtually all combat forces and operations. Some soldiers actually entered combat in Grozny without weapons or ammunition. The Russian transportation system failed. Supply officers were unprepared for the high volume demands for ammunition, hand grenades, smoke grenades, demolition charges, and one-shot antitank weapons.

(2) Russian logistics personnel were ill-prepared in basic soldier skills. Russian infantry often had to be used to conduct logistics operations into Grozny, because the logistics personnel lack the basic skills to defend themselves against the Chechens.

(3) The Russian ground forces required special equipment not normally found in unit T/Es. The T/E for the typical Russian infantry unit did not include such items as ropes, grappling hooks, and ladders, all of which proved essential for urban combat.

(4) Armored vehicle recovery is extremely difficult and dangerous. The Russians found the recovery of vehicles to be extremely difficult due to the terrain (i.e., narrow streets, rubble and debris), and Chechen opposition.

(5) Some Russian equipment was modified in the field to counter Chechen tactics and equipment. The Russians had to adopt add-on armor for their vehicles. They fabricated metal screens and wire mesh guards, which helped to defeat RPG and Molotov cocktail attacks.

(6) Medical support was of great importance. Russian casualty evacuation was difficult due to the lack of vehicular mobility, and the vulnerability of helicopters to Chechen fire. Many casualties were psychological, due in great part to the intense, violent nature of

.3

close quarter combat. The Russians were ill-prepared for the level of mental health problems they encountered. Additionally, logistics units were often unable to provide fresh drinking water, which caused Russian troops to consume contaminated water that resulted in other health problems.

f. Force Protection.

(1) Operations took much longer than expected, and the cost in terms of casualties was much greater than anticipated. The Russians committed unprepared, poorly trained and poorly equipped forces to Chechnya, based on an unrealistic operational time frame. Ultimately, this cost the Russians significant casualties and the war.

(2) Fratricide was a serious and continuing problem throughout the campaign. The situation facing Russian combat forces was ripe for fratricide. Poorly trained units, operating in a confused and uncertain urban environment, often unable to tell friend from foe, and lacking quality leadership and inter-unit coordination, were often as dangerous to themselves as they were to the Chechens .

(3) Chechen use of adhoc air defense systems eliminated Russian air support, particularly helicopters, in Grozny. The Chechens were able to improvise effective anti-air defenses using a mixture of anti-air and ground (i.e., heavy machine guns, RPGs) weapons, and ambush tactics specifically adapted to the urban environment.

(4) The Chechens made extensive and effective use of snipers. Snipers fired from well inside rooms versus near window openings, as well as, from rooftops and basements. The Russians lacked an effective sniper and counter-sniper capability of their own.

(5) The Chechens improvised crude chemical weapons. The Chechens, lacking access to military chemical weapons, Used off-the-shelf materials and equipment to make chemical mines (i.e., chlorine gas), detonated remotely by radio.

(6) The Chechens made extensive and effective use of boobytraps. It seemed that the Chechens mined and boobytrapped 7-8everything. They had a good understanding of the average Russian soldier's actions and reactions, and boobytrapped accordingly. Additionally, the Russians found it difficult to maintain boobytrap awareness among their troops.

Appendix K

Glossary

I. Acronyms

AAA .... antiaircraft artillery
AAAV .... advanced assault amphibious vehicle
AAV .... assault amphibious vehicle
ACE .... aviation combat element
AO .... area of operations
APC .... armored personnel carrier
APDS-T .... armor-piercing, discarding sabot with tracer
APFSDS .... armor-piercing, fin-stabilized, discarding sabot
API .... armor-piercing incendiary
ARVN .... Army of the Republic of Vietnam
ATF .... amphibious task force
ATGM .... antitank guided missile
BDA .... battle damage assessment
BMNT .... .beginning of morning nautical twilight
CA .... civil affairs
CAG .... civil affairs group
CE .... command element
CI .... counterintelligence
COC .... combat operations center
CONAD .... consolidated administration center
CP .... command post
CSS .... combat service support
CSSE .... combat service support element
DOD .... Department of Defense
DS .... direct support
EOD .... explosive ordnance disposal
EPW .... enemy prisoner of war
EW .... electronic warfare
FAC(A) .... forward air controller (airborne)
FASCAM .... family of scatterable mines
FDC .... fire direction center
FEBA .... forward edge of the battle area
FLOT .... forward line of own troops
FM .... U.S. Army field manual
FM .... Frequency modulated
FMFM .... Fleet Marine Force manual
FMFRP .... Fleet Marine Force reference publication
FO .... forward observer
FPL .... final protective line
FRAGO .... fragmentary order
FSC .... fire support coordinator
GCE .... ground combat element
GPS .... global positioning system
GS .... general support
HE .... high explosive
HEAT .... high-explosive antitank
HEAT-MP .... HEAT-multipurpose
HEDP .... high-explosive dual-purpose
HEI-T .... high-explosive incendiary with tracer
HLZ .... helicopter landing zone
HMMWV .... high-mobility, multipurpose wheeled vehicle
HSS .... health services support
HUMINT .... human intelligence
IDF .... Israeli Defense Forces
IFV .... infantry fighting vehicle
IPB . . intelligence preparation of the battlefield/battlespace
ITOW .... improved TOW
JTF .... joint task force
KOCOA .... key terrain, observation and fields of fire, cover and concealment, obstacles, and avenues of approach
LAV .... light armored vehicle
LCAC .... landing craft air cushion
LD .... line of departure
LOS .... line of sight
LZ .... landing zone
MAGTF .... Marine air-ground task force
MCDP .... Marine Corps doctrinal publication
MCRP .... Marine Corps reference publication
MCWP .... Marine Corps warfighting publication
MEDEVAC .... medical evacuation
MEF .... Marine expeditionary force
MET .... meteorological
METT-T ... mission, enemy, terrain and weather, troops and support available-time available
MEU .... Marine expeditionary unit
MOOTW .... military operations other than war
MOPP .... mission-oriented protective posture
MOUT .... military operations on urbanized terrain
MP .... military police
MPF .... maritime prepositioning force
MULE .... modular universal laser equipment
NBC .... nuclear, biological, and chemical
NCA .... National Command Authorities
NCF .... naval construction force
NEF .... naval expeditionary force
NEO .... noncombatant evacuation operation
NGF .... naval gunfire
NSFS .... naval surface fire support
NVA .... North Vietnamese Army
NWP .... naval warfare publication

Appendix J

Glossary

I. Acronyms

OP .... observation post
OPLAN .... operation plan

OPORD
operation order
OPSEC
operations security
PDDA
power-driven decontamination apparatus
PGM
precision-guided munition
PLO
Palestine Liberation Organization
PLRS
position location reporting system
POL
petroleum, oils, and lubricants
PSYOP
psychological operations
ROE
rules of engagement
ROK
Republic of Korea
RP
red phosphorus
RSOP . . . . reconnaissance, selection, and occupation

of

position
SASR
special application sniper rifle
SMAW
shoulder-launched multipurpose assault

weapon
SOP
standing operating procedure
STANAG
NATO standardization agreement
STB
supertropical bleach
TOW
tube-launched, optically tracked, wire-guided

missile
UAV
unmanned aerial vehicle
UCMJ
Uniform Code of Military Justice
U.N.
United Nations
VC
Viet Cong
VHF
very high frequency

WP . . . . white phosphorus

II. Definitions

A

assault—1. The climax of an attack, closing with the enemy in hand-to-hand fighting. 2. To make a short, violent, but well-ordered attack against a local objective, such as a gun emplacement, a fort, or a machine gun nest. (Joint Pub 1-02)

assault—The culmination of an attack which closes with the enemy. (MCRP 5-2A)

attack—An offensive action characterized by movement supported by fire with the objective of defeating or destroying the enemy. (FMFRP 0-14)

B

battlespace—All aspects of air, surface, subsurface, land, space, and electromagnetic spectrum which encompass the area of influence and area of interest. (FMFRP 0-14)

block—1. A tactical task assigned to a unit that requires it to deny the enemy access to a given area or to prevent enemy advance in a given direction or an avenue of approach. It may be for a specified time. Units assigned this mission may have to retain terrain and accept decisive engagement. 2. An obstacle effect that integrates fire planning and obstacle effort to stop an attacker on a specific avenue of approach or to prevent an enemy from exiting an engagement area. (MCRP 5-2A)

breach—A tactical task where any means available are employed to break through or secure a passage through an enemy defense, obstacle, minefield, or fortification. (MCRP 5-2A)

built-up area—A concentration of structures, facilities, and populations, such as villages, cities, and towns, that forms the economic and cultural focus for the surrounding area. (MCRP 5-2A, modified)

bypass—A tactical task that involves maneuvering around an obstacle, position, or enemy force to maintain the momentum of advance. Bypassed obstacles and enemy forces are reported to higher headquarters. (MCRP 5-2A)

C

clear—A tactical task to remove all enemy forces and eliminate organized resistance in an assigned zone, area, or location by destroying, capturing, or forcing the withdrawal of enemy forces such that they cannot interfere with the friendly unit's ability to accomplish its mission. (MCRP 5-2A)

combat arms—Units who close with and destroy enemy forces or provide firepower and destructive capabilities on the battlefield. (MCRP 5-2A, modified)

combat operations center (COC)—The primary operational agency required to control the tactical operations of a command that employs ground and aviation combat, combat support, and combat service support elements or portions thereof. The combat operations center continually monitors, records, and supervises operations in the name of the commander and includes the necessary personnel and communications to do the same. (FMFRP 0-14)

combat service support (CSS)—The essential capabilities, functions, activities, and tasks necessary to sustain all elements of operating forces in theater at all levels of war. Within the national and theater logistic systems, it includes but is not limited to that support rendered by service forces in ensuring the aspects of supply, maintenance, transportation, health services, and other services required by aviation and ground combat troops to permit those units to accomplish their missions in combat. Combat service support encompasses those activities at all levels of war that produce sustainment to all operating forces on the battlefield. (Joint Pub 1-02)

combat support (CS)—Fire support and operational assistance provided to combat elements. (Joint Pub 1-02)

combat support (CS)—Units that provide critical combat functions in conjunction with combat arms units to secure victory (MCRP 5-2A, modified).

command post (CP)—A unit's or subunit's headquarters where the commander and the staff perform their activities. In combat, a unit's or subunit's headquarters is often divided into echelons; the echelon in which the unit or subunit commander is located or from which such commander operates is called a command post. (Joint Pub 1-02)

contain—A tactical task to restrict enemy movement. (MCRP 5-2A)

cover—1. Shelter or protection from enemy observation that reduces the effects of enemy direct and indirect fire. 2. A type of security operation that protects the force from surprise, develops the situation, and gives commanders time and space in which to respond to the enemy's actions. (MCRP 5-2A)

D

defend—A combat operation designed to defeat an attacker and prevent him from achieving his objectives. It employs all means and methods available to prevent, resist, or destroy an enemy attack. Forms of defensive operations are area and mobile. Choices of defensive operations are in-depth and forward. The defensive techniques are defend in sector, defend a battle position, and defend a strong point. (MCRP 5-2A)

destroy—1. A tactical task to physically render an enemy force combat-ineffective unless it is reconstituted. 2. To render a target so damaged that it cannot function as intended nor be restored to a usable condition without being entirely rebuilt. (MCRP 5-2A)

disrupt—A tactical task or obstacle effect that integrates fire planning and obstacle effort to break apart an enemy's formation and tempo, interrupt the enemy's timetable, or cause premature commitment of enemy forces, or the piecemealing of his attack. (MCRP 5-2A)

II. Definitions

F

fire support—Assistance to elements of the Marine air-ground task force engaged with the enemy rendered by other firing units, including (but not limited to) artillery, mortars, naval surface fire support, and offensive air support. (FMFRP 0-14)

fix—1. A tactical task in which actions are taken to prevent the enemy from moving any part of his forces either from a specific location or for a specific period of time by holding or surrounding them to prevent their withdrawal for use elsewhere. 2. A tactical obstacle effect that integrates fire planning and obstacle effort to slow an attacker within a specified area—normally an engagement area. (MCRP 5-2A)

H

helicopterborne operation—A military action in which combat forces and their equipment maneuver about the battlefield by helicopters or vertical-landed aircraft. Aviation activities are under the control of the aviation combat element commander who is assigned in direct or general support of one or more combat element(s). (FMFRP 0-14)

hold—1. To maintain or retain possession of by force, as a position or an area. 2. In an attack, to exert sufficient pressure to prevent movement or redisposition of enemy forces. (Joint Pub 1-02)

I

infiltration—The movement through or into an area or territory occupied by either friendly or enemy troops or organizations. The movement is made, either by small groups or by individuals, at extended or irregular intervals. When used in connection with the enemy, it infers that contact is avoided. (Joint Pub 1-02)

isolate—A tactical task given to a unit to seal off (both physically and psychologically) an enemy from his sources of support, to deny an enemy freedom of movement, and prevent an enemy unit from having contact with other enemy forces. An enemy must not be allowed sanctuary within his present position. (MCRP 5-2A)

L

linkup—An operation wherein two friendly ground forces join together in a hostile area. (FMFRP 0-14)

M

Marine air-ground task force (MAGTF)—A task organization of Marine forces (division, aircraft wing, and service support groups) under a single command and structured to accomplish a specific mission. (Joint Pub 1-02)

military operations on urbanized terrain (MOUT)—All military actions that are planned and conducted on a topographical complex and its adjacent natural terrain where manmade construction is the dominant feature. It includes combat in cities, which is that portion of MOUT involving house-to-house and street-by-street fighting in towns and cities (MCRP 5-2A, modified)

military operations other than war (MOOTW)—Operations that encompass the use of military capabilities across the range of military operations short of war. These military actions can be applied to complement any combination of the other instruments of national power and occur before, during, and after war. (Joint Pub 1-02)

N

neutralize—1. To render enemy personnel or material incapable of interfering with a particular operation. 2. To render safe mines, bombs, missiles, and boobytraps. (MCRP 5-2A)

O

occupy—A tactical task in which a force moves onto an objective, key terrain, or other man-made or natural terrain area without opposition, and controls that entire area. (MCRP 5-2A)

P

peace enforcement—Application of military force, or the threat of its use, normally pursuant to international authorization, to compel compliance with resolutions or sanctions designed to maintain or restore peace and order. (Joint Pub 1-02)

peacekeeping—Military operations undertaken with the consent of all major parties to a dispute, designed to monitor and facilitate implementation of an agreement (ceasefire, truce, or other such agreement) and support diplomatic efforts to reach a long-term political settlement. (Joint Pub 1-02)

protect—A tactical task to prevent observation of or engagement or interference with, a force or location. (MCRP 5-2A)

pursuit—An offensive operation designed to catch or cut off a hostile force attempting to escape, with the aim of destroying it. (Joint Pub 1-02)

R

reduce—1. A tactical task to gain control over an enemy position or objective. 2. A task to create lanes through or over an obstacle sufficient to allow the attacking force to accomplish its mission. (MCRP 5-2A)

retain—A tactical task to occupy and hold a terrain feature to ensure it is free of enemy occupation or use. (MCRP 5-2A)

rules of engagement (ROE)—Directives issued by competent military authority which delineate the circumstances and limitations under which United States forces will initiate and/or continue combat engagement with other forces encountered. (Joint Pub 1-02)

S

screen—A task to maintain surveillance, provide early warning to the main body, or

II. Definitions

impede, destroy, and harass enemy reconnaissance within its capability without becoming decisively engaged. (MCRP 5-2A)

search—A systematic reconnaissance of a defined area, so that all parts of the area have passed within visibility. (Joint Pub 1-02)

secure—A tactical task to gain possession of a position or terrain feature, with or without force, and to deploy in a manner which prevents its destruction or loss to enemy action. The attacking force may or may not have to physically occupy the area. (MCRP 5-2A)

seize—To clear a designated area and obtain control of it. (FMFRP 0-14)

suppression—A tactical task to employ direct or indirect fires, electronic attack, or smoke on enemy personnel, weapons, or equipment to prevent or degrade enemy fires and observation of the friendly forces. (FMFRP 0-14)

T

terrorism—The calculated use of violence or threat of violence to inculcate fear; intended to coerce or to intimidate governments or societies in the pursuit of goals that are generally political, religious, or ideological. (Joint Pub 1-02)

Appendix L

References and Related Publications

Field Manuals (FMs)

3-11	Flame, Riot Control Agent, and Herbicide Operations
5-34	Engineer Field Data
5-102	Countermobility
5-250	Explosives and Demolitions
7-98	Operations in a Low-Intensity Conflict
9-15	Explosive Ordnance Disposal Service and Unit Operations
20-32	Mine/Countermine Operations
21-75	Combat Skills of the Soldier
3-30	Grenades and Pyrotechnic Signals
23-34	TOW Weapon System
27-10	Law of Land Warfare
33-1	Psychological Operations
34-130	Intelligence Preparation of the Battlefield
41-5	Joint Manual for Civil Affairs
41-10	Civil Affairs Operations
90-8	Counterguerrilla Operations
100-20	Military Operations In Low Intensity Conflict

Technical Manual (TM)

9-2350-264-10-2	Operator's Manual for Operation Under Usual and Unusual Conditions for Tank, Combat, Full-Tracked: 120mm Gun, M1A1 General Abrams, with Change 4

Marine Corps Concept Paper

A Concept for Future Military Operations on Urbanized Terrain

Marine Corps Doctrinal Publications (MCDPs)

1	Warfighting
1-3	Tactics

Marine Corps Reference Publications (MCRPs)

5-2A	Operational Terms and Graphics

Marine Corps Warfighting Publications (MCWPs)

3-15.3	Scout Sniping (under development)
3-37.1	Chemical Operations
3-37.4	NBC Reconnaissance

Fleet Marine Force Manuals (FMFMs)

0-9	Field Firing for the M16A2
0-25	The Law of Land Warfare
1-3	Tactics
2-6	MAGTF Rear Area Security
2-11	Antiarmor Operations
3-5	Employment of Military Police in Combat
3-30	Communications
6	Ground Combat Operations
6-7	Scouting and Patrolling for Infantry Units
6-8	Supporting Arms Observer, Spotter, and Controller
6-9	Marine Artillery Support
6-21	Tactical Fundamentals of Helicopterborne Operations
7-14	Combating Terrorism
7-34	MAGTF Civil Affairs
7-34-1	Civil Affairs Guidance for the Small-Unit Leader
7-40	Helicopter Insertion/Extraction
8-2	Counterinsurgency Operations
11-9	NBC Protection
11-10	NBC Decontamination
11-11	Treatment of Chemical Agent Casualties and Conventional Military Chemical Injuries
11-12	Chemical Hazard Assessment Guide
11-13	NBC Handbook
11-17	Chemical and Biological Contamination Avoidance
11-18	Nuclear Contamination Avoidance
11-20	NBC Reconnaissance
13-7	MAGTF Breaching Operations

Fleet Marine Force Reference Publications (FMFRPs)

0-6	Marine Troop Leader's Guide
0-14	Marine Corps Supplement to DOD Dictionary of Military and Associated Terms
4-26	Enemy Prisoners of War and Civilian Internees
7-14A	The Individual's Guide for Understanding and Surviving Terrorism
7-37	Vehicle Bomb Search
12-15	Small Wars Manual

12-18	Mao Tse-Tung on Guerrilla Warfare
12-25	The Guerrilla and How to Fight Him
13-5	Engineer Field Data

Joint Publications

1-02	DOD Dictionary of Military and Associated Terms
3-07	Joint Doctrine for Military Operations Other Than War
3-11	Joint Doctrine for Nuclear, Biological, and Chemical (NBC) Defense
3-15	Joint Doctrine for Barriers, Obstacles, and Mine Warfare
3-53	Doctrine for Joint Psychological Operations
4-06	JTTP for Mortuary Affairs in Joint Operations

Standard NATO Agreements (STANAGs)

2003	Patrol Reports by Army Forces
2014	Operations Orders
2029	Method of Describing Ground Locations, Areas, and Boundaries
2070	Emergency War Burial Procedures
2088	Battlefield Illumination
2101	Establishing Liaison
2868	Land Force Tactical Doctrine (ATP-35(B))

Miscellaneous

Fratricide: Reducing Self-Inflicted Losses, Newsletter No. 92-4, Center for Army Lessons Learned (CALL), U.S. Army Combined Arms Command, Fort Leavenworth, Kansas, April 1992.

HR, 25, Annex to Hague Convention No. IV, *Embodying Regulations Respecting the Laws and Customs of War on Land*, 18 October 1907.

Modern Experience in City Combat, U.S. Army Human Engineering Laboratory, Aberdeen Proving Ground, Maryland, 1987.

Foreword

We remain a nation at war in an era of persistent conflict, but we do not stand alone. Our nation has many multinational partners, equally committed to freedom, rule of law and stability. It is clear that we are stronger when we act with partners in today's operating environment. Therefore, security force assistance is no longer an "additional duty." It is now a core competency of our Army.

As distinctions separating war and peace blur and challenges to security increase, we must seek to help our multinational partners successfully confront their security challenges. Security force assistance builds our multinational partners' capability to defeat regular, irregular, and hybrid threats prevalent in an era of persistent conflict.

The two pillars of security force assistance are the modular brigade and Soldiers acting as advisors. Their tactical efforts build partner capability and improve the security situation. Over time, U.S. forces and partners foreign security forces collectively set the conditions to defeat common threats and ultimately achieve strategic success.

This manual addresses common characteristics and considerations for conducting security force assistance and clarifies what units and individual advisors must understand to work "by, with, and through" their counterparts. Since every situation and foreign security force is unique, units and individuals conducting security force assistance must carefully analyze the operational environment, especially the relationships of foreign security forces and their population. Conducting successful security force assistance requires adaptive units led by well-informed, culturally astute leaders. FM 3-07.1 provides commanders, staffs, and advisors with the common doctrine for achieving this goal.

Conducting foreign security assistance requires great skill in building relationships and "leading from behind." We must all increase our understanding of this important mission.

MARTIN E. DEMPSEY
General, U.S. Army
Commanding General
U.S. Army Training and Doctrine
Command

Field Manual
No. 3-07.1

Headquarters
Department of the Army
Washington, DC, 1 May 2009

Security Force Assistance

Contents

Distribution Restriction: Approved for public release; distribution is unlimited.

Figures

Preface

This field manual (FM) is the Army's doctrinal publication for security force assistance (SFA). It provides doctrinal guidance and direction for how U.S. forces contribute to SFA. It focuses on the brigade combat team (BCT) conducting SFA and advising foreign security forces. It is based on lessons learned from previous advising efforts and recent combat operations with a view to the future. It supports the Army Education System instruction on the theory and conduct of SFA.

The two primary audiences for this manual are leaders in BCTs conducting SFA and Soldiers assigned as advisors. The BCT is the largest fixed tactical organization and the key formation of the Army's modular organization. Army modularity allows commanders to add selective units to assist the BCT as it conducts SFA. The BCT and higher echelons provide the framework for advisors to function and accomplish the mission—developing foreign security forces and, when appropriate, working by, with, and through foreign security forces to achieve the desired end state of the host nation's internal defense and development. Thus, Soldiers assigned as advisors are the key element of SFA.

This FM provides the conceptual framework for conventional forces to conduct SFA within the construct of full spectrum operations, across the spectrum of conflict. It addresses SFA at operational and tactical levels.

Army doctrine is consistent and compatible with joint doctrine. FM 3-07.1 expands on fundamental principles for SFA expressed in FM 3-07. FM 3-07.1 uses text and concepts developed with North Atlantic Treaty Organization partners and ABCA (America, Britain, Canada, Australia, and New Zealand program armies).

Terms that have joint or Army definitions are italicized and the number of the proponent manual follows the definition. FM 3-07.1 will be the proponent manual for foreign security forces as a term.

FM 3-07.1 applies to the Active Army, Army National Guard/U.S. Army National Guard, and U.S. Army Reserve unless otherwise stated.

Headquarters, US Army Training and Doctrine Command is the proponent for this publication. The preparing agency is the Combined Arms Doctrine Directorate, US Army Combined Arms Center. Send written comments and recommendations on Department of the Army (DA) Form 2028 (Recommended Changes to Publications and Blank Forms) directly to: Commander, US Army Combined Arms Center and Fort Leavenworth, ATTN: ATZL-CD (FM 3-07.1), 201 Reynolds Avenue (Building 285), Fort Leavenworth, KS 66027-1352. Send comments and recommendations by e-mail to leav-cadd-web-cadd@conus.army.mil. Follow the DA Form 2028 format or submit an electronic DA Form 2028.

Introduction

Our strategy emphasizes building the capacities of a broad spectrum of partners as the basis for long-term security. We must also seek to strengthen the resiliency of the international system to deal with conflict when it occurs.

National Defense Strategy, June 2008

In an era of persistent conflict, the United States supports the internal defense and development of international partners, regardless of whether those partners are highly developed and stable or less developed and emerging. While many of these partners are nations, they can also include alliances, coalitions, and regional organizations. U.S. support to these partners ranges from providing humanitarian assistance to major combat operations. U.S. support includes conducting conflict transformation, bolstering partner legitimacy, and building partner capacity. A vital part of these three aspects of U.S. support is assisting partner security forces.

Security force assistance (SFA) is not new for Army forces. In fact, General George Washington's Inspector General of the Army acted as an advisor for Army forces. Baron Friedrick Wilhelm von Steuben instilled discipline and professionalism into an army that previously lacked formalized training. His 1779 *Regulations for the Order and Discipline of the Troops of the United States*, adapted from the Prussian army, formed the doctrinal backbone of the Continental Army throughout the Revolutionary War. Additionally, the lineage of the Army's operations field manual, FM 3-0, can be traced to this document. As a benefactor of advisors such as von Steuben, the Army has since undertaken what is called SFA several times throughout its history.

Army doctrine defines *security force assistance* as the unified action to generate, employ, and sustain local, host-nation, or regional security forces in support of a legitimate authority (FM 3-07). SFA is part of the FM 3-0 construct of full spectrum operations. Similarly, it is conducted across the spectrum of conflict or in any of the operational themes. SFA is normally part of a larger security sector reform effort, while in other instances, SFA is not tied to reform but to building partner capacity.

Three general situations exist in which SFA may occur: an internally focused bilateral relationship, an externally focused bilateral relationship, and a multilateral relationship. Consequently, SFA supports the appropriate partner's plans. When SFA supports a host nation, it also supports that host nation's strategy. If SFA supports a host nation's externally focused efforts, it must support the host nation's national security strategy. SFA may support regional security forces, such as those of the African Union or the Organization of American States. In these cases, SFA supports that organization's plans.

ORGANIZATION

This FM primarily focuses on the brigade combat team and advisors. It also provides a wider doctrinal framework for SFA. It is organized in the following manner:

- **Chapter 1 focuses above the brigade combat team**. It provides the overall doctrinal context for SFA.
- **Chapters 2 through 6 focus on the brigade combat team level**. These chapters provide a framework for SFA, its considerations for the brigade combat team operations process, considerations for augmenting the modular brigade for security force assistance, its considerations for the unit employment, and its sustainment considerations.
- **Chapters 7 through 10 focus on the individual advisor**. These chapters discuss the advisor, advisor cultural and communication considerations, the advisor working with counterparts, and cross-cultural influencing and communication.
- The appendices discuss Army special operations forces imperatives, legal considerations, and media considerations.

Chapter 1

The Strategic Context

This chapter places security force assistance in strategic context. First, it explains security force assistance and its relationship to full spectrum operations, spectrum of conflict, and operational themes (see FM 3-0). A discussion of U.S. national strategy and defense policies that impact security force assistance follows. This chapter then examines operations and programs related to security force assistance. Finally, it discusses the relationship of security force assistance and related operations and programs.

SECURITY FORCE ASSISTANCE AND THE STRATEGIC CONTEXT

1-1. *Security force assistance* is the unified action to generate, employ, and sustain local, host-nation or regional security forces in support of a legitimate authority (FM 3-07). Security force assistance (SFA) improves the capability and capacity of host-nation or regional security organization's security forces. These forces are collectively referred to as foreign security forces. ***Foreign security forces*** **are forces—including but not limited to military, paramilitary, police, and intelligence forces; border police, coast guard, and customs officials; and prison guards and correctional personnel—that provide security for a host nation and its relevant population or support a regional security organization's mission.**

1-2. SFA occurs within the framework of full spectrum operations (see FM 3-0). In most situations involving this assistance, there is relatively little weight on offensive and defensive operations from a U.S. perspective. However, when U.S. forces accompany foreign security forces (FSF) in combat, the weight of offensive and defense operations will change to address the situation and align with the foreign security force's efforts. SFA is not just a stability operation, although it is a key contributor to the primary stability tasks of establish civil security and establish civil control.

1-3. SFA can be conducted across the spectrum of conflict, from stable peace to general war. This assistance could focus on improving the security forces of a host nation that is currently under no immediate threat, on paramilitary forces to counter an insurgency, or on advising FSF in major combat operations against an external threat. Figure 1-1 illustrates how SFA spans the spectrum of conflict.

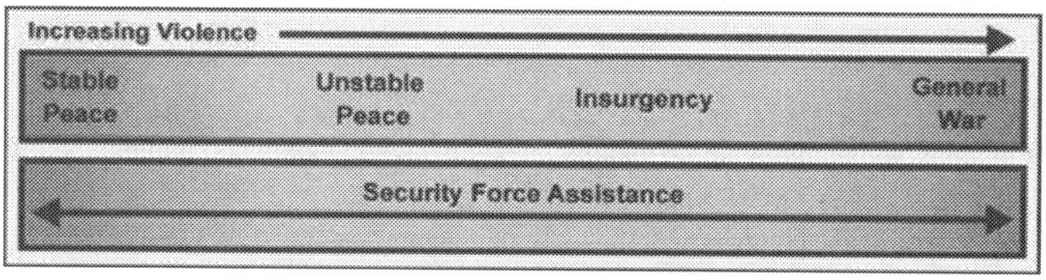

Figure 1-1. The spectrum of conflict and security force assistance

1-4. SFA can occur as part of any operational themes—peacetime limited engagement, limited intervention, peace operations, irregular warfare, and major combat operations. SFA often occurs during peacetime military engagement, peace operations, and irregular warfare. SFA involving paramilitary forces will often be limited to the irregular warfare operational theme, although paramilitary forces should eventually be absorbed, disbanded, or maintained in a long-term relationship with special operations forces.

1-5. While Army units are vital contributors, SFA is part of a comprehensive approach. This comprehensive approach should include close collaboration with military and civilian joint and multinational forces. The host-nation or regional security organization is the key actor within the comprehensive approach. It is vital that SFA is based on an objective, continuous assessment that examines the organization, training, equipment, rebuilding, and advising of the forces involved.

1-6. Law enforcement, military, intelligence, and border forces operate and cooperate within the security sector. U.S. forces understand how these units are intended to operate in the host-nation scheme, not necessarily using a U.S. model. U.S. forces should plan to help develop the respective capabilities so that these units can carry out their security functions.

1-7. Fully developed and institutionalized FSF plan, prepare, and execute at all the appropriate levels. Thus, U.S. forces must integrate SFA into the operations process at all levels. SFA should be designed to work seamlessly with the host-nation government at all levels, from ministries addressing the security sector to initial entry-level FSF training.

NATIONAL STRATEGY AND DEFENSE POLICIES

1-8. U.S. strategy and foreign policy affects all SFA. More importantly, SFA is critical to these policies as it is a primary tool for building partnership capacity with other nations. Key strategy documents for SFA include the National Security Strategy, the National Defense Strategy, the National Military Strategy, the Quadrennial Defense Review, and DODD 3000.05.

1-9. The National Security Strategy outlines the President's vision for providing enduring security for the American people in a volatile, uncertain, complex, and ambiguous strategic environment. The National Security Strategy of 2006 addresses stability operations within the broad engagement strategy for regional conflict. It supports SFA by "tailoring assistance and training of military forces to support civilian control of the military and military respect for human rights in a democratic society." This strategy also states that "[t]he world has found through bitter experience that success often depends on the early establishment of strong local institutions such as effective police forces and a functioning justice and penal system. This governance capacity is critical to establishing the rule of law and a free market economy, which provide long-term stability and prosperity."

1-10. The National Defense Strategy encompasses the Secretary of Defense's vision for the Department of Defense role in protecting the American people and U.S. interests. This strategy stems from the National Security Strategy. The 2008 National Defense Strategy highlights SFA. It states that "[o]ur strategy emphasizes building the capacities of a broad spectrum of partners as the basis for long-term security.... [l]ocal and regional conflicts in particular remain a serious and immediate problem...we will help build the internal capacities of countries at risk...[b]y helping others to police themselves and their regions, we will collectively address threats to the broader international system."

1-11. The National Military Strategy supports the National Security Strategy, National Defense Strategy, and Quadrennial Defense Review. The 2004 National Military Strategy states "[t]he Department will also work to strengthen alliances and partnerships by helping other nations increase their ability to defend themselves and protect common security interests." Working with other nations and strengthening partnerships appears throughout the National Military Strategy, which supports the importance of SFA.

1-12. The Quadrennial Defense Review is a congressionally mandated Department of Defense (DOD) review of strategy, programs, and resources. The 2006 Quadrennial Defense Review directed that joint ground forces "possess the ability to train, mentor and advise FSF and conduct counterinsurgency campaigns." This review also stated that conventional forces need to be capable of training, equipping, and advising local forces; deploying and engaging with partners; and supporting stability operations. Additionally, the review envisioned conventional ground forces continuing to perform more of the tasks traditionally performed by special operations forces. Conventional forces must understand foreign cultures and societies as well as be able to train, mentor, and advise FSF.

1-13. DODD 3000.05 provides guidance on stability operations. DODD 3000.05 establishes that stability operations are a core military mission and equal in importance to combat operations. DODD 3000.05 also reinforces the importance of SFA by stating "[t]he immediate goal often is to provide the local populace with security, restore essential services, and meet humanitarian needs. The long-term goal is to help develop host-nation capacity for securing essential services, a viable market economy, rule of law, democratic institutions, and a robust civil society."

RELATED OPERATIONS AND PROGRAMS

1-14. Several operations or programs directly or potentially relate to SFA. These operations and programs provide the context for SFA. They include security cooperation, security assistance, foreign internal defense, internal defense and development (IDAD), and security sector reform. For more information on operations related to SFA, see DOD 5105.38M, DOD security cooperation guidance, JP 3-07.1, FM 3-07, FM 3-05.137, and AR 11-31.

SECURITY COOPERATION

1-15. *Security cooperation* is all Department of Defense interactions with foreign defense establishments to build defense relationships that promote specific U.S. security interests, develop allied and friendly military capabilities for self-defense and multinational operations, and provide U.S. forces with peacetime and contingency access to a host nation (JP 3-07.1). Finally, security cooperation occurs across the spectrum of conflict and is not exclusively a peacetime activity. Security cooperation includes security assistance programs administered by the DOD as well as activities that enhance interoperability and the collective capability of combined forces (using Title 10 or exercise funding and authority). Security cooperation consists of a focused program of bilateral and multilateral defense activities conducted with foreign countries to serve mutual security interests and build defense partnerships. Security cooperation is governed by various sections of Title 10 and specific public laws addressing DOD interactions with other nations.

1-16. AR 11-31 establishes Department of the Army policy and prescribes responsibilities and procedures for the planning, integration, programming, budgeting, and execution of Army security cooperation activities. It is the policy of the U.S. Army to conduct security cooperation activities in compliance with higher-level guidance and in the execution of Army responsibilities under Title 10 and Title 22 which governs the transfer, exchange, conduct, and development of articles and services via a variety of U.S. Government programs.

1-17. Army security cooperation consists of official, cooperative, and noncombat interactions among any Army elements, Active Army, or Reserve Components. This includes any U.S. Government or nongovernment entity supporting the military and civilian joint and multinational organizations.

1-18. Commanders distinguish security cooperation from SFA based its emphasis in building relationships and capacities by using programmatic activities. Thus, security cooperation may support or be supported by SFA.

SECURITY ASSISTANCE

1-19. *Security assistance* is a group of programs authorized by the Foreign Assistance Act of 1961, as amended, and the Arms Export Control Act of 1976, as amended, or other related statutes by which the United States provides defense articles, military training, and other defense-related services by grant, loan, credit, or cash sales in furtherance of national policies and objectives (JP 3-57). Security assistance is a specific subset of security cooperation and may focus on external or internal threats.

1-20. DOD 5105.38-M describes the scope of security assistance programs in detail. Security assistance allows the transfer of military articles and services to friendly foreign governments. These transfers may be carried out via sales, grants, leases, or loans. If these transfers are essential to the security and economic well-being of allied governments and international organizations, they are equally vital to the security and economic well-being of the United States. U.S. forces can use security assistance to deliver defense weapon systems to foreign governments; to train international students; to advise other governments on improving internal defense capabilities; and to provide guidance and assistance in establishing infrastructures and economic bases for regional stability. Finally, security assistance cannot be conducted using Title 10 or exercise funds.

1-21. The military component of security assistance, implemented by DOD with policies established by the Department of State, has four principal components. They are international military education and training, foreign military sales, foreign military financing program, and peace operations. The foreign military financing program and international military education and training fall inside the military assistance budget process of the Department of State. Security assistance can also include funding peace operations.

International Military Education and Training

1-22. International military education and training contributes to internal and external security of a country by providing training to selected foreign militaries and related civilian personnel on a grant aid basis (see JP 3-07.1). These programs help to strengthen foreign militaries through training for the proper functioning of a civilian-controlled, apolitical, professional military. International military education and training serves as a foreign policy tool where the United States shapes doctrine; promotes self sufficiency in maintaining and operating U.S. acquired defense equipment; encourages the value of rule of law; and occasionally has a marked effect on the policies of the recipient governments. Foreign students, many of who occupy the middle and upper echelons of their country's military and political establishments, are taught U.S. doctrine and weapons systems employment resulting in greater cooperation and interoperability. All international military education and training activities are SFA.

Foreign Military Sales

1-23. Foreign military sales is a nonappropriated program through which foreign governments can purchase defense articles, services, and training from the United States (see JP 3-07.1). Eligible nations use this program to help build national security infrastructures. However, nations that require assistance are often unable to finance their needs. SFA supports foreign military sales when these sales include U.S. forces providing equipment and related training. It does not support foreign military sales when these sales provide equipment without related training.

Foreign Military Financing Program

1-24. The foreign military financing program provides funding to purchase defense articles and services, design and construction services, and training through foreign military sales or commercial channels (see JP 3-07.1). This program can assist nations with weak economies that would otherwise be unable to afford U.S. assistance. The foreign military financing program can fund foreign military sales. SFA and the foreign military financing program may be separate activities, although they can also be complementary.

FOREIGN INTERNAL DEFENSE

1-25. *Foreign internal defense* is the participation by civilian and military agencies of a government in any of the action programs taken by another government or other designated organization to free and protect its society from subversion, lawlessness, and insurgency (JP 3-05). U.S. foreign internal defense efforts involve all instruments of national power to support host-nation IDAD programs. The military instrument of foreign internal defense includes indirect support, direct support (not involving U.S. combat operations), and combat operations. Foreign internal defense can occur across the spectrum of conflict and within any operational theme. Foreign internal defense is often a part of irregular warfare or peacetime military engagement operational theme. SFA may support the military instrument of foreign internal defense. JP 3-07.1 discusses foreign internal defense in detail.

Note: It is a national policy decision for U.S. forces to accompany FSF on combat operations.

Military Instrument of Foreign Internal Defense

1-26. The military instrument of foreign internal defense supports the other instruments of national power through activities ranging from peacetime military engagement to major combat operations. Although foreign internal defense may include capacity building across the host nation, U.S. forces focus on combating internal threats to help the host nation maintain legitimacy and influence over the population. As such, foreign internal defense may be part of irregular warfare.

1-27. SFA and foreign internal defense support the host nation's IDAD. Consequently, forces conducting SFA must have proper mindset—they are part of a developmental and supporting effort. This applies even if SFA is part of working with a potential adversary or an existing enemy. SFA and foreign internal defense focus on working by, with, and through FSF and not conducting unilateral efforts to achieve objectives.

Indirect Support

1-28. Indirect support efforts are foreign internal defense operations that emphasize the principle of host-nation self-sufficiency. Indirect support focuses on building strong national infrastructures through economic and military capabilities that contribute to self-sufficiency. For example, indirect support could include U.S. assistance in improving a host-nation's intelligence infrastructure. Contributions from U.S. forces to indirect support come from security cooperation guidance through security assistance. They are supplemented by multinational exercises, exchange programs, and selected joint exercises. Indirect support that is focused on assisting security forces is SFA.

1-29. Although they indirectly assist those involved, military exchange programs are not SFA. Military exchange programs foster mutual understanding between forces; familiarize each force with the organization, administration, and operations of the other; and enhance cultural awareness. These exchange programs often have long-term implications for strengthening democratic ideals and respect for human rights among supported governments. However, such exchange programs (funded with Title 10 monies) do not themselves become vehicles for security assistance training, as they do not intend to assist security forces but to foster mutual understanding and enhance cultural awareness.

1-30. Joint and multinational exercises strengthen U.S. and host-nation relations and interoperability of forces, but they are not SFA. They complement security assistance and civil-military operations by validating host-nation needs and capabilities and by providing a vehicle for the conduct of humanitarian and civic assistance programs (see JP 3-07.1). Strict legal restrictions clarify the type of support that U.S. forces can provide and on the monetary limits of such support. These efforts are not SFA as they do not intend to directly assist security forces; instead, they intend to improve relations and interoperability. See appendix B.

Direct Support

1-31. Direct support does not involve combat operations but includes the use of U.S. forces providing direct assistance to the host-nation civilians or military. Direct support often focuses on civil-military operations (primarily, providing services to local civilians), psychological operations, intelligence and communications sharing, and logistic support. Sometimes direct support will be SFA, such as psychological operations, military training support, logistic support, and intelligence and communications sharing. SFA is not a civil affairs activity such as foreign humanitarian assistance, humanitarian and civic assistance, or military civic action.

1-32. Direct support differs from security assistance in that direct support is joint- or Service-funded. Direct support does not usually involve transferring arms and equipment and does not often include training local military forces. Direct support operations are normally conducted when the host nation has not attained or regained self-sufficiency and when it still faces social, economic, or military threats beyond its capability.

Combat Operations

1-33. The introduction of U.S. combat forces into foreign internal defense operations requires a Presidential decision. These operations are a temporary solution until host-nation forces can stabilize the situation and provide security for civilians. In all cases, U.S. combat operations support the host-nation IDAD program and remain strategically defensive in nature. When U.S. forces are advising, mentoring, partnering, or augmenting host-nation forces in foreign internal defense combat operations, this is SFA. When Army units conduct independent combat operations alongside FSF, this is not SFA.

1-34. U.S. tactical participation in host-nation internal conflicts requires judicious and prudent rules of engagement and guidelines for applying force. Inappropriate destruction and violence attributed to U.S. forces may easily reduce the legitimacy and sovereignty of the supported government. Adversaries may also use such incidents to fuel anti-American sentiments and assist their cause.

Security Relationships

1-35. A complex relationship exists among security cooperation, security assistance, and the military instrument of foreign internal defense. The left side of figure 1-2 depicts this relationship, including how aspects of foreign internal defense and security assistance overlap. The right side of figure 1-2 illustrates how foreign internal defense focuses on internal threats to a host nation and how security assistance focuses on external threats. The column depicts how security (military, intelligence, and civilian), economic, and governance are considerations common to both foreign internal defense and security assistance.

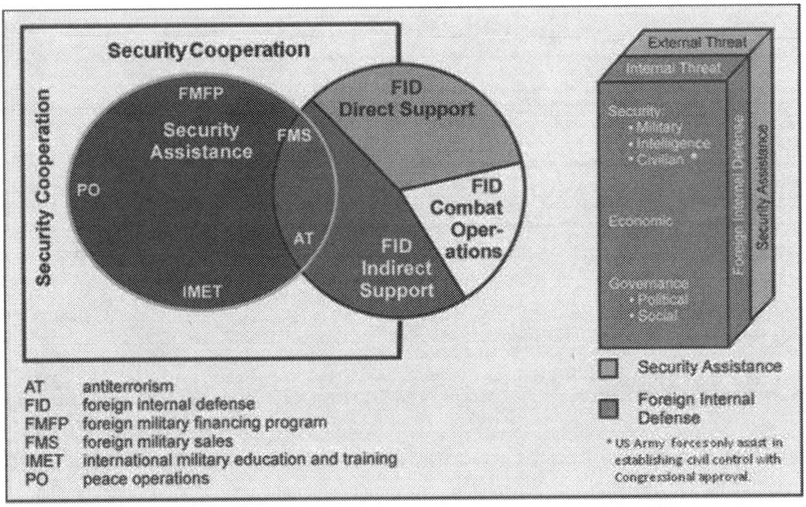

Figure 1-2. Security cooperation, security assistance, and foreign internal defense

Internal Defense and Development

1-36. *Internal defense and development* is the full range of measures taken by a nation to promote its growth and to protect itself from subversion, lawlessness, and insurgency. It focuses on building viable institutions (political, economic, social, and military) that respond to the needs of society (JP 3-07.1). IDAD aims to forestall or defeat the threat and to correct core grievances that prompt violence. Ideally it is a preemptive strategy. However, if an insurgency or other threat develops, it becomes an active strategy to combat that threat. To support the host nation effectively, U.S. forces, especially planners, consider the host-nation's IDAD strategy.

1-37. IDAD focuses on building viable institutions that respond to the needs of society. IDAD is a U.S. doctrinal term; it is not used universally by all countries. Internal defense and development blends four interdependent functions to prevent or counter internal threats: balanced development, security, neutralization, and mobilization. Finally, IDAD involves a cyclic interaction of execution, assessment, and adaptation. JP 3-07.1 discusses IDAD and its functions in detail.

SECURITY SECTOR REFORM

1-38. *Security sector reform* is the set of policies, plans, programs, and activities that a government undertakes to improve the way it provides safety, security, and justice (FM 3-07). Security sector reform aims to provide an effective and legitimate public service that is transparent, accountable to civil authority, and responsive to the needs of the public. It may include integrated activities to support defense and armed forces reform; civilian management and oversight; justice, police, corrections, and intelligence reform; national security planning and strategy support; border management; disarmament, demobilization, and reintegration; and concurrent reduction of armed violence. SFA is a vital component of security sector reform when security sector reform includes U.S. assistance to FSF. See chapter 6 of FM 3-07 for an additional discussion of security sector reform.

SECURITY FORCE ASSISTANCE RELATIONSHIP

1-39. Paragraph 1-35 explained and figure 1-2 supported the complex relationship among security cooperation, security assistance, and the military instrument of foreign internal defense. Figure 1-3 explains the relationship of SFA to security cooperation, security assistance, and the military instrument of foreign internal defense. SFA supports the military instrument of foreign internal defense, much of security assistance efforts, and some security cooperation efforts. The right side of figure 1-3 also shows that SFA addresses security issues, but not economic or governance issues.

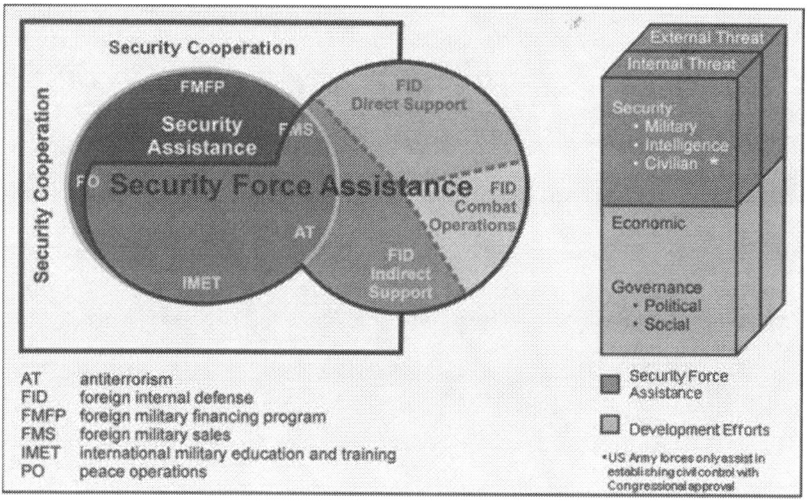

Figure 1-3. Relationship of security force assistance with security cooperation, security assistance, and foreign internal defense

Country Team

1-40. The country team will play a vital role with respect to SFA. The country team is the senior in-country U.S. coordinating and supervising body headed by the U.S. chief of mission. The chief of mission leads the country team and integrates U.S. efforts in support of the host nation. The chief of mission is the highest U.S. civil authority in a foreign country. As the senior U.S. Government official permanently assigned in the host nation, the chief of mission is responsible to the President for policy oversight of all U.S. Government programs. The Foreign Service Act assigns the chief of mission to a foreign country with responsibility for the direction, coordination, and supervision of all government executive branch employees in that country except for Service members and employees under the command of a U.S. area military commander.

1-41. The country team is composed of the senior member of each represented department or agency. As permanently established interagency organizations, country teams represent a priceless resource. They often provide deep reservoirs of local knowledge and interaction with the host-nation government and population. See FM 3-07 and JP 3-07.4 for more information on the country team.

Theater Command Structures

1-42. SFA is inherently multinational. Nations do not relinquish their national interests by participating in multinational operations. As in any multinational operation, commanders conducting SFA prepare to address issues related to legality, mission mandate, and prudence early in the planning process. In multinational operations, consensus often stems from compromise.

1-43. Political considerations heavily influence the ultimate shape of the command structure in which U.S. forces will conduct SFA. However, participating nations should strive to achieve unity of command for the operation to the maximum extent possible. All participants must understand missions, tasks, responsibilities, and authorities. While well defined in U.S. doctrine, command relationships are not necessarily part of the doctrinal lexicon of nations with which the United States may operate in coalition or combined operations.

1-44. Command and control relationships of higher echelons take precedence when conducting SFA. As a result, the brigade combat team (BCT) may be a supported command from a U.S. perspective as well as acting as a supporting command for FSF in the BCT's area of operations. For example, a BCT may support a joint interagency task force conducting international police training. Another example is a joint command supporting a BCT with joint fires; intelligence, surveillance, and reconnaissance; or other capabilities while the BCT supports FSF in training. The basic structures for multinational operations fall into one of three types: lead nation, integrated, or parallel command. Alliance and coalition commands use these structures.

Lead Nation Command Structure

1-45. A lead nation command structure exists when all member nations place their forces under the control of one nation. The lead nation command has a dominant lead nation command and staff arrangement with subordinate elements retaining strict national integrity. This command may also have an integrated staff and multinational subordinate forces. Integrating the staff allows the commander to draw upon the expertise of allied or coalition partners in areas where the lead nation may have less experience. This is the most desirable command structure for SFA.

Integrated Command Structure

1-46. Multinational commands organized under an integrated command structure provide unity of effort in a multinational setting. An integrated, combined command has several distinguishing features. Besides a designated single commander, the staff consists of representatives from all member nations. Subordinate commands and staffs are integrated into the lowest echelon necessary. Although effective for SFA, the integrated command structure works less effectively than the lead nation command structure.

Parallel Command Structure

1-47. Under a parallel command structure, no single force commander is designated. The coalition leadership develops a means for coordination among the participants to attain unity of effort. They can accomplish this by using coordination centers. Nonetheless, because of the absence of a single commander, the use of a parallel command structure should be avoided if at all possible. This is the least desirable command structure for SFA.

Alliance Command Structures

1-48. In combined commands, national political objectives are addressed and generally subsumed within multinational force objectives at the alliance treaty level. Combined command relationships often reflect either an integrated command structure or a lead nation command structure. In combined operations such as SFA, these structures should be used to the maximum practical extent. Combined command and force structures often mirror the degree of allied member participation. Senior military officers from member nations often lead subordinate commands. Effective operations within an alliance, including SFA, require senior political and military authorities to agree on the type of command relationships that will govern the operations of the forces. Potential political sensitivities associated with SFA can impact command relationships and operating procedures.

Coalition Command Structures

1-49. Coalitions often form in response to unforeseen crises that occur outside the area of an established alliance, which is often the case for SFA. Coalition command relationships routinely evolve as a coalition develops. Coalitions often have one of two basic structures: lead nation or parallel. For SFA, all participating coalition members should augment the host-nation headquarters staff with representatives. These representatives might include designated deputies or assistant commanders, planners, and logisticians. Such staffing provides the coalition commander with representative leadership and accessible expertise on the capabilities of the coalition members as well as facilitates the planning process. This integration of the multinational command elements into the coalition planning process should occur as early as practical.

Integration of Conventional and Special Operations Forces

1-50. Many criteria determine when conventional forces, special operations forces, or a combination are appropriate to conduct SFA. Both force levels and force characteristics suggest optimal, acceptable, and undesirable force package options in planning and resourcing SFA. Options for the deployment of a modular brigade augmented for SFA, a select number of conventional military transition teams, or special operations forces depend on conditions of the operational environment, priorities of the IDAD strategy, overall U.S. national policy, and forces available.

1-51. Rarely will U.S. forces conducting SFA be homogenous in terms of conventional or special operations forces, but rather contain both. FM 3-0 and other Army doctrine define command and control relationships and support relationships when both forces operate in the same area. Recent operational experience has shown that, in improving the effectiveness of FSF, the supported commander conducting SFA may be either a conventional force or special operations force commander. Subordinate commanders and staffs of both conventional and special operations forces may act in the supporting role to the U.S. ambassador. Typically this is done through the security assistance officer or the defense attaché at the American Embassy.

1-52. If U.S. forces are to train a large number of FSF in a short time, then the requirement may exceed the capacity of available special operations forces. Large SFA missions occur at the geographic command level or higher headquarters. Planners at these levels determine the number of conventional and special operations forces as part of their mission analysis aimed at training enough FSF in the given time frame. To best prepare for a large SFA, U.S. forces might deploy one or more augmented modular brigades supported by Army special operations forces. These forces conduct complementary foreign internal defense tasks under Title 22 authorities. Another option involves using special operations forces training FSF over several years while using a broad mix of funding and authorities. Another option uses a modular brigade to support a select number of company- or battalion-sized foreign units.

1-53. A force analysis of how best to integrate conventional and special operations forces to accomplish SFA considers host-nation and U.S. national policies. Skill sets among conventional and special operations forces trainers, advisors, and mentors avoid universally excluding or including final, integrated force packages for SFA. Planners leverage skill sets by integrating and coordinating conventional and special operations units with assigned tasks within SFA. Force analysis of the FSF to be trained often suggests a two-phase approach to training. This analysis also suggests a transition to the FSF by either conventional or special operations forces using basic and advanced training. The type of training depends on the nature of the foreign security force being trained. Additionally, integrating the advisor may be a phased process using either conventional or special operations forces as the initial advisor. This method lends itself to the classic crawl-walk-run approach to assisting FSF.

Chapter 2
Framework

This chapter discusses a doctrinal framework for the conduct of security force assistance. This security force assistance framework is composed of the mindset required of units and Soldiers, imperatives for success, inherent tasks and activities, and the three types of security force assistance.

MINDSET

2-1. Conducting successful security force assistance (SFA) requires a specific mindset. This mindset focuses on working by, with, and through foreign security forces (FSF) to support the host nation's internal defense and development (which includes local security requirements) or regional organization's charter. Soldiers conducting SFA must also understand that legitimacy is vital. The relevant population must perceive FSF as legitimate for long-term success. Those conducting SFA must understand that the military instrument of national power is only one part of a comprehensive approach. The imperatives of SFA provide the foundation for proper mindset.

Note: Host nation includes partner organization.

2-2. SFA focuses on how FSF organizations, institutions, systems, capabilities, and limitations—not U.S. organizations, systems, procedures, and equipment—can be enabled to address the host-nation or partner organization problems. Local forces have advantages over outsiders. They inherently understand the local culture and behavior that outsiders simply lack. To tap into those advantages, advisors must resist blatant military solutions. To overcome the temptation to do what they know and do best, whether relevant or not to the situation, advisors must accept that they are bound by unique situations.

IMPERATIVES

2-3. The imperatives of SFA come from the historical record and recent experience. These imperatives do not replace the principles of war or the principles of joint operations. Rather, they provide focus on how to successfully conduct SFA. The six imperatives apply to SFA at every level of war, for any echelon, and for any Soldier.

2-4. Since there is a close relationship between SFA and Army special forces conducting foreign internal defense, SFA planners should also consider the twelve Army special operations forces imperatives, especially at the tactical level (see appendix A).

UNDERSTAND THE OPERATIONAL ENVIRONMENT

2-5. An in-depth understanding of the operational environment—including available FSF, opposing threats, and civil considerations—is critical to planning and conducting effective SFA. Units and Soldiers conducting SFA must clearly understand the theater, population, and FSF with which they are working, especially FSF capabilities. Diplomatic, informational, military, economic, sociological, psychological, and geographic research and understanding are essential prerequisites for successful SFA. Tactically, successful SFA requires identifying the friendly and hostile decisionmakers, their objectives and strategies, and the ways they interact. Further, the conditions of the operational environment can rapidly change, and those conducting SFA must anticipate these changes and exploit potential and possibly fleeting opportunities.

Provide Effective Leadership

2-6. Leadership, a critical aspect of any application of combat power, proves important in the dynamic and complex environments associated with SFA. The operational environment in which SFA occurs places a premium on effective leadership at all levels, from the most junior to the most senior general officer and agency director. Leading FSF or a combined group of U.S. and foreign security forces is inherently more challenging due to differences in culture, language, training, and other aspects. Leaders focus on transitions; their decisions move the foreign security force toward independent operations.

Build Legitimacy

2-7. Ultimately, SFA aims to develop security forces that contribute to the legitimate governance of the local populace. Significant policy and legal considerations may apply to SFA activities. Legitimacy is the most crucial factor in developing and maintaining internal and international support. The United States cannot sustain long-term SFA without legitimacy. Legitimacy is a concept that goes beyond a strict legal definition; it includes the moral and political legitimacy of a host-nation government or partner organization. Local civilians and the international community determine the government's legitimacy based on collective perception of the credibility of its cause and methods.

Manage Information

2-8. Successful SFA disseminates timely and protected relevant information, integrates it during planning, and leverages it appropriately during execution. Effective and efficient information management supports decisionmaking throughout capacity building. Managing information encompasses the collection, analysis, management, application, and preparation of information. Maintaining continuity between rotating SFA elements is critical. For long-term SFA efforts, FSF expect more professional abilities from U.S. advisors. To meet expectations and success in general, units conducting SFA establish and integrate lessons learned. Capturing and incorporating evolving tactics, techniques, and procedures and aspects of the operational environment is vital; with each successive rotation, planners strive to maintain continuity and momentum. Finally, planners synchronize effective information engagement with the entire mission. See appendix C. SFA operations inherently affect the dynamic operational environment.

Ensure Unity of Effort

2-9. SFA often includes many actors, making unity of effort essential for success. SFA will include U.S. and foreign security forces, including conventional forces, special forces, or a combination. Other civilian and military joint and military organizations are often involved in SFA. Planners integrate them into one cohesive effort.

Sustain the Effort

2-10. Sustainability consists of two major components: the ability to sustain SFA effort throughout the operation and the ability of the FSF to sustain their operations independently. While each situation will vary, Army personnel conducting SFA must avoid assisting FSF in techniques and procedures beyond the FSF's capability to sustain. U.S. tactics, techniques, and procedures must be modified to fit the culture, educational level, and technological capability of the FSF. Those involved in SFA must recognize the need for programs that are durable, consistent, and sustainable by both the U.S. and FSF. They must not begin programs that are beyond the economic, technological, or cultural capabilities of the host nation to maintain without U.S. assistance. Such efforts are counterproductive.

TASKS

2-11. SFA aims to establish conditions that support the partner's end state, which includes legitimate, credible, competent, capable, committed, and confident security forces. This requires a force capable of securing borders, protecting the population, holding individuals accountable for criminal activities, regulating the behavior of individuals or groups that pose a security risk, and setting conditions in the operational area that enable the success of other actors.

2-12. This section addresses the six SFA tasks: organize, train, equip, rebuild and build, and advise and assist. When supporting the development of FSF, commanders understand their role with the host-nation or regional security organization. These tasks facilitate SFA planners to assess and allocate resources based on conditions.

ORGANIZE

2-13. Organize is a SFA task that encompasses all measures taken to assist FSF in improving its organizational structure, processes, institutions, and infrastructure. U.S. forces must understand the existing security organizations of FSF to better assist them. Subsequently, SFA personnel may help the host nation organize its security forces to meet the needs of its security environment.

2-14. Organizing a foreign security force depends on the host nation's social and economic conditions, cultural and historical factors, and security threats. SFA aims to create an efficient organization with a command, intelligence, logistic, and operations structure viable for the host nation. Conventional forces with limited special purpose teams (such as explosive ordnance disposal) are preferred. Doctrine or standing operating procedures should apply across the force, as should unit structures. The organization must facilitate the collection, processing, and dissemination of intelligence across and throughout all security forces. As the foreign security force strengthens, U.S. leaders and trainers should expect more independent organizational decisions.

2-15. The host nation determines the structure of its military forces, to include approving all organizational designs. These may include changing the numbers of forces, types of units, and internal organizational designs.

2-16. Organization should address all FSF, from the ministerial level to the patrolling police officer and soldier. Building a competent FSF infrastructure—including civilian information systems—is critical for success. Commanders work with ministries responsible for national and internal security, including the ministry of national defense, the interior ministry, and the justice ministry. Commanders assess strengths and weaknesses of the ministerial organization as well as training requirements of their employees. The joint and multinational advisor team at the ministry level helps the host nation develop a procurement and management system to meet its requirements.

2-17. Organizing a foreign security force requires resolving issues related to—

- Recruiting.
- Promotion screening and selection.
- Pay and benefits.
- Leader recruiting and selection.
- Personnel accountability.
- Demobilization of security force personnel.

Recruiting

2-18. Recruiting is critical when establishing FSF. The recruiting program assimilates local culture and themes that resonate with the populace. The program ensures that FSF have members from all major demographic groups. Joint and multinational partners should encourage and support efforts to recruit from minority populations. Mobile recruiting can target specific areas, ethnic groups, or tribes. Forces should encourage moderate groups and factions within hostile or potentially hostile ethnic groups to join the foreign security force. Often host-nation governments resist recruiting disaffected ethnic groups. However, even moderate success in recruiting from these groups provides enormous payoffs. It builds the security forces' legitimacy and often quiets legitimate fears of such groups regarding their relationship to the host-nation government. Effectively disarming, demobilizing, and reintegrating former enemies or other armed groups should be part of the overall plan. These programs should be mentioned in recruiting efforts.

2-19. A proper recruiting program specifies appropriate behavioral, physical, and moral standards. Ideally, recruits are centrally screened and inducted. Recruiting centers reside in safe areas safe. All recruits undergo a basic security check and are vetted against lists of suspected enemies. Foreign agencies and personnel perform this screening. Membership in illegal organizations needs to be carefully monitored. Past membership need not preclude a person from joining the foreign security force; however, any ongoing relationship with an illegal organization requires constant monitoring. FSF personnel limits the members of an illegal unit, tribal militia, or other militant faction to a single military or police unit.

Promotion Screening and Selection

2-20. Selection for promotion must stem from proven performance and aptitude for increased responsibility. Objective evaluations ensure promotion is by merit and not through influence or family ties. One method of selection identifies the most competent performers, trains them, and recommends them for promotion. The second method identifies those with social or professional status within the training group, then trains and recommends them for promotion. The first method may lead to more competent leaders but could be resisted for cultural reasons. The second method ensures the new leader will be accepted culturally but may sacrifice competence. The most effective solution comes from combining the two methods.

Pay and Benefits

2-21. Appropriate compensation precludes a culture of corruption in FSF. Spending the money needed for adequate wages and producing quality security forces is less costly than ending up with corrupt and abusive forces that alienate the populace. Paying the police adequately is especially important; their duties and contact with civilians often expose them to opportunities for corruption.

2-22. Payroll procedures and systems are also vitally important. They must be transparent and accountable, so FSF members receive their full pay and entitlements. Centralized administration of compensation through secure, automated systems minimizes the risk for corruption and manipulation; however, the supporting infrastructure will have to be available or developed over time.

2-23. Effective FSF can help improve host-nation social and economic development through the benefits each member receives. Every recruit should receive a basic education, job training, and morals and values inculcation.

Leader Recruiting and Selection

2-24. Officer candidate standards should be high. Candidates should have good health and pass an academic test with higher standards than the test for enlisted recruits. Recruiters carefully vet officer candidates to ensure that they do not have close ties to any radical or enemy organization.

2-25. Noncommissioned officers should be selected from the best enlisted FSF members. Objective standards, including proficiency tests, should be established and enforced to ensure that promotion to their ranks comes from merit. Many armies lack a professional noncommissioned officer corps; establishing one for a foreign security force may prove difficult. In the meantime, adjustments will have to be made, placing more responsibility on commissioned officers.

Personnel Accountability

2-26. Host-nation leaders carefully track and account for FSF. Proper personnel accountability reduces corruption, particularly in countries with manual banking systems where soldiers are paid in cash. In addition, many personnel failing to report for duty can indicate possible attacks, low unit morale, or enemy and militia influences on the foreign security force.

Demobilization of Foreign Security Force Personnel

2-27. Host nations develop programs to keep a class of impoverished and disgruntled former officers and soldiers from forming. As the foreign security force mature, officers who perform poorly or fail to meet the standards will need to be removed. Providing some form of government-provided education grants or low-interest business loans enables discharged personnel to earn a living outside the military. All soldiers who serve for several years and are then removed should receive a lump-sum payment or pension to ease their transition to civilian life. These programs should not apply to those guilty of major human rights abuses or corruption. Transition or demobilization planning should start as soon as commanders anticipate the need (it may not be required in all cases). Similar programs may be required when demobilizing nongovernment militias.

TRAIN

2-28. Train is a SFA task to assist FSF by developing programs and institutions to train and educate. These efforts must fit the nature and requirements of their security environment. The imperatives of Army special operations forces for facilitating a comprehensive approach and engaging the threat discriminately apply to the train task (see appendix A).

Training the U.S. Trainers

2-29. Soldiers assigned training missions should receive training on the requirements of developing FSF. The training should emphasize the host nation's cultural background, introduce its language, and provide cultural tips for developing a good rapport. Training should also include protection for troops working with FSF. U.S. trainees must become familiar with FSF organization and equipment, especially weapons not in the U.S. inventory. This training must emphasize the following:

- Sustaining training and reinforcing individual and team skills.
- Using the smallest possible student-to-instructor ratio.
- Developing host-nation trainers.
- Training to standards—not to time.
- Providing immediate feedback; using after action reviews.
- Respecting the culture but learning to distinguish between cultural practices and excuses.
- Learning the language.
- Working with interpreters.

2-30. U.S. forces should show respect for local religions and traditions. They should willingly accept many aspects of the local and national culture, including food (if sanitation standards permit). U.S. forces must stress that they do not intend to undermine or change the local religion or traditions. However, they have to strive to reduce how dysfunctional social practices affect the foreign security force's ability to conduct operations. U.S. trainers and advisors must have enough awareness to identify and stop inappropriate behavior, or at least report it to the appropriate chains of command.

Establishing Training Standards

2-31. Training in SFA involves many of the individual and collective skills performed in conventional military operations. All levels of training for all components should include values training. Metrics for evaluating units should include subjective measures, such as loyalty to the host-nation government, as well as competence in military tasks. Soldiers know how to evaluate military training. However, the acceptance of values, such as ethnic equality or the rejection of corruption, may be a better measure of training effectiveness. Gauging this acceptance is far more difficult than evaluating task performance. While the operational environment varies widely, FSF and trainers can still establish clear measures to evaluate the training of individuals, leaders, and units.

2-32. Effective training programs require clear, detailed individual, leader, and unit performance standards. These standards take into account cultural factors that directly affect the ability of the individual or unit to operate. For example, training a staff or unit to conduct effective operations requires more time in countries where the average soldier is illiterate. Building a security force from the ground up takes far more time than creating one around a trained cadre. Thus, using existing military personnel to form units and cadres for units often proves better than creating novice security forces.

2-33. Often, poorly trained leaders and units commit more human rights violations than well-trained, well-led units. Leaders and units unprepared for the pressure of active operations tend to use indiscriminate force, target civilians, and abuse prisoners. These actions can threaten the popular support and government legitimacy essential for SFA success. Badly disciplined and poorly led FSF can facilitate insurgent recruiting and propaganda efforts.

2-34. Setting realistic measures for FSF and following through on training plans consume time. The pressure is strong to find training shortcuts, employ quick fixes, or train personnel on the job. Trainers should resist such approaches. Such approaches often create more problems than they solve. However, trainers should also avoid the temptation to create long, complex training programs based on unrealistic standards. Effective programs account for the host nation's culture, resources, and short-term security needs. No firm rules exist on how long particular training programs should take. Trainers can use existing and historical training programs to determine how long training should take. To a certain extent, the enemy threat dictates the length of training. As security improves, training programs can expand to facilitate achievement of the long-term end state.

Training Foreign Security Forces

2-35. Members of FSF develop through a systematic training program (individual training and education as well as collective training). The program builds basic skills, then teaches them to work as a team, and finally allows them to function as a unit. Basic military, intelligence, or law enforcement training focuses first on basic skills such as legal considerations, first aid, marksmanship, and fire discipline. Leaders have training in tactics, including patrolling, urban operations, and legal evidence collection. Everyone must master rules of engagement and the law of armed conflict. FSF units should train to standard for conducting the major operations they will face.

Training Leaders

2-36. The effectiveness of FSF directly relates to the quality of their leadership. Building effective leaders requires a comprehensive program of officer, staff, and specialized training. The ultimate success of any SFA effort depends on creating viable FSF leaders able to carry on the fight at all levels and build their nation on their own.

2-37. The standards of leader training reinforce different levels of authority within the foreign security force. Clearly established responsibilities for commissioned and noncommissioned officers specify what is expected of recruits and leaders. To ensure civilian control, subordinate relationships to civilian authorities also are reinforced. In addition, training should establish team dynamics. In some cultures, security forces may need training to understand the vital role of members not in primary leadership positions. For example, noncommissioned officers may be a new or different concept for some FSF.

Employing Newly Trained Forces

2-38. SFA must build the morale and confidence of FSF. Committing poorly trained and badly led forces results in high casualties and invites tactical defeats. While defeat in a small operation may have little strategic consequence in a conventional war, often a small tactical defeat of FSF has serious strategic consequences. If the foreign security force fails, the local populace may begin to lose confidence in the host-nation government's ability to protect them. FSF must prepare for operations so they have every possible advantage. As much as possible, FSF should begin with simpler missions. As their confidence and competence grows, these forces can assume more complex assignments. Collaborating with joint or multinational units can help new forces to adjust to combat stress.

2-39. Newly trained units should enter their first combat operation in support of more experienced foreign units. Operational performance of such inexperienced organizations should be carefully monitored and evaluated to correct weaknesses quickly. The employment plan for FSF should allow enough time for additional training after each operation. Gradually introducing units into combat allows the command to identify poor leaders for retraining or other actions. Competent leaders are also identified and given greater authority and responsibility.

Equip

2-40. Equip is a SFA task encompassing all efforts to assess and assist FSF with the procurement, fielding, and sustainment of equipment. All equipment must fit the nature of the operational environment. The SFA principle of ensuring long-term sustainment is a vital consideration for the equip task.

2-41. The strategic plan for security force development should outline equipment requirements. Appropriate equipment is affordable and suitable against the threat. Forces must be able to train on the equipment. Interoperability may be desired in some cases. A central consideration includes the host nation's long-term ability to support and maintain the equipment.

2-42. The initial FSF development plan should use a comprehensive approach with local, regional, national, and other appropriate actors. Fundamentally, it should support the internal defense and development strategy. It should include phases with objectives for FSF to meet over three to four years. Since potential enemies adapt rapidly and situations change, commanders must continually assess the direction and progress of developing FSF.

2-43. The requirement to provide equipment may be as simple as assisting with maintenance of existing formations or as extensive as providing everything from shoes to communications and investigation kits. If the enemy uses heavy machine guns and rocket-propelled grenades, then FSF need comparable or better equipment. This especially applies to police forces, which are often lightly armed and vulnerable to well-armed enemies.

2-44. Primary considerations should include maintainability, ease of operation, and long-term sustainment costs. Few developing nations can support highly complex equipment. In most operations, having many versatile vehicles that require simple maintenance is often better than having a few highly capable armored vehicles or combat systems that require extensive maintenance. Effective maintenance systems for FSF often begin with major maintenance performed by contractors. The program then progresses to arrangements with U.S. forces as they train foreign personnel to conduct the support mission.

2-45. Sources of materials for FSF include U.S. foreign military sales (not technically SFA), multinational or third-nation resale of property, contracts with internal suppliers, or purchases on the international market. The organizations responsible for equipping FSF require flexibility to obtain equipment that meets their needs for quality, timeliness, and cost. As part of their training, FSF also need to learn property accountability to reduce corruption and ensure proper equipment usage. FSF provide equipment the same level of control and protection that U.S. forces provide for similar equipment. (See AR 12-1 and DOD 5105.38M.)

Rebuild and Build

2-46. Rebuild and build is a SFA task to assess, rebuild, and build the existing capabilities and capacities of FSF and their supporting infrastructure. This task requires an in-depth analysis of the capability, capacity, and structures required to meet the desired end state and operational environment. Some FSF may require assistance in building and rebuilding, while other FSF may only need assistance in building.

2-47. FSF need infrastructure support. People need buildings for storage, training, and shelter. Often requirements include barracks, ranges, motor pools, and other military facilities. Construction takes time; the host nation needs to invest early in such facilities if they are to be available when needed. Forces must plan to protect any infrastructures, including headquarters facilities, since these building are attractive targets for insurgents. (See FM 3-34.400 for information on hardening measures to increase infrastructure survivability and improve protection.)

2-48. During some operations, such as counterinsurgency, foreign military and police forces often operate from local bases. Building training centers and unit garrisons requires a long-term force-basing plan. If possible, garrisons should include housing for the host-nation soldiers and their families; government-provided healthcare for the families; and other attractive benefits.

2-49. The host nation may need to make large investments in time and resources to restore or create the nationwide infrastructure necessary to effectively command and control FSF. The host nation will build functional regional and national headquarters and ministries as well as local bases and police stations.

ADVISE AND ASSIST

2-50. Advise and assist is a SFA task in which U.S. personnel work with FSF to improve their capability and capacity. Advising establishes a personal and a professional relationship where trust and confidence define how well the advisor will be able to influence the foreign security force. Assisting is providing the required supporting or sustaining capabilities so FSF can meet objectives and the end state. The level of advice and assistance is based on conditions and should continue until FSF can establish required systems or until conditions no longer require it. Leaders cannot permit FSF to fail critically at a point that would undermine the overall effort. See chapter 6 for a discussion on advisors.

ACTIVITIES

2-51. SFA is inherently a developmental effort. Success is measured by the increase in the foreign security force's capability, capacity, competency, commitment, and confidence in areas in which U.S. forces are assisting them and as a whole. SFA activities include plan and resource, generate, employ, transition, and sustain. Figure 2-1 depicts the SFA activities. These activities blend over time based on the capability and capacity of FSF.

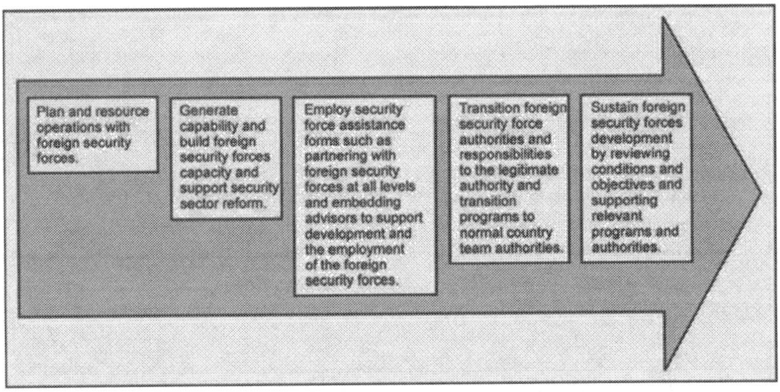

Figure 2-1. Security force assistance activities

PLAN AND RESOURCE

2-52. The plan and resource SFA activity begins as commanders understand the operational environment and determine the requirements of FSF. It also ensures that the United States provides SFA that achieves the objectives and end state of both the host nation and United States. The host nation and United States should then analyze the resource requirements and efforts so that developing FSF have sufficient and appropriate resources. Chapter 4 provides additional detail on planning for facilities and infrastructure.

GENERATE

2-53. The generate SFA activity includes generation of FSF, including the forces, leaders, and warfighting function capabilities (movement and maneuver, intelligence, fires, sustainment, command and control, and protection) based on the desired end state.

Employ

2-54. The employ SFA activity involves FSF transitioning from force generation to mission employment. It does not rely on the maturity of the force or supporting institutions as a whole but is focused on the foreign element in question. Conditions determine when to use an element of FSF. Leaders of U.S. and foreign security forces assess the risk associated with employment and mitigate that risk as much as possible.

Transition

2-55. The transition SFA activity defines the transition between two security forces, when applicable. This could be a transition of authority between U.S. forces to a new foreign security force. Another example could be from a regional foreign security force to a foreign security force with U.S. advisors. It could also be from a host-nation military force to a host-nation police force, with or without U.S. advisors.

Sustain

2-56. The sustain SFA activity occurs when the institutional capacity of the foreign security force has been developed to a point where it is self sustaining. It may continue to have SFA contact through combined exercises, educational opportunity exchange, intelligence sharing, and foreign military sales.

TYPES

2-57. The U.S. or other actors use three types of SFA—augmenting, partnering, and advising—to develop confident, competent, capable, committed, and credible FSF. These types of SFA define the relationship between outside actors and the foreign security force. They may be employed simultaneously, sequentially, or in combination. The progression and types of SFA are determined by the operational environment, the assessment of the FSF, and by resources available. Each of these types requires decidedly different requirements, objectives, and legal considerations.

Advising

2-58. Advising is the primary type of security force assistance. Advising is the use of influence to teach, coach, and advise (see chapter 6) while working by, with, and through FSF. This type of SFA relies on the ability of the advisor to provide relevant and timely advice to FSF. Advising helps FSF conduct independent decisionmaking and operations. Advisors may also provide FSF with direct access to joint and multinational capabilities such as air support, artillery, medical evacuation, and intelligence. However, the advisor and advisor team require proper manning and equipment to perform these secondary support functions while staying focused on advising. Multiple sources will pressure advisors as they perform their missions. For example, nearby maneuver units may have other priorities and may focus on accomplishing their missions with or without FSF.

2-59. Advisor teams require a clearly defined and structured chain of command under which to operate. This is not only for logistics and support, but also keeps the advisor focused on developing FSF. Advisor teams will often find themselves answering to their higher military assistance group, the brigade combat team they are attached to, and the host-nation unit with which they are embedded. A well-defined chain of command alleviates confusion concerning who tasks or monitors the team's progress as well as ensures constant and adequate team sustainment.

2-60. Advisors are not partners; U.S. forces act as partners. Advising and partnering are complementary but inherently different activities. Advising requires relationship building and candid discourse to influence development of a professional security force. Partnering incorporates training with combined operations to achieve the same SFA goals. Advisors perform partnership shaping functions, shape discussions with their counterparts, and create opportunities for the partner units.

Partnering

2-61. Partnering attaches units at various levels to leverage the strengths of both U.S. and foreign security forces. As a foreign security force's capabilities mature, the echelon and degree of partnering decrease. As the foreign security force conducts more autonomous operations, U.S. forces still provide quick reaction forces and other assistance as appropriate.

2-62. Partner units should establish combined cells for intelligence, operations, planning, and sustainment. While effective coordination is always required and initial efforts may require completely fused efforts, FSF should eventually build the capability and capacity to conduct all efforts autonomously. These combined cells have several functions. They support transparent operations and a comprehensive approach. They also enhance the relationships among U.S. and foreign security forces by demonstrating trust. Finally, they develop the capacity of FSF in key staff areas. However, combined cells are not without risk. Operations and plans always at risk of compromise and prudent precautions should be taken.

2-63. Another partnering technique establishes relationships among command and staff elements of U.S. and foreign security forces. It requires less reorganization. It also allows counterparts to understand each other since foreign personnel can observe U.S. personnel performing their duties. This construct works well when the foreign culture is extremely sensitive to maintaining honor. In this case, foreign personnel can observe without having to expose their lack of knowledge or appear subordinate.

2-64. Unit partnerships do not replace advisor roles or functions. If partnering and advising are used in combination, it forms a three-part relationship amongst FSF, advisors, and the partner units. Partner units should look to the advisor to identify, shape, and facilitate operational partnering opportunities and training events. Advisors support U.S., coalition, and partner unit objectives, but, depending on the operational phase, the partner unit may support advisors or advisors may support the partner unit. Therefore, some level of advisor skills training should be included in the partner unit training program if those units will be conducting SFA activities.

Augmenting

2-65. Augmenting is an arrangement where FSF provide individuals or elements to combine with U.S. units, or U.S. individuals or elements combine with FSF. Augmentation improves the interdependence and interoperability of U.S. and foreign security forces. Augmentation can occur at many levels and in many different forms. For example, a U.S. squad can be augmented with host-nation individuals, a U.S. company can be augmented with a host-nation platoon, or a U.S. battalion can be augmented with a company from a foreign security force. Similarly, augmentation can be of short duration for a specific operation or of a longer duration for an enduring mission. Augmenting immerses FSF in a U.S. environment to provide language and cultural awareness to the U.S. unit. U.S. forces can also augment FSF.

Combining Tasks, Activities, and Assessment

2-66. Execution of successful SFA links its activities, tasks, and assessment. Figure 2-2 depicts how to arrange SFA activities, tasks, and assessment in time and purpose. The tasks that occur in one SFA activity will normally continue throughout operations. For example, organize, train, equip, rebuild and build, and advise and assist tasks will continue throughout the SFA activities of employment, transition, and sustain.

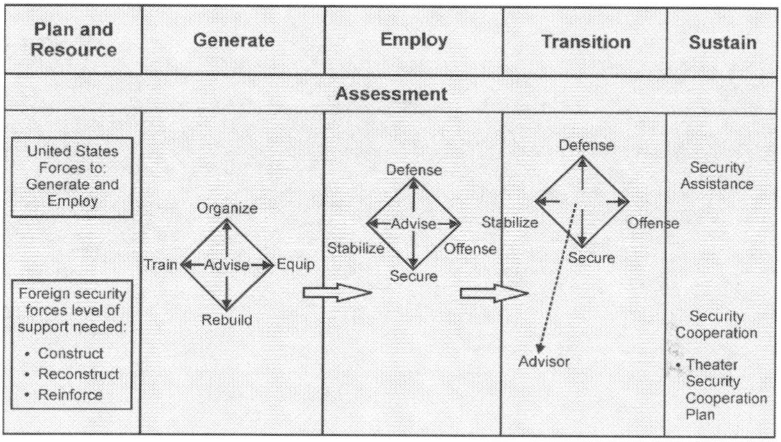

Figure 2-2. Security force assistance tasks, activities, and assessment

Plan and Resource

2-67. The plan and resource SFA activity determines the level of support and number of U.S. forces needed. As all efforts in SFA this activity is part of a comprehensive approach. Planners assess whether they will be constructing a foreign security force from the ground up, reconstructing a foreign security force based on existing capabilities and structure, or merely reinforcing an existing foreign security force. Understanding what capability and capacity the foreign security force should have and what capability and capacity they currently have is vital. Commanders and planners can then determine the size and structure, resources, and funding for the U.S. unit or units that will facilitate or assist the generation of FSF. Chapters 3 and 4 discuss planning and execution considerations for SFA.

Generate

2-68. In the generate SFA activity, U.S. forces focus on organizing, training, equipping, and advising FSF. During generation of FSF, the condition of the existing foreign security force's infrastructure and capabilities determine the assistance needed from U.S. forces. The generate activity adapts to the needs and requirements of the foreign security force. It can be designed to take the individual recruit from initial entry through to an employable unit, or it can augment an existing foreign unit with additional or enhanced capabilities. Like all SFA efforts, generate focuses on bringing a foreign security force's capabilities to a level where it can conduct its mission. With help from advisors in the next phase, FSF can work toward independent and sustained employment. In large SFA efforts, an organization may be required to command and control this activity.

2-69. Once elements of FSF are trained, the focus shifts from organizing, training, equipping, and rebuilding to employing the foreign security force. However, not all FSF units and organizations are ready for employment at the same time. As one unit is employed, others may be part of the generation activity, while still others may be completing the transition activity. FSF may be employed for offensive, defensive, or stability operations while other FSF may be employed against internal or external threats. The primary SFA task for U.S. forces is to advise and assist to facilitate the foreign security force operations.

Transition

2-70. The transition activity focuses on removing the U.S. elements and advisors once FSF are ready to conduct independent operations. All involved actors work in a comprehensive approach to assess when this can be achieved and to coordinate the transition itself. As FSF gradually progress toward the transfer of authority, close relationships forged between the FSF and their partners prove essential to sustainable development and successful transition. Genuine relationships engender trust and confidence, enabling increased responsibility and a well-executed transition process. These relationships also foster a clear understanding of command responsibilities and authorities. Such an understanding ensures FSF approach transition prepared to assume the full weight of their future role in the security sector. Success in developing FSF often depends more on relationships and personalities at the unit level than any other factor.

Chapter 3

Brigade Operations Process

This chapter examines operational considerations for the modular brigade augmented for security force assistance (SFA). It uses battle command and the operations process to examine these considerations.

BATTLE COMMAND

3-1. *Battle command* is the art and science of understanding, visualizing, describing, directing, leading, and assessing forces to impose the commander's will on a hostile, thinking, and adaptive enemy. Battle command applies leadership to translate decisions into actions—by synchronizing forces and warfighting functions in time, space, and purpose—to accomplish missions (FM 3-0). Commanders are central to battle command for brigade combat team (BCT) operations just as they are for any other operation. However, there are key considerations for successful battle command of a BCT conducting SFA. Most importantly, BCT commanders must work with their foreign counterparts, which may require facilitating operations when the foreign security forces (FSF) commander has the lead for the operations process.

UNDERSTAND

3-2. Successful SFA requires that BCT commanders thoroughly understand the operational environment. Conditions of the operational environment affect what SFA commanders provide. Commanders must understand the conditions that make up the current situation, including the relevant population, other actors, and FSF. From this understanding, commanders visualize desired conditions that represent a desired end state. Commanders must understand the capacity and capabilities of foreign security. The host-nation or regional security organization helps to determine desired conditions associated with the foreign security forces. After envisioning a desired end state, commanders then conceptualize how to change the current situation to the desired end state.

3-3. BCT commanders must understand the host-nation's security and political concerns. Security concerns focus on internal and external threats, their capability to affect the operational environment, and the foreign security force's strength and capability to counter those threats. Politically, the considerations associated with the security sector are the legitimacy of the foreign security force, FSF capability and capacity, host-nation or partner organization policy, and the ability to support the host-nation and U.S. Government objectives.

3-4. Commanders use the information obtained from intelligence preparation of the battlefield and intelligence, surveillance, and reconnaissance (ISR) to determine objectives and requirements for assessing and mitigating risk. Risk applies to how well FSF, U.S. Government, host-nation, partner organization, and international organizations can tolerate changes in the operational environment, as well as the challenges and conditions inherent to the operation.

VISUALIZE

3-5. *Commander's visualization* is the mental process of developing situational understanding, determining a desired end state, and envisioning the broad sequence of events by which the force will achieve that end state (FM 3-0). Visualization begins with assignment of the mission, which provides the focus for developing the commander's visualization. However, visualization must be continuous as SFA is dynamic. During planning, commander's visualization provides the basis for developing plans and orders.

3-6. The nature and extent of assistance provided to FSF are determined by this desired end state. SFA is often a slow process that does not lend itself to quick solutions. To ensure long-term success, commanders and host-nation or regional security organizations clarify early what conditions they desire. These conditions include the desired end state for FSF, which often include being—

- **Flexible**. FSF can accomplish broad missions. Flexibility requires an effective command and control organization appropriate for the partner or host-nation. This can include both internal and external threats.
- **Proficient**. The level of proficiency of FSF should include—
  - The capability of working effectively within a comprehensive approach.
  - The capability to integrate their operations with those of multinational partners.
  - Nonmilitary security forces competent in maintaining civil order, enforcing laws, controlling borders, securing key infrastructure, and detaining criminal suspects.
  - Nonmilitary security forces thoroughly trained in modern police ethos and procedures and which understand the basics of investigation, evidence collection, and proper court and legal procedures.
- **Self-sustained**. FSF can manage their own equipment throughout its lifecycle and perform administrative support.
- **Well led**. Leaders at all levels possess sound professional standards and appropriate military values; they are selected and promoted based on competence and merit.
- **Professional**. FSF are—
  - Honest, impartial, and committed to protecting and serving the entire population; they operate under the rule of law and respect human rights.
  - Loyal to the partner organization or host-nation government and serving partner or national interests; they recognize their role as the servants and not masters.
- **Integrated into society**. In the case of a host nation, FSF represent the host nation's major social groups; they are not seen as elements of just one faction.

3-7. These conditions may vary in different countries, but well-trained FSF should—

- Provide reasonable levels of security from external threats without threatening regional security.
- Provide reasonable levels of internal security without infringing on the populace's civil liberties or posing a coup threat to the host-nation government.
- Be founded upon the rule of law, including international principles of the rule of law.
- Be sustainable after joint and multinational forces depart.

3-8. When addressing internal threats, foreign military and police forces often perform unconventional functions. The military may have to fill an internal security role usually reserved for the police, or police may have forces so heavily armed that they would normally be part of the military. In the near term, FSF should—

- Integrate military capabilities with those of local, regional, and national police.
- Maintain the flexibility to transition to more conventional roles of external and internal defense based on long-term requirements.

3-9. To achieve this end state and intermediate objectives, the host nation should develop an internal defense and development (IDAD) strategy (with multinational assistance if required). This strategy should have a section on IDAD that addresses all aspects of SFA.

3-10. U.S. doctrine divides force development into domains: doctrine, organization, training, materiel, leadership and education, personnel, and facilities (DOTMLPF), which may be useful to analyze the IDAD. DOTMLPF elements are tightly linked, simultaneously pursued, and difficult to prioritize. Commanders monitor progress in all domains. Often military forces involved in such programs are tempted to impose their own doctrine and judgment on the host nation. Foreign security force doctrine, like the remaining DOTMLPF domains, must be appropriate to host-nation capabilities and requirements. Paragraphs 3-59 through 3-70 use DOTMLPF as a tool to assess FSF.

DESCRIBE

3-11. After BCT commanders visualize how to conduct SFA, they describe it to their staffs and subordinates to facilitate shared understanding of the mission and commander's intent. Commanders ensure subordinates understand the visualization well enough to begin planning. Commanders express their initial visualization as—

- Initial commander's intent.
- Planning guidance, including an initial concept of operations.
- Commander's critical information requirements.
- Essential elements of friendly information that must be protected.

Commander's Intent

3-12. Like any operation, SFA requires a clear, concise statement of what the force must do and the conditions the force must meet to succeed with respect to the enemy, terrain, and desired end state. In such an operation, the end state should include a description of desired conditions with respect to FSF capability and capacity. The commander's intent links the mission and concept of operations. It describes the key tasks and end state that, along with the mission, form the basis for subordinates' initiative.

3-13. A sample end state could discuss—

- How civilians perceive FSF as credible and legitimate and representing a legitimate authority.
- How FSF have the capability to support stability and security, retain the initiative and freedom of movement, disrupt enemy activities, influence the population, operate with other friendly forces, and sustain themselves.

3-14. Like any operation, how the BCT supports the higher organization's mission is vital. Additionally, the BCT supports FSF. Some sample key tasks are:

- Support the organization of a division of foreign infantry and two battalions of foreign police.
- Train leaders of FSF to create a cadre that can sustain their own training programs.
- Oversee the equipping of FSF and provide new equipment training.
- Support the rebuilding or building of infrastructure required to develop FSF.
- Advise FSF to develop the required capabilities and competencies associated with shared objectives.
- Assist the partner's security forces in meeting their operational requirements.

Planning Guidance

3-15. Commanders develop planning guidance for the staff from the commander's visualization. Planning guidance may be as broad or detailed as circumstances require; however, it must convey to the staff the essence of the commander's visualization. Due to the complex and dynamic nature of FSF, SFA often requires detailed planning guidance. Commanders ensure the staff understands their visualization while still permitting the staff to explore different options. Some considerations are:

- Friendly courses of action (COAs) with differing SFA forms and differing methods on how to augment, partner, or advise.
- Command and control architecture, intelligence architecture, and communications architecture to support dispersed SFA.
- Enemy COAs as they seek to defeat friendly efforts with respect to SFA activities.
- Enemy adaptation given the development of the foreign security force.
- Guidance for security during the conduct of SFA, including the BCT securing its advisor teams. The BCT determines the staging, needs, and ad hoc arrangements of advisor teams.
- ISR guidance to include identifying the focus of effort as it relates to SFA.

Commander's Critical Information Requirements

3-16. Commander's critical information requirements (CCIRs) will be developed for SFA just as for any other operation. These requirements are what commanders need to make decisions or the information they think is important for understanding the operational environment. These CCIRs should be compatible with the FSF commander's information requirements. CCIRs include priority intelligence requirements and friendly force information requirements.

3-17. Priority intelligence requirements are vital to support decisionmaking. During SFA, priority intelligence requirements must be focused on the enemy and the operational environment, not on FSF. U.S. forces do not focus intelligence efforts, such as collection, on the foreign forces they assist.

3-18. Friendly force information requirements provide the information that commanders need to understand the conditions of their own forces, as well as the conditions of the foreign security forces. For example, a BCT commander may direct the staff to better define a foreign security force's critical equipment shortages or maintenance requirements. During execution, commanders may focus on when FSF meet certain objectives or conditions. This information allows the commander and FSF commander to make adjustments, such as changes to the task organization, support requirements, or even the type of SFA.

Essential Elements of Friendly Information

3-19. An *essential element of friendly information* is a critical aspect of a friendly operation that, if known by the enemy, would subsequently compromise, lead to failure, or limit success of the operation, and therefore should be protected from enemy detection (FM 3-0). Essential elements of friendly information should be determined through a comprehensive approach. Information sharing and trust are vital elements of SFA. U.S. commanders must be aware of the inherent risks in this context and employ prudent risk mitigation efforts. Thus, successful SFA requires a balance between sharing information and risk mitigation efforts. Although essential elements of friendly information are not CCIRs, they have the same priority.

Intelligence, Surveillance, and Reconnaissance

3-20. *Intelligence, surveillance, and reconnaissance* is an activity that synchronizes and integrates the planning and operation of sensors, assets, and processing, exploitation, and dissemination systems in direct support of current and future operations. This is an integrated intelligence and operations function. For Army forces, this activity is a combined arms operation that focuses on priority intelligence requirements while answering the commander's critical information requirements (FM 3-0). For the brigade, these efforts will often support FSF. Moreover, the BCT must have means that allow effective intelligence collaboration, such as an intelligence fusion cell with the participation of all appropriate actors.

3-21. ISR efforts focus on the enemy and operational environment within the context of SFA. ISR efforts in SFA are not used on the foreign security force. Rather, U.S. forces assess the foreign security force within a comprehensive approach.

Direct

3-22. BCT commanders direct all aspects of operations, although a comprehensive approach, including FSF and other appropriate actors is fundamental to successful SFA. The commander's direction takes different forms during planning, preparation, and execution. Commanders make decisions and direct actions based on their situational understanding, which they maintain by continuous assessment and coordination with FSF. SFA is directed by—

- Preparing and approving plans and orders with FSF.
- Assigning and adjusting missions, tasks, task organization, and control.
- Positioning units to maximize their support to FSF, anticipate threat actions, or create or preserve options.
- Positioning key leaders to ensure observation and supervision of FSF at critical times and places.

- Adjusting SFA and other support priorities and allocating resources based on FSF priorities, as well as opportunities and threats.
- Changing support arrangements, either internally or with respect to FSF.

ASSESS AND LEAD

3-23. Effective battle command requires commanders to continuously assess and lead. Assessment helps commanders to better understand current conditions and determine how the operation is progressing and how best to support FSF. Based on an assessment that is the product of a comprehensive approach, commanders modify plans and orders to better accomplish the mission. This may include changing the type or combination of SFA, support, sustainment, or task organization. If their assessment reveals a significant variance from the original shared commander's visualization, a comprehensive approach must be used to reframe the visualization and develop a new plan.

3-24. Commanders in a BCT conducting SFA lead by example and presence, as Army commanders do in any operation. However, commanders have the additional challenge of leading by, with, and through FSF. Command posts for BCTs conducting SFA have similar challenges in that they should be integrated headquarters as much as practical.

OPERATIONS PROCESS

3-25. As they would in any other mission, BCT commanders conducting SFA balance their time and the staff's time and resources among four major activities in a continuous learning and adaptive cycle called the operations process. The *operations process* consists of the major command and control activities performed during operations: planning, preparing, executing, and continuously assessing the operation. The commander drives the operations process (FM 3-0). While simple in concept, the operations process is complex in execution. The operations process activities may be sequential or simultaneous.

3-26. Figure 3-1 depicts the operations process activities as a cycle that is driven by battle command. This cyclic process applies equally to a BCT conducting major combat operational theme or operations within an irregular warfare operational theme. During the operations process involving SFA, commanders consider the presence and inclusion of other actors, including FSF. Policy, the operational environment, and the forms of SFA determine how U.S. forces, FSF, and other actors interact in the operations process.

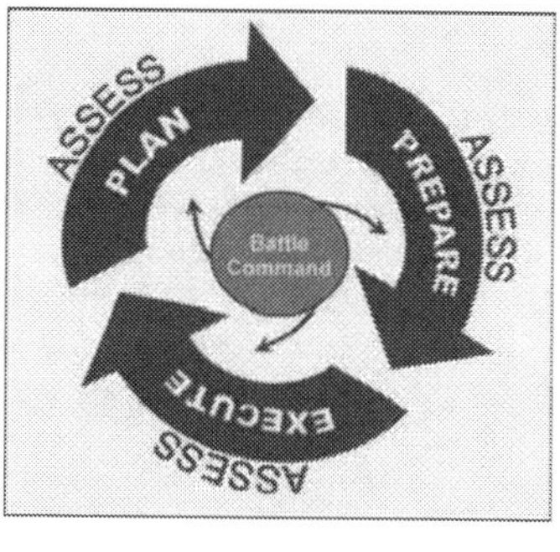

Figure 3-1. The operations process

PLAN

3-27. Planning for SFA, like any other operation, begins either with the anticipation of a new mission or the receipt of mission as part of the military decisionmaking process. *Planning* is the process by which commanders (and the staff, if available) translate the commander's visualization into a specific course of action for preparation and execution, focusing on the expected results (FM 3-0). SFA requires a comprehensive approach to ensure the U.S. commander's visualization synchronizes with policy and the foreign commander's visualization. SFA is the major requirement for a BCT conducting SFA, but the mission may require the brigade to perform other tasks.

3-28. Continuous and open to change, planning for SFA includes identifying how to best assist the FSF and developing a sequence of actions to change the situation. This planning must occur within a comprehensive approach that includes FSF, U.S. forces, and other actors. Similarly, this plan must be nested within policy, IDAD strategy, the campaign plan, and any other higher-echelon plans. It also involves anticipating consequences of actions and developing ways to mitigate them.

3-29. The plan, which includes the commander's intent, provides a common understanding to U.S. and foreign security forces on the actions to take. Plans and orders also provide decision points and branches that anticipate options that enable the force to adapt as the operation unfolds. This is especially important for SFA, as these operations tend to be prolonged efforts. BCTs conducting SFA often rotate before achieving all objectives. As a result, the planning team should establish objectives and milestones that can be achieved during the brigade's mission. These BCT objectives and milestones must support higher-echelon plans, including the campaign plan and IDAD strategy.

Functional Requirements

3-30. The composition of advisor teams is subject to objectives and conditions. For example, security conditions may require three armored vehicles, a crew-served weapon, and nine personnel to conduct independent movement. Another example involves the difference in the number of advisors. Newly generated FSF may require many advisors whereas experienced FSF may only require assistance in a few warfighting functions and consequently need fewer advisors. Understanding the functional requirements provides flexibility to the units assisting FSF. Three examples of determining functional requirements based on the understanding the operational environment include the BCT, a partnering or augmenting unit, and an advisor team.

Brigade Combat Team

3-31. A BCT headquarters has specific functions and requirements. These can include, but are not limited to—

- The functions of—
  - Carrying out SFA tasks and assisting FSF in support of objectives and the end state.
  - Facilitating partnerships between U.S. and foreign security forces.
  - Requesting, coordinating, and synchronizing resources.
  - Providing command and control of subordinate units and assigned advisor teams.
  - Coordinating with other commands and adjacent units.
  - Coordinating with other participants integrated within the comprehensive approach.
  - Developing, implementing, and synchronizing information engagement in assigned areas.
  - Developing relationships and an understanding with appropriate actors involved in SFA.

3-32. The requirements of—

- Having enough staff to handle the communications traffic generated by conditions.
- Being able to analyze information to aid commander and subordinate elements.
- Being able to articulate resource requirements in support of subordinate elements and efforts to conduct SFA tasks.
- Being organized and equipped to conduct operations and support and sustain FSF.
- Including regionally oriented host-nation officers to provide regional expertise.

Partnering or Augmenting Unit

3-33. Functions and requirements depend on the degree and level of partnering or augmenting. No prescriptive way to partner or augment exists; it is based on specific conditions and subject to the needs of policy, higher headquarters guidance, and needs of the FSF. A BCT may designate one battalion to partner or augment while designating other battalions to advise. A partnering or augmenting unit conducting SFA has specific functions and requirements. These can include, but are not limited to—

- The functions of—
  - Partnering with foreign security under the BCT level.
  - Increasing capability, capacity, competency, confidence, and commitment of FSF by conducting combined tactical operations.
  - Providing feedback on performance of FSF and certain shortfalls to embedded advisors and BCT headquarters.
  - Conducting sustainment and medical training with FSF at their home station or on operations.
  - Reporting conditions in the operational environment.
- The requirements of—
  - Having U.S. forces that are capable, competent, committed, and confident.
  - Being interoperable with FSF.
  - Being available for an appropriate period.

Military Advisor Team

3-34. The functions and requirements of a military advisor team are based on conditions—requirements, availability, and footprint. A military advisor team conducting SFA has specific functions and requirements. These can include, but are not limited to—

- The functions of—
  - Under the BCT headquarters, advising, training, and assisting FSF to which it is assigned.
  - Increasing capability, capacity, competency, confidence, and commitment of FSF by providing advice and support during battalion level and higher operations.
  - Assessing partner leaders, staff, and certain shortfalls to BCT headquarters.
  - Conducting sustainment training with FSF at their home or on operations.
  - Reporting on conditions in the operational environment.
- The requirements of—
  - Being manned with trained personnel who are properly equipped so as to be capable of executing their mission in current conditions.
  - Providing personnel with rank, education, and experience corresponding to the echelon and type of unit being advised.

Support and Sustainment Requirements

3-35. Support and sustainment requirements play a vital role in COA development (see chapter 6). The BCT may operate across a broad geographical area when conducting or supporting SFA. Partnering may be on certain levels with FSF and may support numerous embedded advisor teams or other actors. Since FSF often have varied relationships with higher echelons and institutions, the BCT may be involved vertically and spatially. These considerations may require more support and sustainment requirements than an initial analysis might indicate. Support and sustainment considerations include, but are not limited to—

- Distribution of FSF to which U.S. forces are providing SFA.
- The capability of FSF to provide support, sustainment, and medical functions.
- The technical capability and interoperability of U.S. and foreign security forces.
- The infrastructure that supports SFA.
- Overall security conditions where SFA occurs.

- The available contracting support.
- The ability of the environment to accept and absorb the U.S. forces' footprint.

Determining a Course of Action

3-36. COAs for SFA require an understanding of the breadth and width of the security sector. They may have to account for the roles ranging from the individual soldier or policeman to the ministerial level. COA development examines the roles other actors, for example, a gendarme or paramilitary force, in the security sector. Because the United States does not have a gendarme, U.S. forces may not fully appreciate its importance.

3-37. COAs for SFA also need to be flexible and inclusive. Commanders consider the following when examining a partner's security sector:

- The status and challenges of the host-nation security sector or regional security organization.
- The composition and disposition of the partner's security sector.
- Supporting institutions.
- Links among ministries, institutions, and components of the security sector.
- Partner capability and capacity shortfalls and impediments to their development.
- Environmental requirements to enable the building of security capability and capacity.
- Functional requirements to assist building capability and capacity.
- The impact of other actors within the comprehensive approach.

3-38. Once planners analyze the security sector, they can develop COAs that best combine SFA tasks, activities, and types (see chapter 2). While the advising type of SFA can be of any duration, partnering and augmenting tend to be enduring if used.

3-39. All SFA types tend to be more effective if they are long-term efforts. Advising fosters a more personal relationship and infers a stronger commitment to see the partner succeed. Partnering and augmenting include U.S. and foreign security forces conducting operations together. There are levels of partnering. Different advisory relationships use advisor teams embedded within various partner echelons and units, use one advisor team to advise multiple entities, or have a senior leader visit a partner unit to give advice.

3-40. Using any form or combination of forms depends on conditions (such as the level of competency), capability of FSF, and resources available. Policy, guidance, or formal agreements can also help commanders determine which form to use. Historically, a combination of forms appears to be the most effective. However, combining forms often proves more resource intensive and affected by time, geography, and the number of forces. COAs should remain flexible with respect to the form of SFA. Some example changes in conditions are the increased capability of one element in the security sector over another, the desire to increase one element over another based on partner objectives, or the increased targeting of one element over another by the enemy.

3-41. If a desired COA is not feasible given the scope of the security sector or due to other considerations, it may be useful to sequence SFA forms with portions of FSF. For example, a COA that affects smaller portions of the FSF, but in the right area and at the right time, may do more for achieving an objective or end state than a COA that attempts to support all FSF at one time.

3-42. BCTs are task-organized and augmented to accomplish a specific mission. How BCTs are specifically task-organized and augmented is based on the phase and types of SFA that policy and higher headquarters have directed. Chapter 4 describes augmentation in detail.

3-43. Attached units or units under operational control are task-organized under different headquarters as necessary to accomplish assigned and implied SFA missions. Deciding to task-organize specific formations for SFA may reflect previously approved contingency plans or result from current planning efforts. As these missions end, units return to their parent headquarters or are further task-organized by the controlling headquarters.

3-44. As with other operations, gaining commanders task-organize for SFA by designating operational control, attached, or support relationships to another unit or headquarters. Within each BCT, the brigade commander task-organizes all organic, assigned, attached, or operational control units to best suit the operational environment and types of SFA. Commanders may choose to use a support relationship (direct support, reinforcing, or general support reinforcing) among units of the BCT instead of a command relationship. This can also apply to relationships with foreign units.

3-45. SFA is often highly dispersed, as BCTs may have individuals or units assisting FSF over a wide area. This dispersion increases communications and sustainment requirements for the BCT. The nature of SFA also requires BCTs to help FSF meet requests for assets. To do so, the BCT and other units may need to task-organize to provide tactical support or sustainment. Dispersion and the nature of SFA often necessitate a decentralized leadership approach to facilitate effective and rapid support of FSF.

3-46. BCT commanders use the three types of SFA—advising, partnering, or augmenting—to accomplish the mission. The exact nature of a unit's assigned mission, the operational variables, and the mission variables drive modifications to headquarters and maneuver units. For example, a commander may need a unit tasked with advising, a unit tasked with partnering, and a unit tasked with augmenting. In this case, the unit tasked to advise would provide teams for the appropriate FSF while other foreign units continue their traditional missions. The unit that is partnering may provide the host-nation staff to assist the FSF headquarters. The third unit may provide squads to augment foreign platoons. Units may request additional support, sustainment, and medical support. However, BCTs conducting SFA must retain a reserve to conduct full spectrum operations.

3-47. Based on the overall mission analysis, including an analysis of functions and requirements, commanders and staffs begin to determine how to task organize. Due to the level of detail that may be involved, task organization should begin early in the military decisionmaking process, especially during COA development. No standard task organization exists since conditions vary. For example, a squad of infantry and three additional leaders may work as an advisor team for one mission, where another mission may require an advisor team with specialists for different warfighting functions. The art in SFA involves determining where, when, and if to allocate resources; it involves either breaking up or simply moving personnel and units from one foreign unit to another.

3-48. The techniques used by BCT commanders and staffs to allocate other resources and support are just as valid in SFA as in other types of operations. For example, if it is determined that the most important effort is advising, then the BCT commander may weight that effort with additional resources, such as communications equipment or experienced personnel.

3-49. BCTs emphasize unity of effort with FSF. Ideally, interoperability and shared resources extend to the lowest level. This means having one tactical operations center, dining facility, clinic, motor pool, and so on. Not only does this use resources more efficiently, but it also facilitates understanding, trust, and rapport between U.S. and foreign security forces. For Soldiers to effectively conduct SFA work by, with, and through FSF requires a close rapport and high level of trust.

3-50. Part of COA development for SFA includes an analysis of where to focus the brigade's subordinate units. This analysis includes determining which subordinate units support key elements of FSF. The brigade's subordinate units may support provincial reconstruction teams or transition teams that are not organic to the brigade. Another method is for the brigade may provide organic forces to make up transition teams or the military portion of provincial reconstruction teams or transition teams. Transition teams may include, but are not limited to police, border, point of entry, or military transition teams. Figure 3-2 depicts an example for a brigade providing forces as part of a provincial reconstruction team as well as providing the other transition teams. Figure 3-2 is an example of how a brigade may focus its subordinate units. Additionally, the teams are only examples. All of these teams may not be necessary for all brigades, and some brigades may require other types of advisor teams.

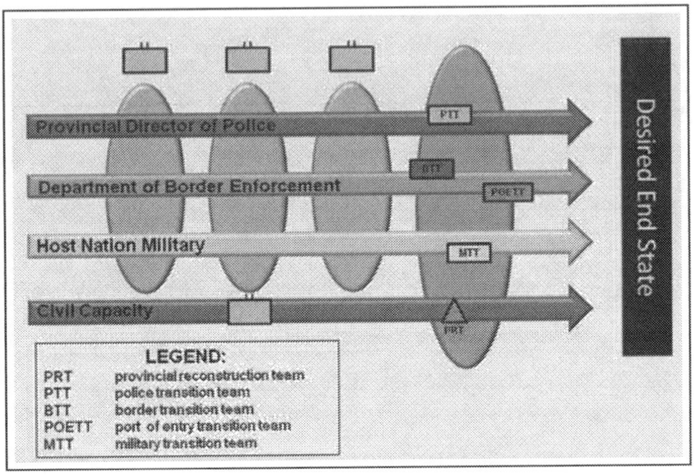

Figure 3-2. Example of brigade focus for subordinate units

PREPARE

3-51. *Preparation* consists of activities performed by units to improve their ability to execute an operation. Preparation includes, but is not limited to, plan refinement; rehearsals; intelligence, surveillance, and reconnaissance; coordination; inspections; and movement (FM 3-0). Like any operation, preparation for SFA creates conditions that improve friendly forces' opportunities for success. It facilitates and sustains transitions, including those to branches and sequels, which are of vital importance for the often dynamic operational environment for SFA. Chapter 5 covers preparation for units conducting SFA, including the BCT.

EXECUTE

3-52. *Execution* is putting a plan into action by applying combat power to accomplish the mission and using situational understanding to assess progress and make execution and adjustment decisions (FM 3-0). Execution of SFA focuses on supporting the needs of the FSF and, like many other operations, is often characterized by phases and transitions between phases. Chapter 4 examines BCT phases for SFA.

Acceptance of Prudent Risk

3-53. SFA, like any other operation, requires prudent, disciplined risk-taking focused on winning rather than preventing defeat. Because uncertainty exists in all operations, every decision involves risk. Among key elements of the art of command are deciding how much risk to accept and minimizing the effects of accepted risk (see FM 6-0). Due to SFA's inherent collaborative nature, determining prudent risk is complex and requires a comprehensive approach.

3-54. Risk reduction measures identified in planning add to the plan's flexibility during execution. A flexible plan can mitigate risk by partially compensating for a lack of intelligence. Again, SFA requires a thorough comprehensive approach to analyzing and agreeing upon risk reduction measures.

Execution Keys to Success

3-55. History and recent operations point to several keys to success for BCTs executing SFA:

- Communication is vital, especially when the BCT shares its operational area with other entities that have cultural differences and lack of communication. It can include FSF or multinational forces, as well as U.S. forces with another mission. For example, an advisor team can develop effective relationships with its foreign security force and local authorities only to have another unit conduct an operation that unhinges those relationships.
- BCTs establish close and continuing relationships with all advisor teams, other actors operating in their operational area, and foreign area officers with local or regional expertise. In addition to advisor teams assigned to military units in direct contact with the BCT, advisors may embed with local police, border, and other forces.
- BCTs establish close and continuing relationships with all foreign units (military, police, and others) operating in their operational areas.
- BCTs establish close and continuing relationships with all political entities and actors in their operational areas.
- Before each operation, as part of the military decisionmaking process, the BCT considers second- and third-order effects of the operation. The team coordinates the commander's intent with advisor teams of FSF and other units in the operational area.
- Unit leaders develop cultural awareness and use this awareness so that operations and relationships achieve the desired end state. In short, the closer a maneuver commander works with advisor teams and the more they interact with local political and cultural leaders, the better the overall chances of mission success.

ASSESS

3-56. The operations process and effective battle command require commanders to continuously assess and lead. Assessment helps commanders to better understand current situations and how SFA is progressing. *Assessment* is the continuous monitoring and evaluation of the current situation, particularly the enemy, and progress of an operation (FM 3-0). Assessment is conducted during planning, preparation, and execution and is evaluated against measures of effectiveness and performances to make decisions and adjustments (see FM 6-0).

3-57. Assessment of SFA requires a comprehensive approach that analyzes the overall operational environment and focuses on FSF and their impact. SFA assessment is performed in close collaboration with the U.S. country team, the host-nation government, and other multinational partners. These partners continually assess the security situation and its influence on other operations. From the assessment, planners develop short-, mid-, and long-range plans. As circumstances change, so do the plans. An insurgency might require employing host-nation forces at earlier stages of development. Some existing security forces may be so dysfunctional or corrupt that the organizations ideally should be disbanded rather than rehabilitated. In some cases, commanders will need to be replaced before their units will become functional. Political realities, however, may make both prospects undesirable. Minimizing and marginalizing particularly corrupt units or leaders while focusing on other units can be effective in this case.

3-58. While every situation differs, BCT leaders assess the following factors throughout planning, preparation, and execution of the operation:

- The operational environment.
- The civil considerations of the operational environment (see appendix B of FM 6-0).
- The core grievances and the threat (possibly including the prerequisites of insurgency).
- FSF elements.
- Methods, successes, and failures of FSF security efforts.
- State of training at all levels, and the specialties and education of leaders.

- Equipment and priority placed on maintenance.
- Logistic and support structure and their ability to meet the force's requirements.
- Extent of acceptance of ethnic and religious minorities.
- Laws and regulations governing the security forces and their relationships to national leaders.

Initial Assessment

3-59. An initial comprehensive assessment offers the BCT a baseline for establishing the scope of effort required. This assessment includes a troop-to-task analysis that determines the type and size of forces needed. FSF may require complete reestablishment or only assistance to increase capacity. They may completely lack a capability (such as internal affairs, corrections, or formal schools for leaders) or they may require temporary reinforcement. As with other military operations, efforts to assist FSF should reinforce success, not failure. For example, instead of building new police stations in every town, improve the good stations and use them as models for other organizations.

3-60. BCT leaders need to make decisions on what shortfalls to address first. The extent of the threats combined with resource limitations inevitably forces commanders to set priorities. However, making efforts to legitimize the host-nation leaders by encouraging them to make the decisions often proves more successful.

3-61. A key effort for SFA is the initial assessment of the DOTMLPF and policy capabilities of a foreign security force. During the early phases of a campaign, a geographic combatant command receives an initial assessment or a unit operating in the area has an assessment of FSF. The incoming BCT updates this assessment as it begins SFA.

3-62. Assessments require objective and subjective reporting, which may require a classification level. At a minimum, a BCT should assess—

- Doctrine.
- Organization.
- Training.
- Materiel and equipment.
- Leadership.
- Personnel.
- Command, control, communications, and intelligence.
- Operational effectiveness.

3-63. The doctrine assessment provides the BCT with an analysis on how the FSF operate. This assessment could focus on formal doctrine or more localized practices.

3-64. The organization portion of the overall assessment provides a perspective on how the FSF are organized. This includes the array of foreign institutions and units.

3-65. The level of training for FSF is a vital consideration. While this assessment is partly subjective, it should objectively examine individual and collective training strategies and mission-essential task lists. In the absence of mission-essential task lists, the BCT should compare FSF to U.S. equivalents.

3-66. The materiel and equipment assessment includes, but is not limited to, weapons, vehicles, equipment, and communications. It should be divided into equipment on hand, equipment readiness, and sustainment. Equipment on hand is quantitative and based on the type of equipment available to security operations. Equipment readiness reflects the operational readiness condition or serviceability of on-hand equipment. Sustainment readiness provides a qualitative and quantitative assessment that reflects how well FSF perform supply, maintenance, transportation, and medical functions. This includes facilities, systems, and structures.

3-67. The leadership assessment analyzes FSF commanders, officers, noncommissioned officers, and other leaders. This includes an assessment of staff members.

3-68. The foreign personnel assessment is quantitative and examines manning, staff manning, and administrative capabilities. It also examines officers, noncommissioned officers, and other personnel.

3-69. The command, control, communications, and intelligence assessment is qualitative and should reflect how well FSF conduct these functions.

3-70. The operational effectiveness assessment examines the foreign security force's ability to execute operations. This portion of the assessment should not be confused with the examination of training, as relatively high levels of training do not necessarily mean that FSF will perform in operations. This assessment is highly subjective.

Assessment During Operations

3-71. Assessment for the BCT conducting SFA should be part of a comprehensive approach. Assessment is directly tied to the commander's decisions throughout planning, preparation, and execution. Critical to effective assessment is developing criteria to evaluate progress toward task accomplishment, achievement of objectives, and attainment of the end state conditions. Assessment criteria are expressed as measures of effectiveness or measures of performance.

3-72. During operations, BCT and FSF commanders and staffs periodically assess the underlying framework of the plan itself. This involves reexamining the logic and assumptions used to develop the original plan, and it requires a deliberately conceived effort to learn as a part of the operation. BCT and FSF collaboration among higher, lower, and adjacent commanders, staffs, and other actors, backed up by qualitative and quantitative assessments, contributes to this learning. Based on this reexamination, commanders update their visualization, direct changes to the order, or develop an entirely new plan to adapt the force to better accomplish the mission.

Chapter 4

The Modular Brigade Augmented for Security Force Assistance

This chapter examines the typical phases for the brigade combat team conducting security force assistance, including a discussion of baseline augmentation by phase. The final section examines some of the subordinate units and elements of the modular brigade augmented for security force assistance.

THE BRIGADE COMBAT TEAM

4-1. The brigade combat team (BCT) is the cornerstone for Army modularity and is designed to operate at the tactical level across the spectrum of conflict. The modular brigade can be augmented, based on the requirements of the operational environment, with enabling assets and capabilities to support distributed security force assistance (SFA). However, the BCT augmented for SFA retains the capability to conduct full spectrum operations—offense, defense, and stability. Any of the three modular BCTs—heavy, infantry, or Stryker—can support SFA. Battalion-sized maneuver, fires, reconnaissance, and sustainment subordinate elements of the BCT support foreign security forces (FSF).

4-2. The gaining theater army force tailors all modular Army forces for use in major combat operations and contingencies. The theater army commander, working for the geographic combatant commander, determines the mix of forces and capabilities required for the BCT. This force tailoring can be done as part of the joint deliberate planning process or as a result of crisis action planning. Force tailoring includes providing BCTs with additional forces, personnel, or capabilities, including the possibility of an embedded provincial reconstruction team. Additional assets and capabilities can include, but are not limited to, command and control, communications, sustainment, engineering, military police, and intelligence.

4-3. In recent operations, the BCT has proven its versatility and capability to learn at the individual and collective levels. It supported other U.S. units conducting SFA and directly increased the effectiveness of FSF. The BCT has operated under varying levels of enemy activity, host-nation capabilities, objectives, and other conditions.

4-4. Because the BCT can operate in nonpermissive and permissive environments, it can conduct SFA across the spectrum of conflict. Thus, it can support civilian and military joint and multinational actors. The support can range from movement security and sustainment to augmentation of reconstruction teams and security for elections.

4-5. BCTs conducting SFA may support FSF development, assist FSF operations, and support and assist the development of host-nation institutions and infrastructure. While providing this support to the host nation, the BCT remains capable of conducting full spectrum operations independently.

BRIGADE COMBAT TEAM PHASES

4-6. If a force lacks the means to accomplish its mission in a single effort, then commanders normally phase the operation. A *phase* is a planning and execution tool used to divide an operation in duration or activity. A change in phase usually involves a change of mission, task organization, or rules of engagement. Phasing helps in planning and controlling and may be indicated by time, distance, terrain, or an event (FM 3-0). Phasing SFA assists in planning and controlling these dispersed and complicated operations. These phases often involve key changes in supporting SFA tasks and types employed.

4-7. BCT phases for SFA are initial response, transformation, and fostering sustainability. These phases mirror the stability operations framework described in FM 3-07 and are based on the operational environment. SFA can start in any phase or may even move to a previous phase due to changes in the conditions of the operational environment.

4-8. Figure 4-1 illustrates an SFA operation that includes all three SFA phases. The horizontal axis depicts an improving security situation, and the vertical axis depicts security capability. The example depicts an ideal situation, although this may not be the case in actual execution. Issues that arise require proper assessment and potentially the use of branches or sequels.

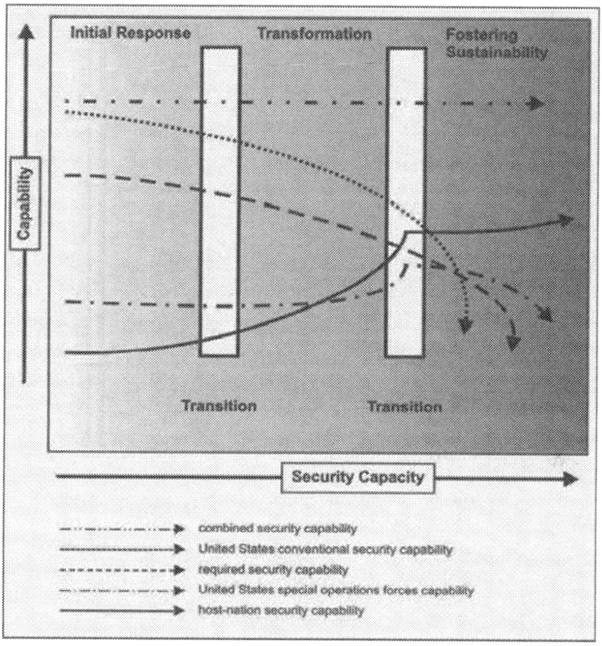

Figure 4-1. Example of phases of security force assistance

4-9. In figure 4-1, the combined security capability represents the overall security capacity of all security forces involved; it is the sum of U.S. conventional security, host-nation security, and the U.S. special operations forces security capabilities. This example assumes that the combined security capability exceeds the required security capability to effectively maintain a secure environment and counter any threat. The U.S. conventional security capability refers to tactical U.S. elements, such as the BCT. It does not refer to higher tactical or operational headquarters or support, sustainment, or medical assets. The U.S. special operations forces capability is depicted as declining in the end of the sustainment phase; as the FSF capability improves, the need for special operations forces declines. The host-nation or regional security organization and U.S. policy determine the end state for special operations.

4-10. For figure 4-1, a major change to enemy resources may require more FSF or additional. Drastic changes to the situation may even require a new operation plan. The figure focuses on capability, not forces, so host-nation security capability may increase without increasing the number or size of host-nation forces.

4-11. Figure 4-2 illustrates how the area of operations for a modular brigade augmented for SFA may change throughout the SFA phases. The differences between the initial response phase and the transformation phase may not change on the surface, but the BCT's relationship with FSF changes drastically. For example, FSF in the north met required conditions so that one of the U.S. brigades conducting SFA was no longer required. The responsibility of the other U.S. brigade thus expanded from providing assistance to one division to two divisions. This brigade's span of control and area of operations similarly expanded. The latter stage of the transformation phase can differ greatly. Areas of operation generally increase in size as they provide SFA to more FSF.

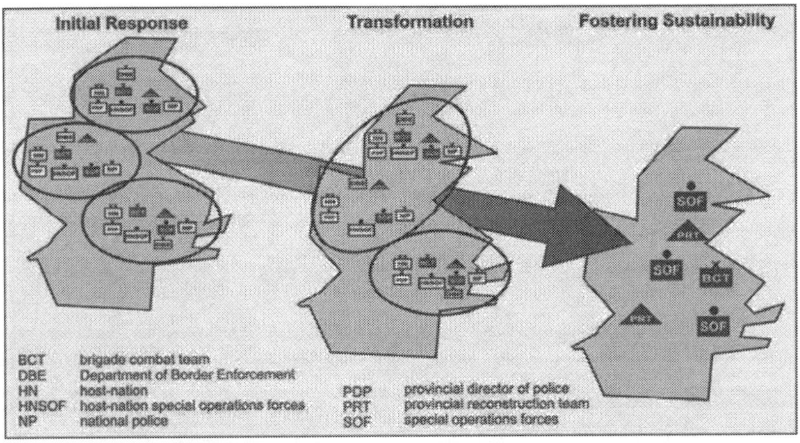

Figure 4-2. Example of changes to a brigade combat team's area of operations

AUGMENTATION BY PHASE SUMMARY

4-12. BCT augmentation is based on the operational environment and mission. As the three phases are based on the operational environment, they provide a baseline for augmentation. Potential augmentation may require military police, legal, public affairs, civil affairs, psychological operations, engineering, sociocultural experts, sustainment, and military transition team personnel. Table 4-1 illustrates a baseline for augmentation by phase for BCTs.

4-13. An embedded provincial reconstruction team is a key element of a BCT conducting security for assistance. The BCT has the lead on the following tasks: establish civil security, establish civil control (when approved by Congress), and develop and enable foreign security forces. An embedded provincial reconstruction team has the lead for the following tasks: support to economic and infrastructure development, restore essential services, and support to governance. Thus, the BCT and an embedded provincial reconstruction team are able to effectively support FSF and execute all five stability tasks.

Table 4-1. Baseline augmentation for modular brigade

CAPABILITY	INITIAL PHASE (1 HN DIV)	TRANSFORMATION PHASE (2 HN DIV)	TRANSFORMATION PHASE (3 HN DIV)
Police Provincial District	171 - MP Company 2 - MP Team (LTC/SGM) 3 - MP MAJ	171 - MP Company 4 – 2 x MP Team (LTC/SGM) 6 - MP MAJ	171 - MP Company 6 – 3 x MP Team (LTC/SGM) 9 - MP MAJ
Legal HN BDE HN DIV	2 - JAG Team (MAJ/SFC) 2 - JAG Team (CPT/SSG)	4 – 2 x JAG Team (MAJ/SFC) 4 – 2 x JAG Team (CPT/SSG)	6 – 3 x JAG Team (MAJ/SFC) 6 – 3 x JAG Team (CPT/SSG)
Public Affairs	2 - PA (Officer/NCO)	2 - PA (Officer/NCO)	2 - PA (Officer/NCO)
Civil Affairs & Psychological Operations	32 - CA Company – 5 CA Teams 70 – Tactical PSYOP Company – 9 TPTs	32 - CA Company – 5 CA Teams 70 – Tactical PSYOP Company – 9 TPTs	197 - CA Battalion (4 CA Companies) – 20 CA Teams 70 – Tactical PSYOP Company – 9 TPTs
Engineer	8 - FEST-A Team 2 - EOD (CPT/SFC)	8 - FEST-A Team 2 - EOD (CPT/SFC)	8 - FEST-A Team 2 - EOD (CPT/SFC)
Sociocultural Experts	8	8	8
Military Transition Team	56 - MTT Augment	4 - MTT Augment	8 - MTT Augment
TOTAL	358	315	493
EPRT	~15-20	~15-20	~15-20
Sustainment Truck Squad Supply Platoon CCT FMST	 21 personnel 32 personnel 4 personnel 4 personnel	 4 personnel 4 personnel	 4 personnel 4 personnel

CA	civil affairs	LTC	lieutenant colonel
CPT	captain	MAJ	major
CCT	contingency contracting team	MP	military police
DIV	division	MTT	military transition team
JAG	Judge Advocate General	NCO	noncommissioned officer
EOD	explosive ordnance disposal	PA	public affairs
EPRT	embedded provincial reconstruction team	SFC	sergeant first class
FEST-A	forward engineering support team – augmenting	PSYOP	psychological operations team
FMST	financial management support team	SGM	sergeant major
HN	host-nation	SSG	staff sergeant
HTT	human terrain team	TPT	tactical psychological operations team

INITIAL RESPONSE PHASE

4-14. The initial response phase seeks to assist FSF stabilize the operational environment in a crisis state. This phase occurs during or immediately after a conflict where the operational environment prevents civilian personnel from operating effectively. The operational environment is typified as nonpermissive.

Thus, the objective of this phase is to improve the security situation, reducing the threat to the populace and creating the conditions that allow civilian personnel to safely operate.

4-15. SFA in the initial response phase is normally required when FSF lack the capability or capacity to provide the required level of security. This phase often requires SFA efforts to help generate and train or assist new and existing FSF. This phase may require a combination of the types of SFA and considerable support, sustainment, and medical resources. BCT activities during the initial response may have to be conducted with multinational major combat operations, to include providing a safe, secure environment for the local populace.

4-16. The BCT's SFA efforts during this phase focus on improving the foreign security force's capability and capacity so all security forces—U.S., other, and foreign security forces—provide a secure environment and reduce the threat. Figure 4-3 illustrates the BCT's span of control. The figure also depicts three BCTs' operational areas in the initial response phase. As security conditions improve, transition to the transformation phase begins.

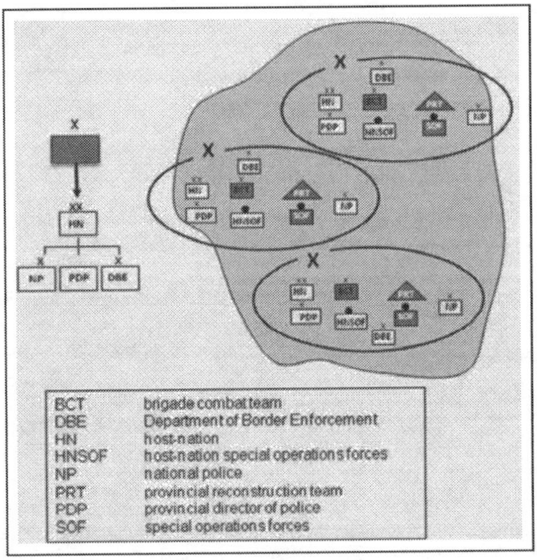

Figure 4-3. Example of initial response span of control and area of operations

4-17. The BCT requires augmentation for the initial response phase. These embedded units provide key enabling capabilities for the brigade, supported FSF, and other supported actors. The embedded provincial reconstruction team is one of the most important of the brigade's augmentation, as it provides interim civilian expertise in developing local institutions to take the lead in national governance, the provision of basic services, fostering economic development, and enforcement of the rule of law. Column two of table 4-1 summarizes augmentation during the initial phase. Although each brigade's mission and conditions in its area of operations are unique, the baseline augmentation for the modular brigade operating in this phase is as follows:

- Embedded provincial reconstruction team.
- Forward engineering support team-augmenting.
- Explosive ordnance disposal team.
- Sociocultural experts.

- Military police company (provincial police transition team).
- Military police team (district police transition team).
- Three military police majors (district police transition team).
- Two judge advocate general teams.
- Civil affairs company.
- Tactical psychological operations company.
- Public affairs team.
- Contingency contracting team.
- Financial management support team.
- Military transition team (specific augmentation personnel include two lieutenant colonels, thirty-one majors, and twenty-three master sergeants).

Transformation Phase

4-18. In the transformation phase, the BCT seeks to assist FSF to stabilize the operational environment in a crisis or vulnerable state. The operational environment in this phase is more permissive than the initial response phase; however, military forces will often be required to provide security to some actors. Activities in this phase normally include a broad range of post-conflict reconstruction, stabilization, and capacity-building efforts, which the embedded provincial reconstruction team is essential for long-term success. Objectives in this phase include continuing efforts to improve the security situation, reducing the threat to the populace, building host-nation capacity across the stability sectors, and facilitating the comprehensive approach to assist FSF.

4-19. The transformation phase represents a broad range of SFA activities that the BCT could perform to support FSF. The initial response phase differs from the transformation phase in the foreign security force's capability to provide for a safe and secure environment. More specifically, FSF have a level of proficiency so they no longer need a permanent U.S. and foreign security forces relationship for tactical operations. However, they may still need full-time advisors and support, sustainment, and medical assistance. The embedded provincial reconstruction team will continue to play a vital role in assisting governance and development efforts throughout this phase.

4-20. Figure 4-4 illustrates the BCT conducting SFA in the transformation phase. In this example, the BCT's span of control expands from assisting one foreign division to assisting two foreign divisions. The figure also depicts how a BCT's operational area has similarly expanded. SFA activities in this phase seek to establish conditions so the host nation's security sector can provide a secure environment with its own security forces.

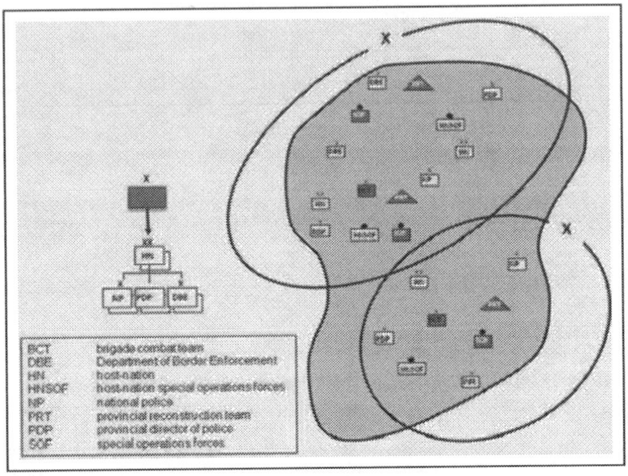

Figure 4-4. Example of transformation phase span of control and area of operations

4-21. The BCT requires augmentation for the transformation phase, like the initial response phase. It often requires similar augmentation due to its greater span of control and area or operations. Column three of table 4-1 summarizes augmentation during the transformation phase with two foreign divisions. The baseline augmentation for the brigade in this phase is as follows:

- Embedded provincial reconstruction team.
- Forward engineering support team-augmenting.
- Explosive ordnance disposal team.
- Sociocultural experts.
- Military police company (provincial police transition team).
- Two military police teams (district police transition team).
- Six military police majors (district police transition team).
- Four judge advocate general teams.
- Civil affairs company.
- Tactical psychological operations company.
- Public affairs team.
- Contingency contracting team.
- Financial management support team.
- Military transition team (specific augmentation personnel include two lieutenant colonels and two master sergeants).

4-22. As conditions of the operational environment improve, the BCT may require a transition to assist three foreign divisions, rather than two divisions as depicted in figure 4-4. This represents a significant expansion of the span of control and area of operations, with a corresponding expansion of required augmentation. Figure 4-5 depicts the BCT supporting three foreign divisions.

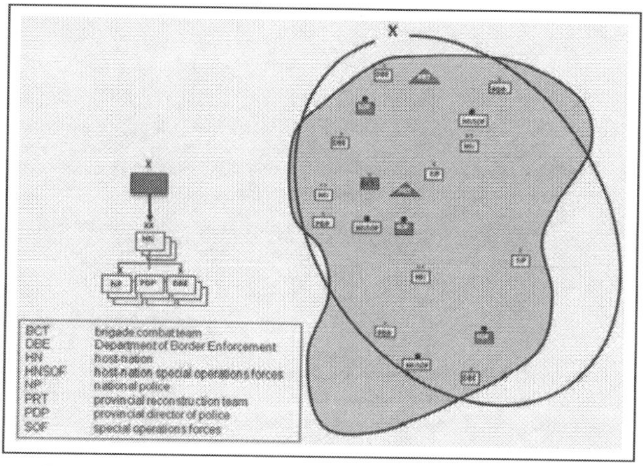

Figure 4-5. Example of a three-division span of control and area of operations

4-23. As in other phases, the BCT requires augmentation to support three divisions. Column four of table 4-1 summarizes augmentation during the transformation phase with three foreign divisions. The baseline augmentation for the brigade in this phase is as follows:

- Embedded provincial reconstruction team.
- Forward engineering support team-augmenting.
- Explosive ordnance disposal team.
- Sociocultural experts.
- Military police company (provincial police transition team).
- Three military police teams (district police transition team).
- Nine military police majors (district police transition team).
- Six judge advocate general teams.
- Civil affairs company.
- Tactical psychological operations company.
- Public affairs team.
- Contingency contracting team.
- Financial management support team.
- Military transition team (specific augmentation personnel include four lieutenant colonels and four master sergeants).

Fostering Sustainability Phase

4-24. In this phase, the focus of SFA continues to shift toward assisting institutions required to sustain FSF. This phase encompasses long-term efforts to assist FSF. In this phase, FSF are capable and viable. They conduct independent operations and can provide a safe, secure internal environment. While the brigade may be initially required during this phase, it is normally withdrawn during this phase. The determination to withdraw the brigade is based on the policy and conditions of the operational environment. Special operations forces, provincial reconstruction teams, and other forces may remain to support a theater security cooperation plan. Thus, remaining SFA missions are normally special operations forces missions, not a BCT mission (special operations forces will normally be involved in all phases of SFA).

Transitions

4-25. SFA can require a vast array of transitions. These transitions are conducted according to the tactical, operational, and strategic conditions identified during planning (which must support the internal defense and development strategy and using an overall comprehensive approach). Transitions can occur simultaneously or sequentially in different levels or war and in separate echelons. This includes potentially having tactical transitions for different units within one brigade's area of operations.

4-26. There are several major transitions. First, the supported unit may be the U.S. BCT in the beginning of an initial response phase but the foreign security forces will transition to be the supported unit. At this point in the transformation phase, the area in which the BCT conducts SFA will expand. This expansion can occur multiple times during the transformation phase, which is based on conditions, especially the capability and capacity of FSF.

4-27. Operations never proceed exactly as planned. To facilitate flexibility, commanders incorporate branches and sequels into the overall plan. Visualizing and planning branches and sequels are important because they involve transitions. Unless planned, prepared for, and executed efficiently, transitions can reduce the tempo of the operation, slow its momentum, and surrender the initiative to the adversary.

SUBORDINATE UNITS

4-28. The modular brigade augmented for SFA has subordinate units whose sole focus is working with FSF. These advisor teams may be formed from brigade organic resources, external augmentation, or a combination. These teams optimally are embedded with the counterpart unit, or they may reside on a U.S. camp and commute to FSF they support. The method depends on U.S. policy, direction from higher headquarters, the conditions of the operational environment, and capacity of the foreign security force's camps to accommodate the U.S. forces.

4-29. The military transition team (MTT) is a key subordinate unit for the brigade conducting SFA. The MTT mission is to assist FSF units. Figure 4-6 illustrates the relationship between a brigade conducting SFA and a MTT assisting a foreign brigade. In this example, the brigade has a battalion and a division MTT as subordinate units, although it would have other subordinate units. The brigade MTT is subordinate to the battalion and has operational control of one of the parent U.S. battalion's company teams.

4-30. When the brigade provides organic forces to form the basis for the brigade MTT, the company team is the foundation of a brigade MTT. The company team will be augmented with additional personnel and assets. The U.S. brigade must facilitate the MTT operations, such as force protection, transportation, communications, and reachback.

4-31. As companies do not have organic intelligence analysts, personnel must be used to form company intelligence support teams (COISTs). A COIST, an ad hoc organization, assists the MTT and its FSF. Ideally, the COIST should consist of six individuals, which will allow for 24 hour operations and the depth in the team to both integrate with the staff and to successfully support the company's mission. The team should consist of an officer, a noncommissioned officer, and four Soldiers, although the cell can be minimally manned with four personnel. The company intelligence cell works most effectively if its work complements the battalion S-2's work.

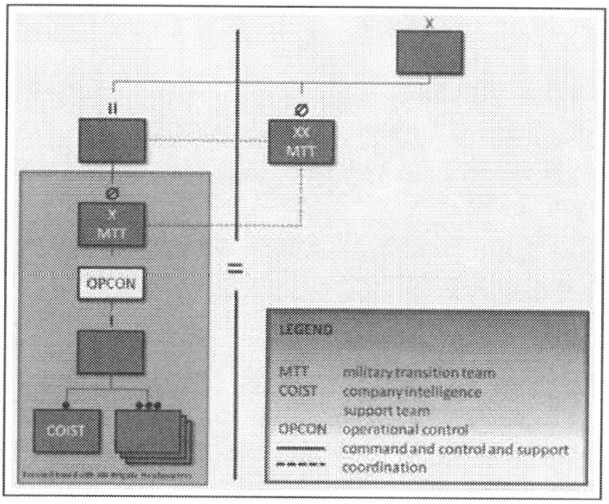

Figure 4-6. Military transition teams

SAMPLE TROOP-TO-TASK MODEL

4-32. The BCT may support assigned to multiple missions, including potentially supporting multiple foreign units in their operational area. Additionally, these FSF organizations may each report through different host-nation government channels and even to different ministries. To synchronize efforts in this case, U.S. forces must achieve unity of command and effort. Similarly, each of the FSF organizational commanders should synchronize their efforts with the host-nation government representatives, as appropriate.

4-33. Figure 4-7 depicts a sample troop-to-task model for a U.S. brigade. It depicts one company team acting as a quick reaction force; thus, the battalion retains the capability to conduct independent full spectrum operations. The figure depicts how U.S. company teams may form the basis of a brigade MTT or a brigade border transition team (and subordinate point of entry transition teams). The figure also depicts how a company team may provide police support. This support is in the form of two platoons supporting police transition teams and a third platoon in a joint security station providing support to a police transition team.

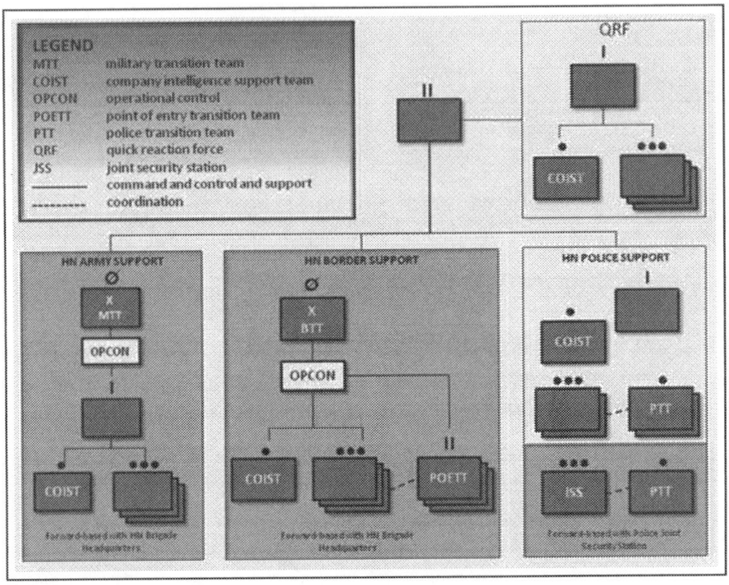

Figure 4-7. Battalions conducting security force assistance

4-34. Subordinate units of BCTs conducting SFA are best located inside a FSF base—not an American base. Co-location facilitates integration with the FSF. It allows the two forces to form mutual understanding and trust. Co-location and the close cooperation it facilitates often improves the population's perception of the legitimacy U.S. and foreign security forces, which can be an essential condition of the overall mission's end state.

4-35. If force protection requires, a U.S. area may be established in the FSF base, although this is not optimal. Key considerations for co-location may include the threat, FSF acceptance, physical space inside the FSF base, sustainment capabilities, medical facilities, and availability of quick reaction forces.

4-36. When U.S. forces are operating out of smaller outposts in an urban environment, the local populace sees the integration and presence of the U.S. and foreign security forces working together. This integration not only enhances overall operational effectiveness and trust, living and working together builds legitimacy of the two forces as well as FSF; it reinforces trust between the FSF and the people they are tasked to protect.

Chapter 5
Unit Operations

Unit operations require continuous activities. Commanders plan for premission, deployment, in-country preparation, employment, redeployment, and postdeployment activities.

PREMISSION

5-1. After receiving a mission, the unit continues detailed preparation activities. The unit obtains supplies, equipment, and training materials and prepares for deployment. Personnel prepare for and rehearse classes to be given in-country. They hone their military skills and conduct extensive area orientation. Paragraphs 5-2 through 5-10 discuss staff actions particular to security force assistance (SFA).

OPERATION PLAN

5-2. The S-3 disseminates the operation plan to unit staff and subordinates. The S-3 ensures predeployment training for Soldiers, to include preparation for training foreign security forces (FSF) and rehearsals for movement. The S-3 reviews the program of instruction for training FSF, to include getting approval from the commander and higher headquarters, if necessary.

CIVIL-MILITARY OPERATIONS

5-3. The civil-military operations section ensures the operation plan minimizes how operations affect the civilian population and addresses ways to mitigate the civilian impact on military operations. Civil-military operations also ensure the operation plan discusses coordination with any augmenting actors. These actors might include civil affairs battalions or brigades, provincial reconstruction teams, or United States Agency for International Development project officers in the operational area. During predeployment training, Soldiers receive training, materials, and briefings on the operational area. This training can cover the history, culture, religion, language, tribal affiliations, local politics, and cultural sensitivities as well as any significant nongovernmental organizations operating in the operational area.

TRAINING

5-4. As discussed in chapter 2, advisors should focus their premission training on the specific requirements of developing FSF. The training emphasizes the host-nation culture and language and provides cultural tips for developing a good rapport with foreign personnel. All premission training for SFA follows the training principles described in FM 7-0.

5-5. Based on the higher commander's training guidance, unit commanders assign missions and approve the draft mission-essential task list that supports SFA. The staff plans, conducts, and evaluates training to support this guidance and the approved mission-essential task list for SFA missions. Commanders prioritize tasks that need training. Since there is never enough time to train in every area, commanders focus on tasks essential for mission accomplishment.

5-6. Once commanders select tasks for training, the staff builds the training schedule and plans on these tasks. The staff provides the training requirements to the commander. After approving the list of tasks to be trained, the commander includes the tasks in the unit training schedule. The staff then coordinates the support and resource requirements with the S-3 and S-4. Finally, the commander ensures standards are enforced during training.

Evaluation

5-7. Evaluations can be either internal or external. Internal evaluations occur at all levels, and they must be inherent in all training. External evaluations are usually more formal and conducted by a headquarters two levels above the unit being evaluated. This subject must be carefully planned and discussed with FSF leaders to account for cultural sensitivities and current capabilities.

5-8. A critical weakness in training is the failure to evaluate each task every time it is executed. Every training exercise provides potential for evaluation feedback. Every evaluation is also a training session. Leaders and trainers must continually evaluate to optimize training. Evaluation must occur as training takes place. Emphasis is on direct, on-the-spot evaluations. However, leaders allow Soldiers to complete the task first. Leaders plan after action reviews at frequent, logical intervals during exercises. This technique allows the correction of shortcomings while they are still fresh in everyone's mind. The after action review eliminates reinforcing bad habits.

Specified Training

5-9. Augmentation elements require area orientation, refresher combat training, field training exercises, and the like. Unit training objectives are for developing capabilities to conduct internal defense and development (IDAD) activities for tactical operations, intelligence operations, psychological operations, populace and resources control operations, and civil affairs and advisory assistance operations in the host-nation language. Units identified for SFA begin intensified training immediately upon deployment notification.

5-10. After deployment to the host nation and before commitment to operations, the unit may receive in-country training at host-nation training centers or at designated training locations. This training helps personnel become psychologically and physically acclimated to the host nation. This training also allows commanders and staff some time to coordinate and plan within their own command and with civilian and military joint and multinational organizations. After commitment, training continues and is stressed between operations, using needed improvements identified in operations as the basis for training.

DEPLOYMENT

5-11. SFA is often conducted in semipermissive environments in which U.S. forces are guests of the host-nation or partner organization. Units move into the operational area by following their deployment standing operating procedures and the operation plan. Paragraphs 5-12 through 5-25 highlight organizations and procedures critical to SFA deployment. This information, however, does not obviate the need for extensive standing operating procedures and an operation plan.

PREDEPLOYMENT SITE SURVEY

5-12. The predeployment site survey unit aims to report accurately to its parent unit the existing FSF mission variables—mission, enemy, terrain and weather, troops and support available, time available, civil considerations (METT-TC). See table 5-1. It also establishes in-country command and control, support, protection, and sustainment relationships for the follow-on unit's mission execution. The site survey coordinates the in-country reception of SFA forces. This coordination wards off most friction during the handoff. This coordination requires all units—outgoing and incoming—to apply a comprehensive approach at all levels of command. Appendix D of FM 3-05.202 discusses site survey procedures in detail.

Table 5-1. Advisor team predeployment site survey checklist

Tasks	*Characteristics*
Area of operations overview	• Map familiarization. • In sector orientation. • Areas of emphasis.
Graphic control measures	• Unit boundaries. • Routes. • Other United States and multinational support locations. • Maintenance support location. • Medical support locations. • Other points of interest.
United States and multinational unit interviews by specialty staff	• Responsibilities. • Battle rhythm. • Required reports. • Standing operating procedures, continuity book. • Stay-behind equipment.
Briefings on advised unit	• Foreign security forces task organization. • Foreign security forces key leadership. • Intelligence assessment. • Unit operational assessment.
Advised unit introductions	• Among commanders. • Among staffs.

Advance Party Deployment

5-13. An advance party deploys to the host-nation or partner organization area of operations after the unit completes its mission planning and the proper commander representative declares it deployable. The advance party conducts predeployment activities and coordinates necessary travel documentation and country clearances. The advance party reviews the rules of engagement and any status-of-forces agreements before deploying. If applicable, the proper staff provides an updated threat briefing to the unit. The S-4 section and support company help with out-loading in the unit area. The unit places all supplies and materials on pallets beforehand and prepares the load manifest.

5-14. Upon arrival, the advance party processes through customs and immediately notifies the higher headquarters of its arrival and status. Often the higher headquarters provides point of contacts to the advance party. These contacts help the advance party obtain the proper identification, documentation, and weapons permits.

5-15. The advance party establishes contact with all U.S. and foreign security forces tactical unit commanders and conducts final premission planning activities. The advance party procures secure working, storage, and living areas for the unit and reconnoiters and prepares the training sites. The advance party coordinates for special support equipment to unload heavy supplies and transport the supplies and personnel to the training site.

Main Body Deployment

5-16. The unit deploys after completing its planning and the appropriate commander declares the unit ready. The unit follows the same procedures performed by the advance party. Members of the advance party meet the main body on its arrival and brief the main body on any required changes to the operation order.

5-17. After arriving in-country, the unit processes through customs and immediately notifies the higher headquarters of its arrival and status. The unit obtains the proper identification, documentation and weapons permits from U.S. or other sources, if applicable. Often the security assistance organization provides to the unit the necessary point of contacts that expedite this process.

5-18. The unit off-loads equipment and personnel onto vehicles for movement to the training site, or it stores the equipment in a secure area until the processing is completed. FSF secure the equipment but the unit members remember the commander is ultimately accountable for the advisor team's physical and personal security. Supply personnel stay with the equipment and help to guard it.

IN-COUNTRY PREPARATION

5-19. Upon arrival, the commander and S-3 brief the higher headquarters on the planned execution of the mission and reconfirm the required command relationship. Local conditions may require the unit to confirm or establish its in-country and external command and control, support, and sustainment relationships from outside its operational area upon arrival. The unit establishes direct working relationships with its next higher in- or out-of-country supporting element to—

- Determine the limits of the available support and expected reaction time between the initiation of the support request and its fulfillment.
- Confirm or establish communications procedures between the supporting element and the unit, to include alternative and emergency procedures for command and control, all available support, and medical evacuation.

5-20. The unit establishes procedures to promote interagency cooperation and synchronization. The unit—

- Identifies the location of the concerned host-nation, U.S., or other agencies.
- Contacts the concerned agency to establish initial coordination.
- Exchanges information or intelligence.
- Confirms or establishes other coordination protocols as necessary.
- Incorporates the newly established or changed procedures into the plans for mission execution.

5-21. The unit immediately establishes operations security procedures to support its mission execution and identifies rally points incorporated into its defensive, evasion, and personnel recovery plans.

5-22. After receiving a detailed briefing and further guidance from the advance party, unit personnel continue to develop effective rapport with the FSF commander and counterparts. They also assess their working, storage, and living areas for security and verify the location of the training site, communications center, dispensary area, and FSF troop area. With the FSF commander, the unit commander—

- Establishes rapport.
- Conducts introductions in a businesslike, congenial manner.
- Briefs on the unit's mission, its capabilities, and the restrictions and limits imposed on the detachment by the higher U.S. commander.
- Ensures all unit personnel fully support FSF and firmly believes a joint U.S.-FSF effort will succeed. Requests counterpart linkup be made under the mutual supervision of the FSF commander and the unit commander.
- Ensures all current unit plans are tentative and that assistance is needed to finalize them.
- Deduces or solicits the actual estimate of unit capabilities and perceived advisory assistance and material requirements.
- Recommends the most desirable courses of action (COAs) while emphasizing how they satisfy present conditions, achieve the desired training, and meet advisory assistance goals.
- Informs the higher in-country U.S. commander of any significant changes in the unit's plan to assist FSF.

5-23. The S-2 supervises the dissemination of intelligence and other operationally pertinent information within the unit and, as applicable, to higher, lower, or adjacent units or agencies. The S-2 also—

- Monitors the implementation of the intelligence collection plans to include updating the commander's priority intelligence requirements (PIRs), conducting area assessment, and coordinating for additional intelligence support.
- Establishes liaison with FSF intelligence and security agencies (within the guidelines provided by applicable higher authority). Assesses the intelligence threat and resulting security requirements, including coordination with the S-3 on specific security and operations security measures.

5-24. Through the S-2, the commander's PIRs are based on the latest information available and requirements for additional PIRs that arise from modified estimates and plans. The S-2 also—

- Analyzes the foreign unit's status to finalize unit plans for advisory assistance. These plans can include task organization of unit with counterparts, staff functions for planning SFA, and advisory assistance for executing SFA.
- Explains analysis to counterparts and encourages them to help with—and participate in—analyzing, preparing, and briefing the analysis to the foreign unit commander.
- Prepares and briefs the plans for training and advisory assistance.
- Helps the foreign unit inspect the available facilities to identify deficiencies. If the unit finds deficiencies, the S-2 prepares estimates of COAs for the FSF commander to correct them.
- Supervises the preparation of the facilities with their counterparts and informs unit and FSF commanders on the status of the facilities.

5-25. The unit ensures its security is based on the present or anticipated threat. Some recommended actions the unit may take include—

- Hardening its positions based on available means and requirements to maintain low visibility.
- Maintaining unit internal guard system with at least one Soldier who is awake and knows the locations of all other unit personnel. The guard reacts to an emergency by following an internal alert plan and starting defensive actions.
- Maintaining communications with all subordinate unit personnel deployed outside the immediate area controlled by the main body.
- Establishing plans for immediate defensive actions in the event of an attack or a loss of rapport with hostile reaction.
- Discussing visible security measures with foreign counterparts to ensure understanding and to maintain effective rapport. Unit personnel do not divulge sensitive information for the sake of possible rapport benefits.
- Encouraging the foreign unit, through counterparts, to adopt additional security measures identified when analyzing the foreign unit's status and inspecting its facilities.
- Coordinating defensive measures with the foreign unit to develop a mutual defensive plan. Unit personnel obtain from the unit's present reaction and defensive plans for attack. They encourage the foreign unit to conduct mutual full-force rehearsals of defensive plans; if unsuccessful, the unit conducts internal rehearsals of the plans.

EMPLOYMENT

5-26. Once the unit has deployed and arrived in-country, it begins employment. Employment is the conduct of SFA to support the FSF commander. Employment occurs within the advisor team itself. It includes foreign counterparts in the mission planning (preparing the FSF for the mission itself) to increase the capability and capacity of their planning processes, as well as increase the probability of success.

PERSONNEL

5-27. During employment, individual unit personnel perform functions described in paragraphs 5-28 through 5-38.

Commander

5-28. Before the mission, the U.S. commander advises and assists the FSF commander. The latter issues planning guidance for planning the execution of the mission and clarifies commander's intent. The U.S. commander advises and assists the FSF unit commander in the command and control process of tactical operations. By accompanying the FSF unit commander when the mission is received from higher headquarters, the U.S. commander assists any subsequent missions. The U.S. commander monitors how FSF understand the commander's intent and all specified or implied tasks.

5-29. During execution of the mission, the U.S. commander helps the foreign unit commander provide command and control during operations. After monitoring the tactical situation, the U.S. commander recommends changes to the COAs to exploit the situation. After monitoring the flow of information, the U.S. commander recommends improvements to use intelligence collection assets and to keep subordinates reporting required information.

Staff

5-30. Before the mission, the advisor staff advises and assists the foreign counterparts in preparing estimated COAs for essential tasks. The staff helps write tentative plans based on the planning guidance and unit's standing operating procedures. These plans include primary, alternate, contingency, and emergency plans.

5-31. During execution, the staff helps foreign counterparts coordinate the execution of the tasks. They disseminate portions of their plans to other personnel, senior and subordinate staff sections, and supporting elements. They help notify higher, lower, or adjacent staff sections of modified estimates and plans. Finally, the staff—with the S-2 and the S-2 counterpart—helps update the commander's critical information requirements with the latest information and requirements that arise from operations.

Executive Officer

5-32. The executive officer performs the organizational analysis of the unit's coordinating staff sections to ensure efficiency during the planning process according to initial planning guidance. With the foreign counterpart, this officer directs foreign staff sections as they develop estimates and plans. The executive officer monitors the liaison and coordination with FSF higher headquarters, recommending changes to improve efficiency.

Adjutant

5-33. The adjutant provides advice, assists, and makes recommendations to the foreign counterpart for conducting human resources. This includes monitoring the maintenance of foreign unit strength, accountability of casualties, unit morale, and postal activities. This may also include concerns with the foreign pay system, leave procedures, and casualty pay procedures.

Intelligence Officer

5-34. The intelligence officer advises and assists the monitoring of FSF operations security to protect classified and sensitive material and operations and recommends improvements. By helping the foreign counterpart update the situation map, the intelligence officer helps to keep the map current. The intelligence officer recommends improvements to the standing operating procedures of the tactical operations center communications so the intelligence section receives situation reports. This officer helps the counterpart monitor the collection, evaluation, interpretation, and dissemination of information. The intelligence officer assists in the examination of captured insurgent documents and material. This officer helps gather and disseminate intelligence reports from available sources to ensure the exploitation of all assets. The intelligence officer helps the counterpart to brief and debrief patrols operating as a part of reconnaissance and surveillance activities. Finally, the intelligence officer assists, with the advisor operations officer, developing reconnaissance and surveillance plans.

Operations Officer

5-35. The operations officer helps the foreign counterpart to prepare tactical plans using estimates, predictions, and information. This officer monitors command and communications nets, assists in preparing all orders and plans, and helps to supervise the training and preparation for operations. Finally, the operations officer monitors the planning process and makes recommendations for consistency with IDAD goals.

Supply Officer

5-36. The supply officer advises and assists the foreign counterpart in maintaining equipment readiness; monitoring the support provided to the foreign unit, its subunits, and attachments; and recommending improvements. Finally, this officer helps to supervise the use of transportation assets.

Civil-Military Team

5-37. Upon deployment, civil-military teams advise the commander and staff on civil-military considerations and coordinate efforts of any civil affairs units supporting the unit. The civil-military team mentors the counterpart team on the supported foreign element staff on civil-military operations, the legal and moral obligations of military commanders to civilian populations under their control, and the importance of respecting human rights. The civil-military team may introduce the counterpart to relevant nongovernmental organizations, United States Agency for International Development project officers, and provincial reconstruction team staff.

Company-Level Advisors

5-38. Advisors sent to elements below battalion level help foreign counterparts to analyze the mission and commander's intent from higher headquarters. They assist FSF leaders restate the mission, conduct an initial risk assessment, identify a tentative decisive point, and define their own intent. They also assist their foreign counterparts to analyze the mission variables. From these variables, advisors help their foreign counterparts to develop a COA that meets the higher headquarters concept of operations and commander's intent. Finally, they advise and assist in the conduct of operations and the flow of information to the FSF higher commander.

TRAINING

5-39. Training FSF during SFA parallels the training of U.S. forces. Training follows the principles of training described in FM 7-0. This training tailors the objectives to account for cultural differences and focus on the FSF mission and problem set.

Training Assessment

5-40. The commander of the advising unit begins with a training assessment of training plans designed during predeployment and passed on from the preceding U.S. units conducted in coordination with the FSF commander. This assessment is important to both evaluate the foreign security force but also as an exercise to establish and enhance the working relationship between U.S. and foreign security forces. The training assessment should cover all aspects of leadership, training, sustainment, and professionalization. To support an assessment, the advising unit should analyze the following specific foreign unit considerations:

- The unit's mission and mission-essential task list and capability to execute them.
- Staff capabilities.
- Personnel and equipment authorization.
- Physical condition.

- Any past or present foreign influence on training and combat operations.
- Operational deficiencies identified during recent operations or exercises with U.S. personnel.
- Sustainment capabilities, to include sustainment training programs.
- Internal training programs and personnel.
- Training facilities.

5-41. The level of professionalism of FSF, both units and individuals, should also be assessed. Adhering to established rules of engagement, ethics that meet the established laws and regulations of the commanding authority, laws for land warfare, and human rights, promotions based on merit and the security forces' support of civilian leaders and political goals all fall within this assessment.

5-42. The advisor unit, working with FSF leaders, also evaluates current members of the security forces for past military skills and positions. Often military reorganizations arbitrarily shift personnel to fill vacancies outside their knowledge and experience.

Training Plan

5-43. After completing the training assessment, advisors analyze the prepared training plan and determine if changes are necessary. The training plan stresses the deficiencies identified in the training assessment. It also identifies those in the host nation able to help train FSF and to strengthen the legitimacy of the process. Finally, the training plan should consider the unit's eventual self-sustainment. As the foreign security force gains sufficient capacity and capabilities to perform independently, advisors transition from a leading role to a mentoring role.

5-44. The plan for training FSF uses a comprehensive approach, supports the IDAD strategy, and has many echelons. Working with other echelons and actors within the area of operations can provide support and expertise that enhance the training and operations process as well as limit redundancy.

Program of Instruction

5-45. In coordination with the FSF commander and training staff, the advising unit develops a program of instruction. This program incorporates all training objectives that satisfy the training requirements identified during the assessment. The training program must support these requirements. The FSF commander approves the program of instruction prior to execution. When executing the program of instruction, advisors adhere to a training schedule consistent with changes in the mission variables. Advisors ensure through their counterparts and the FSF commander that all personnel receive training. Foreign counterpart trainers rehearse all classes approved on the program of instruction.

Presentation of Instruction

5-46. To present the training material properly, trainers follow the lesson outlines approved in the program of instruction. All training clearly states the task, conditions, and standards desired during each lesson, ensuring the foreign students understand them. Trainers state all warning and safety instructions in the host-nation language. The training to reinforce the concepts includes demonstrations of the execution of each task, stressing the execution as a step-by-step process. Trainers monitor students' progress during instruction and practical exercises, correcting mistakes as they are made.

Training Methods

5-47. An effective method of training used by U.S. forces is the crawl-walk-run method of teaching individual tasks, battle drills, collective tasks, and field exercises. (See FM 7-0.) This method is employed to develop well-trained leaders and units. During all phases, the training must include the mission of the unit in the context of the higher unit's mission to assist with the practical application of the training. Identifying the higher commander's mission and intent, as well as the tasks and purposes of other units in the area, would also add context to the training. This method can also be expanded to include the role of other actors.

5-48. At a minimum, units continue individual training to improve and sustain individual task proficiency while training on collective tasks. Collective training requires interaction among individuals or organizations to perform tasks, actions, and activities that contribute to achieving mission-essential task proficiency. Collective training includes performing collective, individual, and leader tasks associated with each training objective, action, or activity.

Collective Training

5-49. Collective training starts at the squad level. Squad battle drills provide key building blocks that support operations. Trainers can link battle drills and collective tasks through a logical, tactical scenario in a situational training exercise. Although this exercise is mission-oriented, it results in more than mission proficiency. Battle drills and collective tasks support situational training exercises, while these exercises support operations. Advisors must understand the operational environment when training FSF; training incorporates how internal and external threats and civilians affect the environment. Flexibility in using U.S. doctrine in training enhances efforts to make training realistic. Trainers modify U.S. doctrine to fit the FSF level of expertise, their command and control, the tactical situation, and sustainment base. Often the structure and capabilities of FSF differ from that required by U.S. doctrine. When FSF counter an insurgency, these exercises emphasize interplay among psychological and tactical, populace and resources control, intelligence, and civil affairs operations. Similarly, scenarios for each tactical exercise must include interplay among the various possible aspects of IDAD.

Individual Training

5-50. Training individuals of FSF emphasizes physical and mental conditioning, tactical training, basic rifle marksmanship, first aid, combatives, and the operational environment. Individual training also includes general tactics and techniques of security operations and the motivation, operations, and objectives of internal and external threats. Tough and realistic training conditions troops to mentally and physically withstand the strain of continuous offensive operations.

5-51. Foreign personnel cross-train on all types of weapons, communications and other equipment, and skills particular to their unit. Personnel losses must never cause weapons, communications equipment, or essential skills to be lost due to a lack of fully trained replacement personnel.

Small-Unit Leader Training

5-52. SFA frequently entail rapidly changing circumstances; thus, FSF small-unit leaders must be able to plan and execute operations with little guidance. Therefore, trainers stress small-unit leadership training which can occur concurrently with individual training. Tools the trainer uses to train leaders are manuals, previously established training, tactical exercises without troops, battle simulations, and unit missions. This training develops aggressiveness, tactical proficiency, and initiative. Small-unit leader training should include combined arms technical training such as procedures for forward observer and close air support. Leadership training must also include land navigation in difficult terrain and under conditions of limited visibility. Mission readiness and the health and welfare of subordinates are also continuous parts of training.

After Action Review

5-53. In any training, the after action review provides the critical link between training and evaluation. It is a professional discussion that includes the training participants and focuses directly on the training goals. An after action review occurs after all collective foreign unit training. Effective after action reviews review training goals with the concerned FSF unit commander. Advisor personnel ask leading questions, surface important tactical lessons, explore alternative COAs, assist the retention teaching points, and keep the after action review positive.

5-54. The advisor encourages the FSF unit commander to review the training with the entire unit or key subordinate leaders. At this review, the FSF commander stresses how to strengthen the chain of command. During the after action review, advisors avoid criticizing or embarrassing the FSF unit commander or subordinates.

5-55. If possible, after action reviews occur during the field portion of the training when the unit assembles at logical stopping points. These reviews provide feedback from instructors and leaders that increases and reinforces learning. This feedback provides a richer database for key points. During the after action review, the senior evaluator draws information from unit leaders to form possible alternative COAs.

Assessment

5-56. In all operations, including SFA, success is generally measured for three periods: short-, mid-, and long-term. Success in SFA is normally defined within the context of these three periods. In the short-term period, FSF make steady progress in fighting threats, meeting political milestones, building democratic institutions, and standing up security forces. In the mid-term period, FSF lead fighting threats and provide security, have a functioning government, and work towards achieving economic potential. In the long-term period, FSF are peaceful, united, stable, and secure; integrated into the international community; and a full partner in international security concerns.

Measures of Effectiveness

5-57. A *measure of effectiveness* is a criterion used to assess changes in system behavior, capability, or operational environment that is tied to measuring the attainment of an end state, achievement of an objective, or creation of an effect (JP 3-0). A measure of effectiveness provides a benchmark against which the commander assesses progress toward accomplishing the mission.

5-58. The following are tools to assess the effects of operations in meeting expectations defined for the short, mid, and long term. Although all SFA is not focused on a host nation, these tools are based on the four IDAD pillars:

- **Balanced development** attempts to achieve national goals through political, social, and economic programs.
- **Security** includes all activities implemented to protect the populace from the threat and to provide a safe environment for national development.
- **Neutralization** is a political concept that encompasses the physical and psychological separation of the threatening elements from the population.
- **Mobilization** provides organized manpower and resources and includes all activities to motivate and organize popular support of the government.

Measures of Performance

5-59. A *measure of performance* is a criterion used to assess friendly actions that is tied to measuring task accomplishment (JP 3-0). This quantitative analysis determines whether the task or action was performed to standard. The standard may be derived from a procedure or may be time-sensitive. Examples of measures of performance are listed in the appropriate Army tasks in FM 7-15.

5-60. The advisor helps the foreign counterpart to monitor the current situation for unanticipated successes, failures, or enemy actions. As the commanders assess the progress of foreign security operations, they look for opportunities, threats, and acceptable progress. Throughout the operation, the commanders visualize, describe, and direct changes to the operation, feeding the assessments of progress or regression back into the planning process.

REDEPLOYMENT

5-61. Redeployment is the return of forces and materiel to the home or mobilization station. Before redeployment, a unit conducting SFA inventories all supplies and equipment (jointly, if possible) to be passed on to the foreign unit or the relieving U.S. or multinational unit. These actions ensure all items are accounted for and the custody chain for property and equipment is unbroken. The outgoing unit also—

- Prepares a final evaluation report showing its estimate of the foreign unit's capabilities and giving an opinion of its future employment.
- Discusses the foreign unit's performance with its commander. The unit submits a copy of the foreign unit's final evaluation to the next higher FSF commander.

- Passes custody of training schedules, lesson plans, foreign operational records, and the foreign unit's final evaluation to the foreign unit and relieving U.S. or coalition unit.
- Presents debriefings, after action reviews, and the foreign unit's final evaluation to the higher in-country U.S. commander.

5-62. The outgoing unit commander supervises the redeployment. This commander approves prepared redeployment plans or, as necessary, directs changes to the redeployment plans. The commander and operations officer assess and select alternatives or changes to the present redeployment COAs based on the intelligence officer's intelligence estimate. They develop and disseminate a fragmentary order for selected alternatives or changes.

5-63. Unit personnel prepare all accompanying supplies and equipment for shipment. They brief ground support personnel on equipment handling procedures, routes, convoy procedures, and actions to take if a terrorist or insurgent incident occurs. They load personnel and equipment for movement to the departure airfield or airport following the load plan in the unit's operation order. They maintain accountability for all their personnel, supplies, and equipment.

Mission Handoff Procedures

5-64. During long-term SFA, conditions determine the rotation of in-theater units. Time is not the only governing factor. Changes in the operational environment may require reshaping force packages as situations change. In addition, internal administrative concerns might prompt or support a commander's decision to rotate units. Regardless, mission handoff is necessary and defined as the process of passing an ongoing mission from one unit to another with no discernible loss of continuity.

5-65. The overall authority for the handoff and the subsequent transfer of authority lies with the commander ordering the change. The authority for determining the handoff process lies with the incoming commander assuming responsibility for the mission. This changeover process may affect conditions under which the mission will continue.

Redeployment Considerations

5-66. Although intended for a direct handoff between U.S. units, commanders must make specific considerations when making a handoff to a multinational force. The considerations include mission, enemy, terrain and weather, troops and support available, time available, and civil considerations. For units relieved of a function by a government agency, procedures typically entail longer handoff times and more complex coordination. However, the other areas of consideration still apply and may in fact be a greater issue for an agency. Outgoing units that have past, present, or future projects planned with agencies prepare to transfer these projects to responsible agents in the incoming unit.

Mission

5-67. The outgoing unit must prepare a detailed study of their unit's mission statement and detail the current and implied mission tasks. The mission may also require an incoming unit with additional skill sets, such as advanced special operations, near real-time connectivity, civil affairs functional specialists, or complex media production ability. Knowing the mission, commander's concept of the mission, commander's critical information requirement, PIRs, and information requirements helps the incoming unit understand the mission. After a complete in-depth study of the operational area, the outgoing unit should help the incoming staff complete the handoff in a manner that allows for continued, uninterrupted mission accomplishment. The changeover must not allow the adversary to gain any operational advantages.

Enemy

5-68. The outgoing unit must provide the latest available intelligence on all threats that affect the mission. This intelligence includes comprehensive data on terrorists and terrorist-related incidents over the previous several months. In addition to the normal intelligence provided to the incoming unit on a regular basis, the situation may call for a liaison from the outgoing unit. Operations security is critical to prevent the enemy from discovering the impending handoff and then exploiting the fluidity of the change and the concentration of U.S. or multinational forces.

5-69. The in-country unit provides continuous information updates to the incoming unit. PIRs and information requirements were established for the original mission along with operational, strategic, and tactical information. The incoming unit must become familiar with the ongoing and upcoming mission PIRs and information requirements along with their links to planned nonlethal effects.

Troops and Support Available

5-70. To the incoming unit, learning about the friendly forces is as important as knowing the enemy situation. The unit must be familiarized with the command and control structure with which it will work daily. The incoming unit must know all friendly units in adjacent operational environments. It also must be aware of joint and multinational units and the capabilities of their mission support base. The incoming unit must also be aware of other operations, units, and their capabilities.

5-71. Mission handoff is the time for the outgoing unit leadership to introduce the incoming unit leadership to leaders of FSF. This practice helps prevent breaks in intelligence networks already developed by the outgoing unit and reduces the time the incoming unit needs to build trust. The outgoing unit supports the incoming unit's plans to prepare for the transition in counterpart relations. However, potential or anticipated friction between a foreign security force and the incoming unit may cause the relief to occur more slowly. Therefore, incoming and outgoing units need time for overlap to allow for in-country, face-to-face contact with their counterparts before the mission handoff. If possible, outgoing unit members should provide biographical data on their counterparts to include photographs. This information allows unit members to become familiar with their counterparts before deployment and to determine which advisor techniques may need more emphasis. Execution of the mission must continue within the capabilities of the incoming unit, the foreign unit, and the available supporting assets.

Time

5-72. The depth and dispersion of units and the number of operations conducted determines the time required to exchange units. Ideally, an overlap occurs to allow the incoming unit to become familiar with the operational environment and to establish rapport with their foreign counterparts. However, the handoff must occur as quickly as possible. The longer it takes, the more vulnerable become personnel in the operational environment. However, both outgoing and incoming units should not sacrifice continued and uninterrupted execution of ongoing operations for speed. The incoming unit needs enough time to observe training techniques and procedures and to conduct debriefing on lessons learned. Outgoing unit personnel need sufficient time to put incoming unit personnel in contact with agency counterparts.

Civil Considerations

5-73. The outgoing unit provides an in-depth area study, giving close attention to local problems. Often sister units can provide general demographic data that can be expanded upon for unit-specific needs. Popular support for U.S. activities occurring in the operational environment may directly influence changes in the mission statement. The outgoing unit provides this critical information and describes all completed and progressing civic action projects. The incoming unit considers how the host-nation government functions and international civilian or government agencies are involved in or influence the situation in its operational environment.

POSTDEPLOYMENT

5-74. The unit commander debriefs to provide an overview of the mission and all relevant information subsets. Topics can range from military geography, political parties, and military forces to insurgents, security forces, and ongoing agency operations. Concerns specific to a foreign security force and relevant in one SFA operation may not be relevant in another operation. The commander should codify and debrief any information subset that may affect mission success.

POSTMISSION DEBRIEFING

5-75. Redeployment is not the end of the mission. Upon arrival at the redeployment location, the unit undergoes an extensive debriefing. The intelligence staff section officer typically organizes and conducts the debriefing. The intelligence officer coordinates with higher-level intelligence organizations to take part in the returning unit's debriefing, particularly if other organizations tasked the unit to obtain information. All deployed personnel, to include attachments, must be available for the debriefing.

DOCUMENTATION

5-76. After the debriefing, the unit commander and staff prepare two documents—an after action review and a report of lessons learned.

5-77. The after action review states the who, what, when, where, and how of the operation. It is a permanent record of the major activities of the unit from receipt of mission to debriefing. As such, it is an extremely important template on which past missions may be compared and future missions planned. Within 48 hours of being debriefed, the unit normally submits an after action review through command channels to the higher command. The intelligence and operations officers at each echelon keep copies of unit after action reviews.

5-78. Shortly after completion of the after action review, or simultaneously with its submission, the unit submits a report of lessons learned. This report documents the commander and staff's reflection of the operation and recommendations for the future. Units often prepare the lessons according to the six warfighting functions: movement and maneuver, intelligence, fire support, sustainment, command and control, and protection. This method addresses what worked and did not work during the operation, why it did or did not work, and what changes or substitutions are needed for existing tactics, techniques, and procedures in the unit.

5-79. As a minimum, the lessons learned report is forwarded to appropriate Service- and joint-level lessons learned agencies to allow integration into subsequent unit training and leader education. These agencies might include the Center for Army Lessons Learned and Joint Center for Operational Analysis.

Chapter 6

Sustainment

A brigade combat team may be required to execute operations across the spectrum of conflict. Hence, providing security force assistance (SFA) may require the sustainment staff to analyze the mission to determine support requirements and available capabilities. Since brigade combat teams conducting SFA may be responsible for a large operational area, an accurate and thorough assessment of U.S. assets as well as the foreign security force's assets is critical to mission success. The sustainment planner understands that though the mission can change, methods by which logistics are distributed within a theater of operations do not. Army sustainment enables the commander to execute a mission and sustain the force through full spectrum operations. Sustainment assets are allocated based on those requirements; no standard arrangement fits all situations.

As discussed in chapter 2, one of the principles of SFA is to ensure long-term sustainment. This principle dictates that all policies, strategies, and programs must be durable, consistent, and sustainable by foreign security forces (FSF) and their government once U.S. forces have completed assistance. This can be accomplished by developing the foreign security force's capability and capacity for asset production, management, performance, and maintenance.

SUSTAINMENT IN FOREIGN SECURITY FORCES CAPACITY

6-1. Sustainment is the provision of the logistics, personnel services, and medical support necessary to maintain operations until mission accomplishment. The endurance of SFA forces is primarily a function of their sustainment. Sustainment determines the depth to which the forces can conduct decisive operations, allowing the commander to seize, retain, and exploit the initiative. Endurance is the ability to employ combat power anywhere for protracted periods. Endurance stems from the ability to generate, protect, and sustain a force. It involves anticipating requirements and making the most effective, efficient use of available resources. Sustainment also enables strategic and operational reach. SFA forces require strategic and operational reach to deploy and immediately conduct operations anywhere with little or no advanced notice. FM 4-0 discusses sustainment in detail.

CORE LOGISTIC CAPABILITIES

6-2. Similar to any operation, the core logistic capabilities in SFA provide a framework to facilitate integrated decisionmaking, synchronize and allocate resources, and optimize logistic processes. The challenges associated with support cut across all the core logistic capabilities: supply, maintenance, deployment and distribution, engineering, medical support, logistic services, and operational contract support.

Supply

6-3. Supply operations include identifying requirements; selecting supply sources; scheduling deliveries; receiving, verifying, and transferring products; and authorizing supplier payments. To provide responsive supply operations in SFA to both the advising organization and FSF, logistic planning requires a collaborative environment that includes operations planning, maintenance operations, and the distribution system.

Maintenance

6-4. In SFA, maintenance operations aim to reduce repair cycle times by replacing components, modules, and assemblies as far forward as possible as well as maximizing reliance on parts distribution, visibility, and replacement. Advisors and their foreign counterparts require maintenance capacity for their freedom of action. Force movement and maneuver depends on sufficient readiness levels of transportation equipment and weapons systems. In SFA, maintenance encompasses both intermediate and organizational levels and consists of shop maintenance supporting the supply chain, transportation assets, and weapon systems.

Deployment and Distribution

6-5. The deployment and distribution capability moves forces and logistic support on time by meeting the required delivery date and providing time-definite delivery. By sharing critical information, units create unity of effort among diverse distribution organizations. This unity of effort enhances foreign support efforts and provides end-to-end support to satisfy deployment execution and sustainment operations for advisors and host-nation forces.

Engineering

6-6. Engineering provides capabilities to maximize freedom of action. These include capabilities and activities that support the maneuver of land combat forces; modify, maintain, or protect the physical environment; and contribute to a clear understanding of the physical environment by providing geospatial information and services to commanders and staffs.

Medical Support

6-7. A component of Army health systems oversee operational management of the health service support and force health protection missions for training, predeployment, deployment, and postdeployment operations. Army health system includes all mission support services performed, provided, or arranged by Army Medical Department to support health service support and force health protection mission requirements for civilian and military joint and multinational forces. For more information on Army health systems, see FM 4-02.

Health Service Support

6-8. All support and services performed, provided, and arranged by Army Medical Department promote, improve, conserve, or restore the mental and physical well-being of Soldiers and personnel in other Services, agencies, and organizations as directed. This includes casualty care, medical evacuation, and medical logistics. The health service support mission is a function of the sustainment warfighting function.

Force Health Protection

6-9. Force health protection measures promote, improve, or conserve the mental and physical well-being of Soldiers. These measures enable a healthy and fit force, prevent injury and illness, and protect the force from health hazards. They include some prevention aspects of Army Medical Department functions, combat and operational stress control, dental services, and laboratory services. The force health protection mission is a function of the protection warfighting function.

Logistic Services

6-10. Logistic services comprise the support capabilities that collectively enable the United States to rapidly provide global sustainment for U.S. forces. Logistic services include many disparate activities that are highly scalable capabilities. Included in this area are food, water and ice, base camp, and hygiene services.

Contract Support

6-11. Operational contract support provides acquisition functions to obtain materials, services, and construction of facilities for U.S. and host-nation forces in the local operating areas. The Department of Defense increasingly relies on contractors to perform many functions and tasks and then provides technical oversight. The expeditionary nature of SFA creates an even greater need for this capability.

Principles of Sustainment

6-12. The principles of sustainment are critical to maintaining combat power, maintaining strategic and operational reach, and providing Army forces with endurance. The principles of sustainment are anticipation, responsiveness, simplicity, economy, survivability, continuity, improvisation, and integration. While these principles are independent, they are also interrelated when used in planning and executing operations. For more information on sustainment principles, see FM 4-0.

Missions

6-13. All brigade support battalions have the same general capability to manage distribution and sustainment operations. The brigade support battalion provides support within an assigned operational area. It is force tailored and task-organized and uses subordinate companies, platoons, and teams to perform specific functions. In the sustainment role, the brigade support battalion is primarily concerned with the continuous management and distribution of stocks and allocation of maintenance in the operational area. Such assistance provides operational reach to maneuver commanders.

6-14. The brigade support battalion contains the following sustainment capabilities:

- Provides supplies, field services, field maintenance, recovery, and field feeding for itself and its assigned subordinates.
- Plans and conducts base and base cluster self-defense; defends against level I threats, assists in destruction of level II threats, and escapes or evades against level III threats.
- Has ability to operate as part of an Army or joint force.
- Coordinates host-nation support.
- Has ability to deploy an advance party to support early entry operations.
- Provides opening, distribution, and sustainment management information and advice to commanders and staff within its operational area.
- Exercises technical supervision over operations for all assigned units.
- Provides limited materiel management for internal stocks—classes I, II, III, IV, V, VII, and IX and maintenance management of internal assets.
- Provides a liaison team to augment other headquarters as necessary.
- Manages and maintains the its property records.
- Maintains data in support of Army equipment status reporting database and Army equipment status reporting system.
- Appoints contracting officer representatives to oversee contractor performance, certify receipt of services, and act as liaisons between the requiring activity and the contracting officer.

6-15. A brigade combat team often has attached companies such as an engineering company, a military intelligence company, an antitank company, or a network support company. Similar to the brigade headquarters and headquarters company, these attached companies are supported by the brigade support battalion regardless of where they are located on the battlefield. If one or part of these companies is task-organized to a maneuver battalion, it remains under the support of the brigade support battalion.

CRITICAL CONSIDERATIONS IN BUILDING HOST-NATION SUSTAINMENT CAPACITY

6-16. Effective SFA planning begins with accurately assessing and understanding U.S. policy objectives, determining the end state, and identifying resources available. Some possible logistic planning considerations—

- Have a holistic perspective.
- Address the causes, not the symptoms.
- Understand the desired end state.
- Understand the level of FSF senior leader support.

6-17. Logistic planning during SFA uses a holistic perspective. Operational environments can range from peacetime cooperation to counterinsurgency, stabilization to reconstruction. SFA forces determine the foreign security force's logistic and reconstruction requirements based on their internal defense and development strategy; they develop a sustainment end state to support the plan. Rebuilding the host nation's infrastructure and industrial capacity not only provides long-term job opportunities, but also builds trust and legitimacy in FSF, the host-nation government, and partner organizations.

6-18. Logistic planning during SFA addresses the causes, not the symptoms. Operational logistic support problems are the symptoms, not the causes. They result from problems that exist in the supporting systems or in the society's industrial capacity and economic strength. Teaching unit-level inventory management alone will not result in any sustainment improvement.

6-19. Logistic planning during SFA understands the desired end state. Depending on the U.S. strategic interest in a host nation and its security forces, levels of engagement can vary significantly to achieve the desired end state. Sustainment capacity-building activities can be force tailored and scaled to assist in any of the sustainment capability areas.

6-20. Logistic planning during SFA understands the level of FSF senior leader support. These leaders develop and set the policies for resource allocation. They support the various development projects and skills enhancement assistance activities. These senior-level advisors must fully understand the problems identified and an assistance strategy developed by SFA planners. From the beginning, SFA planners work with their counterparts to decide the plans and strategy to promote the vision and ensure continued effort.

6-21. SFA depends on capable logistic support for maneuver and sustainment. Without an improvement in sustainment capacity, FSF cannot significantly improve regardless of how much effort is devoted to advising and training. When planning for sustainment capacity building, planners thoroughly understand the logistic systems to identify causes that lead to functional deficiencies. They must have a holistic perspective and understand the various goals and resources required to formulate an operational approach that effectively and efficiently achieves U.S. objectives.

ASSET MANAGEMENT AND PERFORMANCE

6-22. One tool at the disposal of the commander and planners during an evaluation of foreign infrastructure and industrial capacity is called an asset management plan. This plan, commonly employed in the industrial community, manages an organization's infrastructure and assets to deliver a standard of service. An asset management plan provides information to FSF so they can analyze the capacity, condition, and current and future requirements of their facilities and infrastructure. Physical development should include facilities, transportation networks, area development, utilities, and other infrastructure components.

6-23. A description of the perceived problems of current foreign facilities, or problems created by missing assets, is also important. The description includes how the advisor intends to address deficiencies in foreign facilities and infrastructure. Each organization needs to determine each facility's function and value to their organizational mission by analyzing the current asset performance, costs, benefits, and improvements.

CURRENT ASSET PERFORMANCE

6-24. Asset performance is the most difficult portion of the asset management plan process to capture while assessing conditions of facilities and infrastructure. While historical records may exist, combat operations and possible instability of the host-nation government likely render these records irrelevant. Remaining structures, when measured against U.S. construction standards (designed to protect occupants from man-made or natural disasters), often are deficient, even if untouched by hostilities. Most existing facilities require significant modification to be considered for future use. However, FSF may use them in the interim until new facilities are completed.

6-25. Asset performance may also use gap analysis. Gap analysis defines the present state, the desired end state, and the gap between them. Once a gap has been identified, the commander and staff can develop courses of action to rectify or bridge the gap. Gap analysis alone is not adequate for all problem sets, as goals may evolve and emerge during the course of problem solving. Most problems have many alternative solutions.

COSTS

6-26. Establishing the short, medium, and long-term costs for facilities and infrastructure, from construction to control, is a critical step in prioritizing which facilities get built or refurbished first, and which get support at all. Units develop a forward-looking cost-profile for operating, maintaining, refurbishing, and replacing assets. This profile explains to civilian and military officials the implications and continued costs to maintain assets. SFA requires a plan for U.S. forces to hand off assets to FSF.

6-27. In constructing these models, the ideal cost-profile extends to cover the life of the longest-lived asset in the system. By doing so, the profile estimates the whole-life cost as well as determines the average annual costs. Furthermore, this cost evaluation may need to be revisited and updated annually, forming the basis for an annual bid for funding.

BENEFITS

6-28. Assets should provide some measure of benefit to U.S. and foreign security forces that can be measured or explained during and after SFA. This usually involves translating asset costs and projected performance into a monetary figure. Other benefits may be social or environmental, which may be difficult to quantify in monetary terms. Regardless, some attempt to record all the relevant benefits is important; the more quantitative, the better.

6-29. A benefit evaluation is vital to demonstrating that the ongoing cost of an asset system is reasonable given the scale of the benefits. The expenditure incurred by asset management needs to be justified in some way.

IMPROVEMENTS

6-30. How assets of FSF perform may be improved by acquisition, enhancement, or some other means. An improvement may address the potential reduction of performance and its effect on the rest of the asset system. The improvements represent a potential change to the performance and are typically managed as a project. Such management comes with an appraisal of the additional expenditure, comparing different options and selecting a preferred option based on the commander's guidance and host-nation requirements.

OPERATIONAL CONTRACT SUPPORT

6-31. Contracting is a key source of support for deployed forces across the spectrum of conflict. Because of the importance and unique challenges of operational contract support, the commander and staff need to fully understand their role in planning for and managing contracted support in the operational area. This includes understanding the types of contracted support, contract management, and contract close out.

Types of Contract Support

6-32. Current doctrine describes three broad types of contracted support: theater of operations support, external support, and systems support.

Theater of Operations Support

6-33. Contracts for theater of operations support provide support to deployed forces under prearranged contracts or contracts awarded from the mission area. They do so by contracting officers under the command and control of the combat sustainment support battalion. Contractors for theater of operations support acquire goods, services, and minor construction support, usually from local commercial sources, to meet the immediate needs of operational commanders. Theater of operations support contracts are the type of contract typically associated with contingency contracting. Sustainment brigades often are the requiring activity for theater of operations support contract actions.

External Support

6-34. Contracts for external support provide a variety of support to deployed forces. External support contracts may be prearranged contracts or contracts awarded during the contingency. These contracts support the mission and may include a mix of personnel including U.S. citizens, third-country nationals, and host-nation subcontractor employees. The largest and most commonly used external support contract is the logistics civilian augmentation program. This Army program is commonly used to provide life support, transportation support, and other support functions to deployed Army and joint forces.

System Support

6-35. System support contracts are prearranged and include, but are not limited to, newly fielded weapons systems, command and control infrastructure, and communications equipment. System support contractors, made up mostly of U.S. citizens, provide support in garrison and may deploy with the force to both training and real-world operations. The unit does not normally have a significant role to play in planning for or coordinating system support contracts other than coordinating and executing support of system support contract-related personnel.

Contract Management

6-36. Contract management requires nominating and tracking contracting officer representatives for every service contract. It will also nominate a receiving official for all supply contracts. To ensure contractors provide the service or item per the contract, they need a quality contracting officer representative and official support. SFA units must identify and coordinate funding requirements in advance with the supporting resource management staff element to secure the correct amount and type of contract funding associated with the mission.

Contract Close Out

6-37. In long-term operations like SFA, the advising unit ensures direct coordination and transfer of information related to operational contract support before mission handoff. Additionally, when the advance party arrives in the operational area, a designated Soldier actively seeks out current information on local contract support capabilities, policies, and procedures. This individual coordinates the formal handover of existing contract management responsibilities from the redeploying unit. This Soldier must know when recurring service contracts end since it takes 30 to 60 days to obtain funding approval. If the unit waits until the contract is about to expire before requesting additional funds, the sustainment brigade could lose the contracted service until funds are available.

EXTERNAL AGENCIES SUPPORT MANAGEMENT

6-38. Some external agencies help support the management of FSF. External support comes from interagencies, the host nation, and multinational organizations.

INTERAGENCY SUPPORT

6-39. The Department of Defense performs both supported and supporting roles with other agencies. It often is the lead agency for logistic activities in SFA. When tasked to provide military support to civil agencies, Army forces perform a supporting role. Sustainment forces may be tasked to support the U.S. ambassador that, as the President's representative, serves as senior U.S. Government officer in the country, or may employ resources of other U.S. Government agencies or private firms. Whether supported or supporting, close coordination is the key to efficient and effective interagency operations. In the absence of an American Embassy, or during operations in austere environments or remote locations, the advising unit may be the only organization able to provide supplies, services (including force protection), and life support. More typically, during the early stages of the advising unit deployment, the unit may find the American Embassy to be the primary source of supplies, services, and life support.

6-40. Nongovernmental organizations do not operate in either the military or governmental hierarchy. Their relationship with an advisor organization is neither supported nor supporting. An associate or partnership relationship exists between military forces and engaged nongovernmental organizations, ideally focused on unity of effort. If formed, the focal point where U.S. forces provide coordinated support to nongovernmental organizations would be the civil-military operations center of a joint task force headquarters.

HOST-NATION SUPPORT

6-41. Host-nation support and local procurement may provide sustainment, operational support, and tactical support. Host-nation support agreements fulfilling the command requirements for support need to be prenegotiated. Such support arrangements must be integrated into the distribution plan and coordinated with joint, allied, and multinational partners to prevent competition for resources and ensure high priority requirements are met. Host-nation support may include functional or area support and may use host-nation facilities, government agencies, civilians, or military units. Preestablished arrangements for host-nation support can reduce the requirement for early deployment of U.S. assets and can offset requirements for early strategic lift by reducing requirements for moving resources to the theater of operations.

MULTINATIONAL SUPPORT

6-42. Multinational support may consist of support provided from one multinational partner to another. One or more of the following organizational and management options facilitates multinational support:

- National support elements provide national support.
- Individual acquisition and cross-servicing agreements provide limited support.
- A lead nation provides specific support to other contributing nation forces.
- A role-specialist nation provides a specific common supply item or service.
- A multinational integrated sustainment unit provides limited common supply and support.
- A multinational joint sustainment center manages common-user logistic support.

6-43. In all cases, the multinational force commander directs specific multinational support within the applicable laws and regulations of the host nation. When operating within a formal alliance, the advising unit executes support following applicable standardization agreements.

FACILITATING THE REDEPLOYMENT PROCESS

6-44. Sustainment assets facilitate the redeployment and the retrograde process through its command and control of the distribution system. Many of the procedures are used to deploy forces, draw pre-positioned stocks, conduct reception staging and onward movement, and distribute supplies within the theater of operations apply to the redeployment process. Two factors in particular complicate redeployment operations: supporting operations and meeting requirements.

6-45. The same elements that operate and manage the theater of operations distribution system during deployment and sustaining operations perform similar roles during redeployment. When redeployment,

deployment, and sustaining operations occur simultaneously, the theater sustainment command may find it necessary to rebalance its forces or change the missions of subordinate units to effectively support ongoing operations and redeployment.

6-46. Requirements vary widely depending on the nature and scale of redeployment operations, infrastructure, and other factors. For example, redeployment operations could range from personnel only to entire units and their equipment. Depending on the political and military strategy, unit rotations may occur while decisive operations continue unabated or during operational transitions. Key considerations include, but are not limited to, several characteristics. These characteristics include the size of the force redeploying and deploying; infrastructure requirements and limitations; security requirements; traffic circulation; staging areas; distribution system capacity; quantity of supplies and materials to be redistributed; the amount and kinds of materials that require disposal agricultural inspections; and establishing accountability of retrograde cargo. The challenge is effective coordination and synchronization, vertically and horizontally, to ensure responsive simultaneous support to not only ongoing distributed operations, but also redeployment.

Chapter 7

The Advisor

Advisors are the most prominent group of U.S. personnel that serve with foreign security forces (FSF). They live, work, and fight with their FSF. The relationship between advisors and FSF is vital. Advisors are not liaison officers, nor do they command FSF units.

ROLES OF THE ADVISOR

7-1. The military advisor has three roles involving different responsibilities. First and foremost, advisors are members of a U.S. military organization with a well-defined chain of command and familiar responsibilities. Second, advisors embed themselves with their counterparts. Third, advisors are interpreters and communicators between U.S. forces and their foreign counterparts.

7-2. As members of military organizations, advisors receive and execute the orders of superiors. These orders may conflict with the orders their counterparts receive. Among other duties, advisors must act unobtrusively, but nonetheless positively, as inspector general—often observing, evaluating, and reporting on the performance of counterparts and their assigned unit.

7-3. Secondly, advisors live, eat, and work with the officers and men of their host units. Often, advisors soon regard themselves as one of them. The sharing of common hardships and dangers forges potent emotional ties. The success and good name of their units become matters of prime and personal importance to the advisor.

7-4. Finally, advisors are interpreters and communicators between U.S. superiors and foreign counterparts. Advisors must introduce and explain one to the other; they help resolve the myriad of problems, misunderstandings, and suspicions which arise in any human organization, particularly when people of starkly different cultures approach difficult tasks together. Advisors with quick and easy access to influential counterparts can sometimes be the best possible means of communicating.

7-5. To be effective, advisors obviously must gain their counterparts' trust and confidence. This relationship, however, is only a prelude to the advisor's major objective: inspiring and influencing a counterpart to effective action. In pursuing this goal—constantly, relentlessly, and forcefully, yet patiently, persuasively, and diplomatically—advisors must recognize conditions which can benefit or handicap their cause.

CONSIDERATIONS OF THE ADVISOR

7-6. An advisor follows ten considerations when serving with FSF.

BY, WITH, AND THROUGH

7-7. All planned operations of a combat advisory mission must be conducted by, with, and through the FSF. Not counting immediate action battle drill responses, the mark of an effective advisory effort is the amount of personal involvement the FSF take in their own operations. Civilians must see that they are secured by their own security forces to promote the legitimacy of the host nation and their capacity and to build trust and confidence. In a word, combat advisors are shadows whose presence is felt at all times.

Empathy Leads to Competency

7-8. Empathy can be defined as identification with and understanding of another's situation, feelings, and motives. This is tough for experienced U.S. leaders and often harder to explain, but it is the key to the success of an advisory mission. Truly understanding other human beings and where they come from allows for honest relationships to develop. In most cultures, the place to begin understanding another person's feelings and experiences is by understanding the other person's narrative. The narrative is a collective group's identity as an interpretation of both ancient history and recent collective experiences. Delving into the narrative, understanding it and how it affects people, is the beginning of empathy.

Success Is Built on Personal Relationships

7-9. Developing a sense of empathy allows advisors to begin nurturing relationships with those they are tasked to advise. No amount of resources and firepower can compensate for the lack of a relationship between advisors and their counterparts. It must be honest, genuine, and heartfelt. Mutual respect, trust, and understanding create success. Both parties rely on each other for mission accomplishment and often for survival. This relationship is likely to be challenged on numerous occasions; only one built on mutual trust can survive and ensure mission success. Furthermore, advisors learn to pick their battles. Camaraderie and rapport with FSF can vanish in an instant if one does not prioritize and define critical lines in the relationship.

Advisors Are Between U.S. and Foreign Security Forces

7-10. Living and fighting with FSF allows for the development of true bonds, empathy, and trust. These desired results have a price to pay. When addressing their own Army, advisors soon realize that they are also not one of them, increasing their level of frustration. Advisors are often alone, navigating between two military systems and two cultures, never quite fitting in with either one.

Advisors Will Never Win, Nor Should They

7-11. Often advisors find it difficult to understand what victory looks like in this environment. Progress tends to occur at a glacial pace and cannot usually be tracked on a day-to-day basis. The advisor attaining a tactical objective does not achieve success; success is achieved by the foreign security force achieving the objective. Advisors find it difficult to satisfy their own units, and they never fully satisfy the demands of their FSF. They are figuratively and literally caught in the middle. Advising is the art of striving to make a win-win situation for all parties. The advisor and foreign security force relationship is one in which the advisor removes the obstacles in the way. Only a plan by FSF will succeed—and it will only succeed if it is, in fact, their plan.

Advisors Are Not Commanders

7-12. Advisors do not command FSF personnel or units. Advisors provide advice, training, and access to coalition resources to FSF. They are not intended to lead FSF in combat; they are responsible for the command and control of their own team of advisors. However, they can and should influence the commander of the foreign unit.

Advisors Are Honest Brokers

7-13. Advisors are advocates for FSF with conventional forces. They transparently assess capabilities and reveal limitations of foreign units to the FSF's and advisors' higher command. They also provide access to U.S. resources—from close air support to medical evacuation—to FSF they advise.

Advisors Live with Shades of Gray

7-14. Black and white is not available to the advisor. Caught between two cultures, systems, and narratives, the advisor works within a gray area. Unlike operations in a conventional unit, many of the checks and balances are not present. Advisors often find themselves isolated with great autonomy and no supervision. These conditions present moral and ethical dilemmas daily. Advisors need to be comfortable in this environment. Often they lose sleep at night and question their actions and those of their FSF. Thus, this mission calls for Soldiers of enormous character, moral courage, and intellect.

Talent Is Everything, But Advisors Must Understand Rank

7-15. Military forces around the world approach rank and prestige in completely different ways. In some cultures, rank equals experience; in others, it is nobility or education. Often many militaries have rewarded loyalty with promotions, which results in rank-heavy armies. Some FSF recognize talent, while others recognize rank. Advisors likely advise counterparts much more senior in rank. They must understand that rank on the uniform is important to many armies.

Make Do

7-16. Advisors will never have everything they believe they need to succeed. However, their creativity often distinguishes between success and failure. Austerity and economy of force can fail to meet the prevalent expectations that FSF have of a U.S. unit in regards to resources. It is up to the advisor team to close the expectation gap by its creativity and imagination. Scrounging, bartering, and negotiating are daily activities of advisors. An enormous amount of energy must be devoted to these activities to achieve mission success and endear them to their counterparts.

PERSONALITY TRAITS OF THE ADVISOR

7-17. Not every Soldier is well suited to perform advisory functions; even those considered to be the best and most experienced have failed at being an advisor. Effective advisors are only the most capable individuals. Advisors are Soldiers known to take the initiative and who set the standards for others; however, they are also patient and personable enough to work effectively with FSF. Recognizing that not all Soldiers are capable of performing as advisors, leaders should immediately remove advisors who do not exhibit these qualities.

7-18. Because advisors operate in very subjective environments, it is difficult to establish objective criteria by which to assess potential advisors. However, research and experience indicate that several personality traits greatly enhance the advisor's ability to adapt and thrive in a foreign culture. These traits include—

- Tolerance for ambiguity.
- Realistic when setting goals and tasks.
- Open-mindedness.
- Ability to withhold judgment.
- Empathy.
- Communicativeness.
- Flexibility.
- Curiosity.
- Warmth in human relations.
- Motivation of self and others.
- Self-reliance.
- Strong sense of self.
- Tolerance for differences.
- Perceptiveness.
- Ability to accept and learn from failure.
- Sense of humor.

7-19. Of the traits listed above, no single trait is paramount. Advisor selection and training programs seek to develop an understanding of the contributions of each of them.

SKILLS OF THE ADVISOR

7-20. Advisor-specific skills fall into two subcategories: enabling skills and developing skills. Enabling skills build on the individual and collective skills of advisors. Developing skills are the main advisor tasks of teaching, coaching, and advising. Advisors use these skills to develop the capabilities of the foreign security force.

ENABLING SKILLS

7-21. Advisors use enabling skills to work individually and in a group. Enabling skills help advisors communicate across cultures, build rapport, influence, and negotiate. Successfully employing these enabling skills sets the conditions for advisors to move forward with their mission.

Individual Skills

7-22. Individual skills help advisors properly perform their mission. Initially advisors should be selected based on subject matter expertise and not just rank. However, tactical proficiency does not necessarily equate to proficiency as an advisor. Modern-day military advisors must be able to teach and advise FSF. To do this, they must be knowledgeable and proficient in procedures and able to impart this knowledge effectively. They may also be able to advise FSF in the areas of intelligence, communications, operations, and logistics. Some advisors may be required to call in U.S. supporting arms, principally artillery and air support, and coordinate both air and ground casualty evacuation. Finally, advisors should be proficient in basic combat skills. These common skills, warrior skills, and survival skills are often required by advisors. Since these skills are critical to an advisor due to the isolated and independent nature of the mission, advisors should refresh them during predeployment training.

Collective Skills

7-23. As a group of advisors form a team, they must practice their warrior tasks, organize to cover the functions associated with any team, and familiarize its members with their duties and responsibilities. As with individual advisor skills, some are refined before deploying and others are learned in country. Figure 7-1 depicts this graphically.

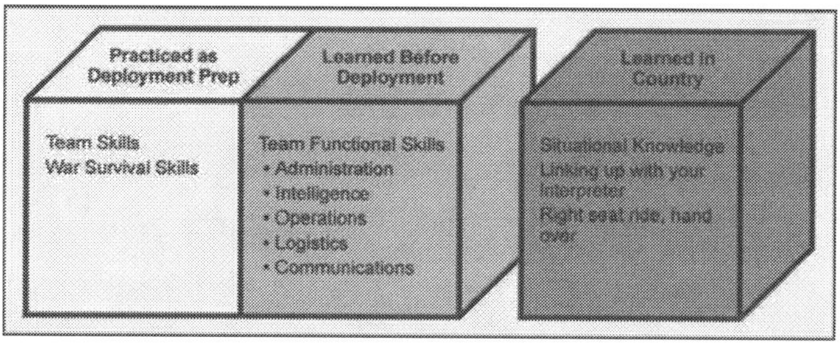

Figure 7-1. Collective advisor skills

DEVELOPING SKILLS

7-24. Developing skills—teaching, coaching, and advising—are required for every advisor. The advisor will make an initial assessment to determine the foreign unit's proficiency to determine the appropriate developing skill with which to start. Figure 7-2 illustrates the appropriate level of advice that corresponds to the foreign unit's capabilities.

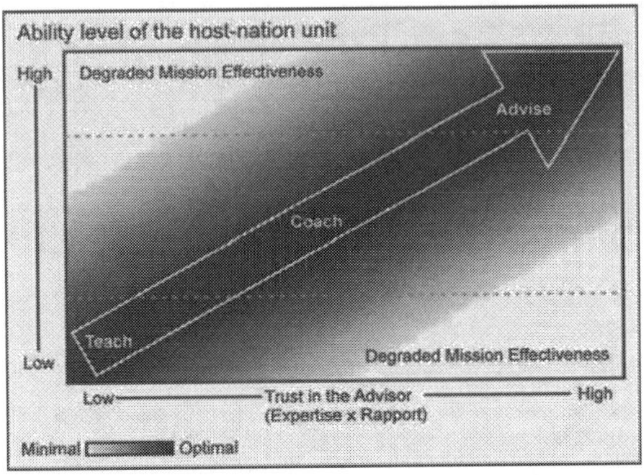

Figure 7-2. Appropriate advising level

7-25. Advisor teams have limited positional power and depend on personal influence to positively affect FSF efforts through teaching, coaching, and advising. Advisor team members often work with people of higher rank or grade than themselves. Advisors should remember the foreign unit's experience and capabilities and carefully choose opportunities to inject or impart knowledge. Foreign units are most receptive to advisor teams that teach unobtrusively. Foreign units most value those who are subtle in their teaching, coaching, and advising. Those advisors who master the ability to create a climate in which FSF personnel feel they are teaching themselves often prove the most effective.

Teaching

7-26. Teaching includes training and education. Methods of teaching can include classroom lectures, seminars, hands-on training, training exercises, and simulations.

Coaching

7-27. Coaching relies on guiding to bring out and enhance capabilities already present. Coaching refers to the function of helping someone through a set of tasks. Those being coached may or may not have appreciated their potential. The coach helps them understand their current level of performance and instructs them how to reach the next level of knowledge and skill. Coaching requires identifying short- and long-term goals and devising a plan to achieve those goals. The coach and the person being coached discuss strengths, weaknesses, and a course of action to sustain or improve the goals.

Advising

7-28. Advising is a combat multiplier that boosts supported unit capability. Advisors have experience in particular areas in which they are advising, but are not required to have similar backgrounds. Advisor relationships are not based on superior to subordinate relationships.

7-29. Advisors provide an expert opinion, advice, and counsel by focusing on both personal development (interpersonal and communication skills) and professional development (technical and tactical knowledge). Advising develops mutual trust and respect.

Advisor Versus Trainer

7-30. Advisors teach, coach, and advise work in a cycle. As FSF master one skill, the advisor can move on to other skills and initiate the process for the new skills. If FSF require additional teaching or coaching, the advisor can take them to that part of the cycle. Eventually, the teaching and coaching should decrease, and most of the advisor's time will be spent as an advisor providing an expert second opinion.

7-31. Ultimately, the question of what is the difference between being a trainer or an advisor will arise. Figure 7-3 shows that every advisor will have to be able to be a trainer, but all trainers are not expected to perform the more sophisticated task of advising. Hence, competent junior officers and enlisted Soldiers can be trainers, but only more experienced individuals can be advisors.

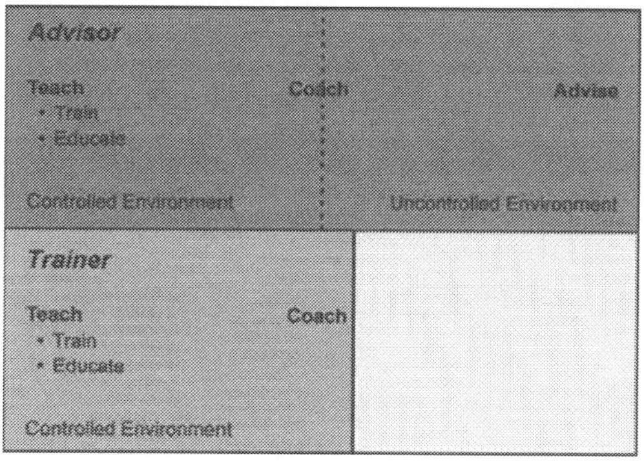

Figure 7-3. Advisor versus trainer

Chapter 8

Culture and Communication

As an advisor, understanding the host-nation population is a crucial element of premission planning and the development of foreign security forces (FSF). Prior knowledge of sociocultural differences aids in building effective relationships and prevents embarrassment, loss of rapport, and compromise of the mission. This chapter discusses three aspects of the sociocultural environment to provide a framework for analysis and comparison of foreign cultures for operational purposes. These aspects are society, culture, and communications.

SOCIETY

8-1. People who have a common culture, share a group identity, occupy a common territory, and are subject to the same political authority can be identified as a society. A society is not easily created or destroyed. Though they share the above factors, societies are not homogeneous nor static. A society usually has a dominant culture but can also have a vast number of secondary cultures, each influencing the other.

8-2. Advisors derive much of their effectiveness from their ability to understand and work with counterparts from a foreign society. Advisors must also consider societies outside their immediate operational environment whose actions, opinions, or political influence can affect the mission.

SOCIAL STRUCTURE

8-3. Society is composed of both social structure and culture. Social structure refers to the relations among groups of persons within a system of groups and is persistent over time. For the purposes of security force assistance, social structure includes groups, institutions, organizations, roles and statuses, and social norms. FM 3-24 contains a more in-depth discussion of social structure.

Groups

8-4. Within societies, groups are composed of two or more people that interact based on mutual interests and interrelated statuses and roles. These groups can be based on race, ethnicity, religion, or kinship. A racial group is defined by its members or by outside groups as different based on physical characteristics or descent. Race is a social category that has no biological basis. An ethnic group is a community whose learned cultural practices, language, history, ancestry, or religion distinguish them from others. Similar to racial groups, ethnic groups can be defined by their members or by outside groups. Religious groups may be subsets of larger ethnic groups and an ethnic group may contain members of different religions. Kinship groups are based on a link by blood relation. It may also be used to indicate a group of persons who live and work together to satisfy basic collective needs and goals. The role of a kinship group is generally seen to care for and educate offspring. This group provides for such needs as food, shelter, and clothing and prepares offspring to be functional members of the society in which they live.

Institutions

8-5. Institutions are the basic building blocks of societies. They are made up of groups that are organized to achieve a common task. Institutions are continuous through many generations and are built to continue to function even as individuals are replaced. An example of an institution from FM 3-24 is a military institution that brings together groups and individuals whose statuses and roles concern defense and security.

Organizations

8-6. Organizations are institutions that have a bounded membership with defined goals and established operations. They include fixed facilities from which they conduct business or operations and a set means of financial or logistic support. The four types of organizations are communicating, religious, economic, and social. Communicating organizations are designed to influence perceptions. Religious organizations are created to regulate societal norms, reaffirm beliefs and values, and provide social support. Economic organizations provide employment and assist in economic development. Social organizations provide support to groups in society, create social networks, and can influence ideologies. Organizations may belong to more than one category and exist to affect the behavior, values, and beliefs of society. As such, advisors must identify influential organizations both inside and outside of the FSF with which they are partnered.

Roles and Statuses

8-7. As members of a group, individuals interact based on assumed or given social positions. These social positions are referred to as statuses. Societies generally associate particular statuses with particular social groups (such as family, lineage, ethnicity, or religion). Each status inherently contains a set of expected behaviors known as a role, which includes how a person of that status should act, feel, and think. A status includes expectations about how others in a society should treat a person of that status.

Social Norms

8-8. Violation of a role derived from a given status results in social disapproval. The standard of conduct for social roles is a social norm. Norms may be either moral (incest prohibition, homicide prohibition) or customary (prayer before a meal, removing shoes before entering a house). Advisors must understand the roles, statuses, and social norms of groups they are advising. This knowledge will clarify the expected behavior of their FSF partners and their role as advisors.

Language

8-9. Language is a learned element of culture that is a system of symbols that people use to communicate with one another. Successful communication within a society requires an understanding of the social setting, appropriate behaviors towards different statuses and roles, and nonverbal cues. The languages used in the advisors' operational environment have a major impact on the advising mission. Languages must be identified to facilitate language training during predeployment and employment, as well as in the requisitioning of translators. Translators are critical for communicating with FSF and interacting with local citizens and leaders.

CULTURE

8-10. Social structure comprises the relationships among groups, institutions, and individuals in a society; in contrast, culture provides meaning to individuals within the society. Social structure can be thought of as a skeleton with culture being the muscle on the bones. The two are mutually dependent and reinforcing. A change in one results in a change in the other.

8-11. Culture influences an individual's range of action and ideas, including what to do and not do, how to do or not do it, and whom to do it with or not to do it with. It is neither a random jumble of customs nor a laundry list of accepted practices and taboos that an advisor should know. It influences judgments about what is right and wrong and what is important and unimportant. Cultural rules are flexible in practice; the customs and actions that an individual practices are based largely on that particular environment.

Cultural Characteristics

8-12. Culture is shaped by history, language, religion, and customs. Members of society use these shared beliefs, values, customs, and behaviors to cope with their world and with one another. It is learned behavior that is shared by members within a society. However, behaviors do not always define a culture.

Frequently, there are more telling cultural signs in the meaning attached to a behavior than in the actual behavior. Culture influences its members' identity, values, belief systems, and cultural forms. As such, it is arbitrary; advisors should make no assumptions regarding what a society considers right and wrong, good and bad. Culture is changeable through the social interactions between people and groups, both inside and outside of that society. TC 31-73 discusses cultural characteristics in great detail.

Identity

8-13. Each individual belongs to multiple groups through birth, assimilation, or achievement. Each group to which individuals belong influences their values and beliefs. Individuals consciously or unconsciously rank their identities into primary and secondary identities. Primary identities are those seen as core to their being; frequently these are national, racial, and religious. In contrast, secondary identities are more peripheral and may include such things as an individual's occupation or hobby.

Values

8-14. Values guide how individuals interact with the world around them. As deeply held beliefs about rightness, values shape how individuals think things should be. Values may never be articulated—in fact, they may not even be consciously held. At times, the beliefs of an individual may even contradict one another. Advisors should have some understanding of a counterpart's value system. Usually values are nonnegotiable. In working with a counterpart with conflicting values, advisors should persuade the individual that the advocated action or behavior really does not conflict with their values.

Belief Systems

8-15. Beliefs primarily take two forms: values (core) and opinions (peripheral). Values, or core beliefs, are not easily changed. Opinions, or peripheral beliefs, stem from thought and decisions. The totality of the identities, values, and beliefs that an individual holds compose that person's belief system.

Forms of Beliefs

8-16. Values are broad moral conclusions about the way to live life. These core beliefs are unstated, taken for granted, resistant to change, and not consciously considered. They reflect strong personal certainties drawn from faith, childhood environments, and life experiences. They lend shape and order to people's lives, helping them to decide if they settled on the right actions, friends, leaders, religions, and careers.

8-17. Opinions, or peripheral beliefs, are immediate conclusions about some aspect of the environment. They are usually quite specific and are open to debate, consciously considered, and easiest to change.

Types of Belief Systems

8-18. Belief systems involve cultural narratives and myths. The interpretation of these narratives and myths provide people with insight into how they should feel, think, and behave. The most prominent systems of beliefs tend to be those associated with religions and ideologies. However, any system of belief in which the interpretation of narratives impacts people's behavior (such as superstitions) can be a contributing component of a society's culture.

8-19. A belief system acts as a filter for new information. It is the lens through which people perceive the world. Beliefs of one group as to what is rational, normal, or true may appear strange, irrational, or illogical to outsiders. Understanding belief systems of various groups in an operational environment allows advisors to work with their foreign counterparts and influence civilians more effectively.

8-20. Belief systems play a fundamental role in the development of a society's values. Careful analysis of a region's belief systems is crucial to understanding the local value system. These systems influence religious doctrine, narratives, and myths and provide enormous insight into the culture of the community. Ultimately, the values which are most prized within a belief system are religious. These beliefs affect that culture's systems of social organization, learned behaviors, and relationship to the physical environment.

Religion as a Belief System

8-21. At its most basic level, religion is both the individual and communal expression of contact with the sacred. Notions of the sacred can vary widely in societies and may include deities and sacred items. In most cases, religions also involve symbols, rituals, and ethics or dogma. Religion can be both an individual practice and a group phenomenon.

8-22. Most societies have some variety of religion—typically a set of sacred beliefs and rituals that provide a common understanding of moral codes and proper conduct. Expressing a common religious understanding helps people to make sense of their world. Often religion provides explanations for human suffering, natural disasters, broken relationships, inequalities, and death. Some societies also have secular equivalents of religion. For instance, America is often remarked to have a civil or civic religion that can be seen in the images, symbols, ideas, and behavior of the political realm.

8-23. Because religion is such an integral part of culture, when advisors carefully prepare and analyze the mission, they must examine the religions and religious groups in the operational environment. Religious beliefs, leaders, and institutions are central to how many societies view the world. When planning any advising operation, planners consider how religion affects FSF and civilians.

Broad Impact of Belief Systems

8-24. In almost every society, followers of dominant religions have a strong societal impact. This influence may be exerted at varying degrees upon the individual citizen, sections of society, the economy, the military, and the political structure. TC 31-73 discusses these in detail.

THE IMPORTANCE OF UNDERSTANDING CULTURE

8-25. Advisors derive their effectiveness from their ability to understand and work with foreign counterparts from another culture. They must understand enough of their own culture and their counterpart's culture to accurately convey ideas, concepts, and purpose without causing counterproductive consequences.

8-26. Advisors must be aware of aspects of the local culture and history that influence behavior in their operational environment. Advisor team members must understand the reasons and motivations underlying personal interaction and must practice patience when working with their counterparts. Group norms guide individual behavior, and advisors must understand how individuals in a society tend to interact as members of a group, whether a race, ethnic, or kinship group.

8-27. Cultural understanding is not derived from demographic information provided to the military through country briefs prior to deployment. It is gained from studying, interacting, and understanding the people, religion, history, customs, and social and political structures within an area. For true understanding, it is necessary to live among the people, gradually understanding the subtleties and nuances of their culture.

CULTURAL EDUCATION

8-28. Culture influences a person's perception of the world. People from one society tend to view people from other societies through the prism of their own culture. Often, their perceptions are biased, are inaccurate, or involve assumptions of superiority. For example, Americans view their society and its culture very differently from the rest of the world. Likewise, as foreigners, advisors view local cultures through the prism of their own idealized culture. This creates a large gap between the culture the advisor observes and the host nation's real culture.

8-29. To narrow the gap of cultural understanding an advisor will encounter, a program of cultural learning should be undertaken. The end state of this cultural learning should be to gain the trust of foreign partners and civilians. Advisors will do so by demonstrating their understanding and sensitivity towards the host nation's culture and social structures.

8-30. An awareness of the cultural aspects of the operational environment can significantly enhance the effectiveness of an advisor. An extensive study program is best, but even a concise program is worthwhile if it provides advisors with the necessary cultural information. The best programs involve credible teachers who intimately understand the culture and have personal experience in the society. Training is dynamic and moves beyond the standard lecture and handout format. Small groups or a discussion panel work quite well. Cultural behavior and cultural familiarization handbooks can also be used. Since handbooks vary greatly in accuracy and quality, credible subject matter experts should be consulted to determine which documents to incorporate into the training. Table 8-1 discusses the training recommendations and tools.

Table 8-1. Cultural education recommendation

	Training Recommendations
Cultural Education	• Comparison of cultural values and social structures (United States compared to those of the operational area). • Local customs and traditions (for example, greetings do's and don'ts). • Geopolitical history (pre-colonial to contemporary and the orientation of each faction or party). • The role of religion in daily life.
	Training Tools
Resources	• Guest speakers native to the country of interest (for example, nongovernmental organization staff, foreign students, recent immigrants, or selected refugees). • Others who have worked in or studied the mission area (for example, special forces personnel, diplomats, and scholars). • Cultural familiarization handbooks.
Format	• Combination of briefings, small group discussions, and question and answer periods. • Handouts to augment—not replace—speakers. • Visual media, specifically slides and videos of the mission area.

8-31. Advisors with previous operational experience have addressed the importance of cultural education. Many civilian agencies have developed intercultural effectiveness programs designed for their personnel working overseas. Often these programs can be revised and tailored for military and paramilitary groups. If available, these programs are an excellent resource for predeployment preparation.

CULTURAL GAP

8-32. Each culture has its own societal rules regarding who a person may speak to, how and when the person may speak, and what topics the person may speak about. Many cultures rely on nonverbal signals to communicate. In such cultures, posture, expression, and actions convey more than spoken or written words. Advisors work to bridge the cultural gap by learning the language as well as nonverbal communication cues. To bridge the gap, advisors develop a sense of cultural awareness, recognize differences, learn to adapt, develop tolerance, and maintain personal contact.

Develop a Sense of Cultural Awareness

8-33. Advisors must recognize they are a product of their own culture. They must learn as much as possible about the culture of the people with whom they need to communicate. When communicating with people across cultures, advisors abandon any sentiments of ethnocentrism—the tendency of individuals to judge all other groups according to their own group's standards, behaviors, and customs. Such notions lead an individual to see other groups as inferior by comparison.

Recognize Differences

8-34. Each culture has its own way of accomplishing required daily tasks. Advisors must understand that each particular society may approach things differently; it does not mean that they are inefficient or less valid. Differences should not be seen as negative. Respect for counterparts must be maintained at all times.

Learn to Adapt

8-35. Advisors must be flexible and ready to adapt and adjust their behavior; however, they must be careful not to overdo their adjustment. Individuals who are overly flexible are often perceived as being insincere. The successful advisor must strive to act in a way that is appropriate to the target culture. Above all, advisors should be themselves and show sincerity.

Develop Tolerance

8-36. Advisors must develop a tolerance for deviations from accepted norms. Events or activities that may seem extraordinary to newcomers may be common practice in the culture. Advisors must be aware that members of the foreign culture may be astounded by that which is commonplace in the United States. Careful observation should be made before judgments are rendered about seemingly peculiar behavior.

Maintain Personal Contact

8-37. Personal contact is the most effective way to bridge organizational barriers. Interorganizational stumbling blocks are very real and the prejudices that arise from them are exacerbated by misunderstanding and ignorance. Often it is too easy to attribute negative attitudes and hostile motives to faceless groups. The advisor must employ superior interpersonal skills and work directly and closely with individual members of other organizations. This type of contact can effectively reinforce commonality and diminish the impact of disparity. These personal relationships are the key to effective interorganizational relationships.

Culture Shock and Adaptation

8-38. The term culture shock describes the anxiety experienced by persons in a completely new environment. It is not the result of a single event. Rather, the condition develops slowly from a series of minor events or conditions. Culture shock expresses the feelings of not having a set direction, not knowing what to do or how to do things in a new environment, and not knowing what is appropriate or inappropriate.

8-39. Culture shock occurs because the mind and body have to go through a period of psychological and physiological adjustment when individuals move from a familiar environment to an unfamiliar one. The cues received by all of the senses suddenly change. During the day the foreigner is bombarded with unfamiliar sights, sounds, smells, tastes, languages, gestures, rules, requirements, interactions, demands, systems, and expectations.

8-40. Some differences that advisors experience between their lives at their home station and their lives when deployed to a foreign location are obvious. These differences include language, climate, religion, food, educational system, and the absence of family and friends. Other differences may not be as obvious. These differences include how people make decisions, spend their leisure time, resolve conflicts, express their emotions, and use their hands, faces, and bodies to express meaning.

Stages of Shock

8-41. Five distinct stages occur in the culture shock process. These include enthusiasm, withdrawal, reemergence, achievement and reentry. Figure 8-1 depicts the stages of shock.

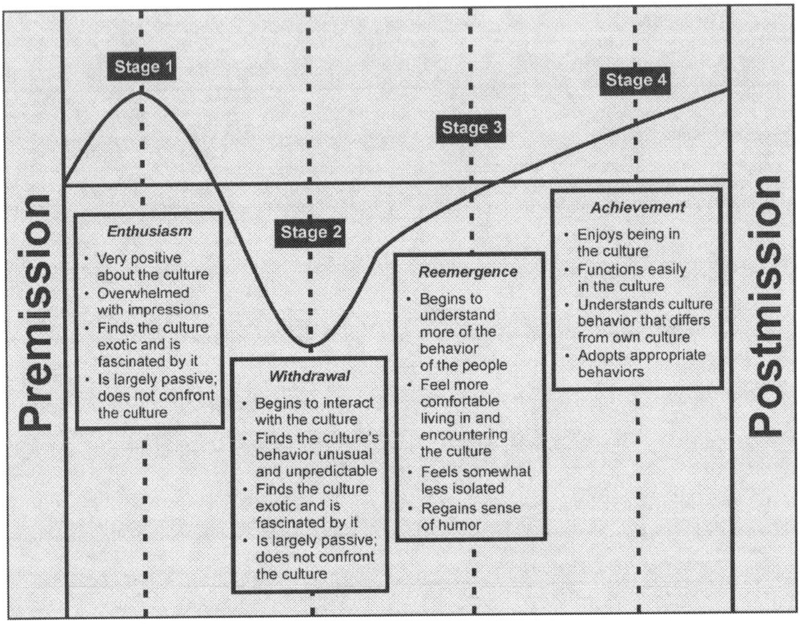

Figure 8-1. Stages of culture shock

Stage 1 – Enthusiasm

8-42. In this stage, a person may feel self-confident and pleasantly challenged. The person may be delighted by all of the new things encountered.

Stage 2 – Withdrawal

8-43. In this stage, a person begins to encounter difficulties and minor crises in daily life. It may be difficult to make oneself understood and foreign counterparts may prove more difficult to work with than anticipated. In this stage, one may have feelings of discontent, impatience, anger, sadness, and even incompetence.

Stage 3 – Reemergence

8-44. This stage is characterized by gaining some understanding of the new culture. A person may not feel as isolated, and a feeling of direction emerges. The individual is more familiar with the environment and is better able to adjust to the circumstances.

Stage 4 – Achievement

8-45. In this stage, the person realizes that the new culture has both good and bad things to offer. This stage is one of integration and the person is increasingly able to function in the new setting. A sense of accomplishment, a reduction of routine annoyances, and a more solid sense of belonging accompany this integration.

Stage 5 – Reentry

8-46. This stage occurs when people return to their own culture. They may find that things are not as they once were. Changes that occurred in their absence—their family, their friends, their communities, and themselves—combine to present a distorted image of home that differs greatly from the one imagined or remembered.

Culture Shock Symptoms

8-47. Being immersed in a society with extreme cultural differences tends to cause feelings of uncertainty and anxiety. The body and mind may react in unusual ways. Some persons may experience more pronounced physical symptoms of stress, such as chronic headaches or upset stomachs. Although uncomfortable, some degree of culture shock is a normal part of the adjustment process. Some common reactions include:

- Feeling irritable with (or even anger toward) one's own group or organization.
- Feeling isolated or alone.
- Tiring easily.
- Changing normal sleep patterns (either too much sleep or not enough).
- Suffering minor (but persistent) body pains, especially in the head, neck, back, and stomach.
- Experiencing feelings of hostility and contempt toward local people.
- Withdrawing from the local population (that is, spending excessive amounts of time alone reading or listening to music).

Overcoming Culture Shock

8-48. An important tool to overcome the obstacles of a new cultural environment is familiarity with the language. An ability to communicate in the new culture, even at the most basic level, reduces the effect and shortens the period needed for adjustment. Other means of overcoming culture shock can include having previous experience in the area (including a regular form of physical activity in personal routines) and acknowledging even slight progress in adjusting to the new culture.

8-49. TC 31-73 discusses culture shock in depth. It identifies the symptoms, cues, and treatment.

CROSS-CULTURAL COMMUNICATIONS

8-50. Communication is the transfer of messages from one person to another. These messages may be passed along verbally, in writing, or by signals (gestures, body language). The sender encodes the message and the receiver decodes it. The type and style of encoding used is based upon the sender's history, beliefs, values, opinions, and belief systems. Communication is determined by culture; as a result it can be either high context or low context. High context communication places an emphasis on clearly stating issues and problems and working towards resolving them in a straightforward manner. Low context communication places a premium on building relationships; as such, problem sets tend to be worked in a less formal matter in societies that utilize this type of communication. Quickly and effectively identifying whether a society uses high or low context communication is a key to effectively conveying material between the advisor and the host-nation population.

8-51. The receiver decodes messages based upon lifestyle, group membership, status and role, worldview, language, and social norms. Communication is a two-way process in which the encoding and decoding methods can affect both sending and receiving. Effective communication occurs when the message is perceived and responded to in the manner the sender intended. Ineffective communication can occur from poorly chosen words, flawed timing, a confused mixture of verbal and nonverbal signals, poor listening skills, and a failure to take culture into account. Communications can be broken into both levels and factors.

LEVELS OF COMMUNICATIONS

8-52. The cross-cultural communications capabilities required by an effective advisor to FSF can be described in three levels. At the lowest level is awareness followed by knowledge. The highest level is reached when these two are combined with well-trained and refined skills. TC 31-73 discusses communications in detail.

Awareness

8-53. Awareness is the basic level of cross-cultural capability. Awareness of cultural differences and their impact is the first prerequisite for successful work with a counterpart. Simply being sensitive to the fact that differences exist and carefully observing actions and reactions can assist the advisor in adjusting behavior and modifying actions to achieve greater influence with the counterpart. Awareness is not region specific and can be instilled in the advisor with relatively little training.

Knowledge

8-54. Knowledge of the details and nuances of a specific culture is the next level of cross-cultural capability. Advisors attain this level through academic study and immersion. Such knowledge is inherently area-specific and rarely transfers from one target area (or culture) to another. Developing in-depth regional knowledge necessary for effective cross-cultural communications requires extensive and time-consuming training. Appropriate personnel assignment policies must support this training.

Skills

8-55. Skills fundamental to effective cross-cultural communications, when combined with awareness and knowledge, form the highest level of cross-cultural capability. These skills include professional competence, language, nonverbal communication, negotiation, interpersonal skills, observation, problem solving, leadership, instruction, and fitness. Although some individuals show greater natural talent for these skills than others, all advisors require continual training in each to achieve and retain their full potential as advisors to foreign counterparts. These skills are discussed in more detail in TC 31-73.

FACTORS AFFECTING COMMUNICATION ACROSS CULTURES

8-56. There are four aspects that can impact communication across cultures. These factors are level of direct communication, perception of time, perception of the individual versus the group, and show of emotion.

Level of Direct Communication

8-57. Cultures fall along a spectrum of formality, with some cultures being direct and explicit in their communication and others less direct and more ambiguous. Cultures that rely on direct communication place more emphasis on verbal communications that are precise and explicit. Context and nonverbal cues are secondary to verbal communication. Cultures that rely on indirect communication rely less on verbal communication and more on context and nonverbal cues. Their verbal communication can seem vague and noncommittal to those from a culture that is more direct.

Perception of Time

8-58. Cultures can differ greatly in their perception of time. This difference divides into two cultural concepts of time: monochronic-time and polychronic-time. In monochronic-time cultures, members emphasize schedules, precise reckoning of time, and promptness. In such cultures, the schedule takes precedence over the interpersonal relation. In polychronic-time cultures, time is viewed as fluid. Members of polychronic-time societies do not observe strict schedules—agendas are subordinate to interpersonal relations.

Perception of the Individual Versus the Group

8-59. Cultures may be individualistic or collectivist in their orientation. An individualist culture is one in which the ties between individuals are loose—where people are expected to take care of themselves and their immediate families. In a collectivist culture, people are raised from birth into strong, cohesive groups. These groups offer a lifetime of protection in exchange for unquestioning loyalty. Advisors must understand from which culture their counterparts come to identify motivations accurately and communicate effectively.

Show of Emotion

8-60. Each culture has its own system of expressing emotion. Some cultures tend to express their emotions and show their feelings openly. Other cultures tend to repress those feelings more. Rather than showing their feelings openly, members of these cultures keep their emotions controlled and subdued. Those from the more expressive culture may view people from the repressive culture as cold or unfeeling. Similarly, those from the more repressive culture may view their more expressive colleagues as eccentric. To communicate effectively, advisors must be able to understand their foreign counterparts' emotional communication and display their own emotions carefully.

Chapter 9
Working with Counterparts

Rapport is a relationship marked by cooperation, conformity, harmony, or affinity. When people discuss good rapport, they describe a relationship founded on mutual trust, understanding, and respect. Relationships characterized by personal dislike, animosity, and other forms of friction often lack rapport. For the advisor, rapport describes the degree of effectiveness.

ESTABLISHMENT OF RAPPORT

9-1. Since advisors are in a unique military position, they establish rapport with their foreign counterparts. That position is one in which the advisor has no positional authority over the actions of their foreign counterparts. This lack of authority means that the doctrinal view of leadership is modified to emphasize interpersonal relationships and deemphasize authoritarian roles. Advisors instead use their people skills to influence the actions and decisions of their counterparts. Advisors use their people skills to building rapport. They have a genuine interest in other people, smile, remember and use people's names, encourage others to talk about themselves, listen to others, discuss what the other person is interested in, and make the other person feel important.

9-2. Advisors obtain certain knowledge before establishing effective rapport. First, they study FM 6-22 to gain basic leadership knowledge for understanding human nature and motivation. Advisors then incorporate information specific to the culture and society of their potential counterparts. This information may take the form of thorough area studies, operational area studies, and other research materials. Knowledge includes the components and techniques of rapport.

COMPONENTS OF RAPPORT

9-3. Three components compose rapport: understanding, respect, and trust. All the components of rapport are two-way streets; to have counterparts share about their culture, language, and experiences, advisors have to be willing and able to share also. Building this rapport may require the advisor to establish a personal level of understanding with their counterparts. Understanding is the first component of rapport. It begins before deployment and may include foreign cultural studies, language training, and equipment and doctrine familiarization. Once in country, advisors continue to broaden their understanding by observing and asking questions.

9-4. Respect is the next component of rapport. It is a reciprocal commodity. The foreign counterpart should grow to respect advisors, what they know, and how they perform. Advisors should look for those things that they respect in their counterparts. This may prove challenging. The counterpart may not fit the mold of U.S. or other coalition officers. Lacking formal training or education, the foreign counterpart may even be illiterate. The uniform standards of the host nation often may differ from the advisors' standards. Start with the basics and expand over time. Additional factors may affect the counterpart's willingness and motivation. For example, by accepting the duties, the counterparts (and their family) may be in mortal danger. The counterpart may have proven to be a fearless warrior, even without formal training. In short, traits deserving respect will exist even before the relationship matures. Mutual respect grows through shared experiences and shared dangers. Advisors should live, eat, and if authorized, fight with their counterparts. By sharing their hardships and dangers, advisors build respect.

9-5. Trust is the final critical component for building rapport. Trust grows gradually from understanding and respect. Building trust needs to begin on day one, but it will not mature until later in the relationship.

Advisors should begin by showing they are reliable and should do everything they say they will do. They should be in the right place at the right time. Two things can enhance this growth. First, the advisor should start out with confidence-building missions. Early success builds self-confidence and trust. Second, advisors should not promise any support they might not be able to deliver.

TECHNIQUES OF RAPPORT

9-6. Foreign security forces (FSF) respect advisors who actually show them the right way to do a task rather than just talk. Experience has shown that foreign units will not hesitate to go into danger if their advisors are right there with them. To further contribute to building rapport, advisors should maintain enthusiasm and a positive attitude; learn the language or how to use an interpreter; respect a counterpart's rank, age, status, and experience; develop negotiation skills; and improve interpersonal skills.

9-7. Other rapport-building techniques can include being confident, competent, and capable. Advisors never expect or demand their counterparts to do something that they are unable or unwilling to do. They demonstrate everything. Advisors learn their counterparts' names and spend time with them at meals and during holidays; they prepare mentally to interact with counterparts at all times. Advisors avoid creating an American-only enclave; they avoid giving the impression of favoring any one group. They frequently assess their counterparts' perceptions. Advisors recognize threats to discipline and enforce the chain of command.

ACCOMPLISHMENT OF RAPPORT

9-8. Advisors measure rapport by how well they can motivate their counterparts to act. The basic techniques of motivation (in the absence of authority) are advising, setting the example, seeking compromise, and coercing. TC 31-73 discusses rapport in detail.

ADVISING

9-9. Advising the counterpart to select a particular course of action is only effective if the counterpart trusts the advisor is professionally competent. If the counterpart does not perceive the proposed solution as realistic, the advisor's competence is questioned. Advisors carefully explain to their counterparts that recommended courses of action are realistic and will be effective.

SETTING THE EXAMPLE

9-10. Setting the example for the counterpart must be an ongoing effort for the advisor to avoid the appearance of a "do as I say, not as I do" attitude. In setting the example, advisors explain that what they are doing is the most effective action for the situation. This is particularly true when the counterpart really does not understand the behavior (or purpose).

SEEKING COMPROMISE

9-11. When seeking compromise with the foreign counterpart, advisors may create a situation in which the counterpart has a personal interest in successful execution. In some cultures, seeking a compromise may allow the counterpart to save face. Furthermore, in certain situations the counterpart—because of practical experience—may have a better solution to the problem at hand.

9-12. When seeking a compromise in certain cultures, the perceived competence of the advisors may suffer. This may be mitigated somewhat by approaching the compromise as two professionals (the advisor and the counterpart) reaching a mutual conclusion. To reach an effective compromise, advisors may have to conduct negotiations.

Note: Two areas of concern must never be compromised for the sake of maintaining rapport—operations security and human rights.

COERCING

9-13. Coercion is the least desirable method of motivation because it can cause irreparable damage to the relationship. Advisors should use coercion only in extreme circumstances. Advisors must avoid unintentionally forcing their counterparts into action. Instead, they establish and maintain good rapport by conveying they are sincerely interested in their counterparts, their nation, and its cause. Advisors will not belittle their counterparts' efforts. Advisors must demonstrate that they have come to help because they believe the counterparts' goals are just, fair, and deserving of success.

POLICIES AND AGREEMENTS

9-14. Advisors must remember their primary aim is to forward U.S. policy. Therefore, they establish relationships with foreign personnel, other U.S. forces, and other government agencies. Advisors promote U.S. policy through agreements, command relationships, and operations. TC 31-73 discusses policy in greater detail.

UNITED STATES AND HOST-NATION POLICIES

9-15. Advisors not only support U.S. national policy in the areas where they operate, they forward the policies of the host-nation government (unless otherwise directed by higher authority). Advisors maintain communications with higher headquarters to ensure activities align with U.S. objectives.

9-16. Host-nation policies may dictate practices and procedures that appear inefficient or uneconomical. Advisors should avoid criticizing or condemning such practices and procedures until they thoroughly understand them. Advisors should make private recommendations critical or contrary to host-nation policy.

AGREEMENTS

9-17. The status of advisors is normally specified by a status-of-forces agreement or other arrangement between the U.S. and host-nation governments. These agreements vary widely; they may offer little political protection or may provide for full diplomatic immunity. If no such agreement exists, advisors are subject to local laws and customs. Even if a status-of-forces agreement or other agreement exists that provides immunity, advisors are still expected to observe local laws.

COMMAND RELATIONSHIPS

9-18. Advisors must use the U.S. chain of command to obtain and disseminate guidance and assistance. They carefully distinguish between the U.S. and foreign security forces' chain of command. In particular, advisors prevent their counterparts from attempting to control subordinates through the U.S. chain of command. Advisors provide recommendations—not orders—to their counterparts. Only the counterpart should issue orders to subordinates.

9-19. Nongovernmental organizations and other government agencies may be co-located in the operational area. Advisors integrate the efforts of all organizations into their mission planning and should seek opportunities for integrated efforts with their counterparts. This type of unified action at the tactical level can be achieved through the planning process to integrate nonlethal actions. Holistically, nonlethal actions include development and diplomatic efforts.

CONDUCT OF OPERATIONS

9-20. Advisors can expect to participate in stability operations. These operations use military and paramilitary forces, as part of the interdepartmental team, to support internal defense and development operations. U.S. assistance may include advice on military organization, training, operations, doctrine, and materiel. In addition, the United States may provide and control U.S. logistics and sustainment for FSF. The objective of this assistance is to increase the capability of foreign organizations to perform their missions and operate efficiently in the given operational environment.

LEVELS OF ADVISORS

9-21. Organizations and individuals possessing greater skills and resources assist by imparting their knowledge through assistance efforts. The success of assistance depends largely on effective interaction between advisors and their foreign counterparts. Advisors may operate in the national, operational, and tactical levels. The national level is considered the largest national subdivision. This may be a nation's combined military headquarters, Service components, national-level government officials, or a combination. Operational-level advisors work at a level below the national-level advisor. This may be with military corps or division officers, with regional government officials, or a combination. Depending upon the size of the nation and its military, operational-level advisors may be assigned as low as brigades. Tactical-level advisors normally advise military units at or below the brigade level and provide counsel and assistance across the spectrum of conflict. They advise and assist counterparts in developing unit combat effectiveness and serve as liaisons between FSF and U.S. combat, support, sustainment and protection forces. Chapter 3 of TC 31-73 discusses these levels in detail.

CONSIDERATIONS OF RAPPORT

9-22. When building and maintaining rapport, advisors consider intelligence, human rights, role shock, ambiguity, personal relationships, bureaucracy, responsiveness, and diversity. These considerations are discussed in greater detail in TC 31-73.

INTELLIGENCE

9-23. To promote national policy and attain overall objectives, advisors may need to plan and conduct intelligence training. The advisor should assist the counterpart in developing a local intelligence collection program, training intelligence personnel in their respective specialties, and utilizing trained intelligence personnel properly.

9-24. The advisor should be familiar with area studies and area assessments. Pertinent documents should be compared to detect trends or changes. The advisor also should evaluate the foreign intelligence staff, its standing operating procedures, its chain of command, intelligence projects, and reference material available from other intelligence agencies. Advisors should prepare and maintain a list of essential elements of information.

HUMAN RIGHTS

9-25. Advisors stress the consequences of mistreating suspects, prisoners, or other persons taken into custody. These persons must be treated per Article 3 of the Third Geneva Convention. Article 3 requires care for the sick and wounded, protection of prisoners and detainees of all types from abuse or other harm. Murder, mutilation, and torture are expressly forbidden, as is humiliating or degrading treatment. Advisors must never be active participants in the conduct of such punishment. Furthermore, they must explain to their counterparts that they are obliged to report any atrocities of which they have knowledge.

ROLE SHOCK

9-26. Role shock results from the discrepancy between the roles individuals expect to play and the roles they actually do play. Role shock may also be a product of the tension created by individuals trying to do jobs themselves versus advising others how to do the jobs. Role shock tends to increase until about the anticipated midpoint of the tour and seldom disappears completely until redeployment. Symptoms of role shock resemble those produced by other stressful life events.

9-27. Many advisors become entangled in complexities of bureaucracy—both foreign and American—and become increasingly frustrated when seemingly familiar things fail to respond in expected ways. This continuing frustration can be a key contributor to role shock.

9-28. Although most advisors undergo some degree of role shock, over time they learn to take the personal, living, and social conditions in stride. More challenging are the problems that arise in connection with their jobs—their professional roles, relationships, personal achievement, self-development, self-determination, and self-image. Often advisors find their duties and activities during deployments differ somewhat from those expected. The actual duties and responsibilities are greater in scope, involve technical work outside of their assigned specialties, and require honed administrative (rather than military) skills.

AMBIGUITY

9-29. Most advisory assignments are ambiguous. What an advisor is expected to achieve is not always made clear. Furthermore, assignments may conflict with what they have trained to do, what others expect of them, or what they want or expect to do.

9-30. U.S. and foreign security forces may spend months—even years—negotiating a project. Still, the overall objectives or definition of the mission to be accomplished might be expressed in general terms. Various interpretations of what is to be accomplished arise. Even more views surface regarding the most appropriate means of reaching those goals.

RELATIONSHIPS

9-31. Ideally, foreign counterparts provide a working context for advisors and help them to understand and adjust to the new culture. Historically, those with good working relationships with their counterparts make better adjustments than those who have no counterparts or bad relations with their counterpart. Good working relationships may be difficult to establish and maintain. Most advisors approach their assignments as teachers and mentors. Foreign counterparts may resist such a relationship, particularly if they were expecting someone to take over and do all or part of the job. In addition, many counterparts want their advisors to share in at least some of the blame if something goes wrong.

9-32. The United States has little control over the availability or capability of counterparts. The assigned individual may be technically incompetent, lack professional commitment, have other interests, or be unavailable when needed. In some cases, numerous counterparts are assigned and rotate in their workings with the advisor. A single counterpart may not be assigned until the advisor has been on the scene for some time. In some scenarios, a specific counterpart is never assigned.

BUREAUCRACY

9-33. Although some Americans might disagree, the United States maintains a system that is—by world standards—relatively permissive and nonbureaucratic. As such, advisors may become aggravated when confronted with or obstructed by host-nation bureaucracy, red tape, administrative centralization, and a cumbersome decisionmaking process. Cultures that emphasize consensus in the decisionmaking process often prove quite frustrating for advisors. To be successful, advisors strive to understand the bureaucracy in which they operate.

9-34. Advisors should recognize the value of being or knowing the middleman. This role is sometimes one of the key contributions an outsider can make. Middlemen can move in, around, and out of organizations somewhat independently of the local hierarchy, protocol, and customs. They can link persons and offices that might otherwise have difficulty communicating. Many advisors dislike this role and refuse to practice it; however, the middleman can be a significant catalyst in any advisory operation.

RESPONSIVENESS

9-35. As frustrating as foreign bureaucracies may be, advisors often get even more irritated with the responsiveness of U.S. forces. Advisors frequently fail to account for distances, lead times, diplomatic issues, and other challenges when conducting security force assistance. Moreover, few Soldiers in the unit conducting security force assistance have had substantive experience in working in an interagency environment. These issues are further complicated by the usual tensions between headquarters and field staffs and between administrators and technical specialists. Expectations, values, and styles of behavior vary from group to group.

9-36. Deploying personnel may not be familiar with the tasks of development and the vast range of problems typical of underdeveloped areas. They come in eager to contribute their knowledge and skill, and anxious to make maximum progress in the limited span of their assignment. In so doing, they may fail to recognize that the factors that frustrate them—such as the lack of professional skills and facilities—are actually their reason for being there.

Diversity

9-37. Organizations attract individuals who differ in abilities, knowledge, skills, work ethics, social behaviors, values, and attitudes. Advisors must be able to recognize personality types and develop techniques to cooperate effectively with each type.

RECOMMENDED PRACTICES IN BUILDING RAPPORT

9-38. Advisors may adopt many practices to help work with counterparts. Practices that help advisors build rapport are discussed in paragraphs 9-39 through 9-51.

Ensure That the Advisor Presence Is Understood

9-39. Advisors entering an operational area work under the sponsorship of a military commander, mayor, village head, or some other recognized local leader. Prior to this, the advisor's arrival may need to be coordinated through the district, sector, or other appropriate local office. Clearances from distant national, state, province, or sector governments cannot compensate for a clear, local explanation of why an American advisor is in the area. This particularly applies to small, isolated communities where it is unusual for a stranger to appear for even an hour without being acknowledged and accepted by local leaders. Without explanations from locally respected persons, civilians often arrive at their own explanations, often to the detriment of the advisor effort.

Find a Basis For Common Interest with the Local People

9-40. If advisors show appreciation for civilians as individuals, cultural gaps and language barriers can be overcome and common ground can usually be found. Advisors should listen and show interest when the locals speak. Initial conversations usually center on universal matters, such as food, shelter, clothing, health, and education. Over time, discussions may naturally move to the matter the advisor wants them to consider. Advisors are better received if they know something about earlier contributions in such matters as agriculture, folk art, religion, and architecture. Naturally, advisors prove more effective and appreciated if they can speak the language.

Try To Understand Why Things Are Done the Way They Are

9-41. Although some local practices may seem strange, they generally have good reasons behind them. Advisors can discover these reasons with careful observation and a creative imagination. Food habits, family traditions, folk cures, and festive celebrations almost always have a great deal of human experience at their root. The advisor also needs to be aware that many villages contain rival subgroups and factions. This tension needs to be accounted for when working with the local people.

Start with Where the People Are And What They Want

9-42. Advisors find out what the local people really want and work with them to achieve this aim. The need felt by the local people often is the best starting point. Once advisors address this aim, people become more appreciative and cooperative—they begin to raise their expectations and become interested in working for other improvements. To address the initial desires of the people, the advisor may need to call in personnel with the specialized skills needed for the particular project. Although this can result in a delay, it will help the advisor to achieve better cooperation.

Work Within the Local, Cultural Framework

9-43. Advisors understand basic cultural matters such as ethnic backgrounds of the people, family relationships, leadership patterns, and value systems. They also need some knowledge of local services such as health, education, communications, and transportation.

Help People Believe They Can Improve Their Situation

9-44. Most traditional peoples of South America, Asia, and Africa live in a largely static environment. Through their experience with change, they are more fearful of losing status than they are hopeful of bettering their condition. Therefore, changes suggested by advisors are often viewed with fear. Concrete local projects that yield easily observed benefits help convince villagers that they can improve their situation and make them more willing to cooperate in other projects.

Be Content with Small Beginnings

9-45. Change tends to come slowly in areas where there have been few in recent times. Advisors remember that knowledge—technical or otherwise—is cumulative. Once a small beginning has been made, greater activity and additional changes soon follow. It is easier to achieve momentum than to maintain it. Regardless, advisors start within as promising a framework as possible and with the support needed to sustain the momentum achieved.

Utilize Local Organizations and Recognize Their Leaders

9-46. The surest way for an activity to be continued after the advisor leaves is for it to have been launched and carried forward within the local organizational and leadership framework. People respond best when their local organizations are recognized as important and useful. Advisors consult and encourage recognized local leaders to contribute as they can. A well-conceived technical activity reflects credit on the local leaders associated with it. Advisors give attention not only to officials and family heads of local groups, but also to the quiet, behind-the-scenes leaders.

Help the Government Get Organized To Serve The People

9-47. For advisors to be most effective, they must understand the organization of the local government and how their activities fit into the overall scheme. Usually a set of agreements exists between various local agencies and the national government (usually through some sort of interministerial council). This agreement provides for a coordinated effort in servicing the varied needs of civilians. The advisor may need to work with appropriate agencies to assist in getting such agreements made. If such agreements already exist, advisors should be careful to recognize and strengthen them.

Train and Use Subprofessional, Multipurpose Local Workers

9-48. Selected young people in the villages can be trained and used as subprofessional, multipurpose village workers. Otherwise, advisors' influence is restricted to where they are standing and the immediate vicinity. Furthermore, advisors may spend so much time establishing and maintaining enough rapport with the villagers that they may become incapable of rendering any real service at all. Often the gap between civilians and the advisor is formidable because of their educational and cultural differences. Often the advisor works with villagers who are poor, illiterate, and devoid of outside contacts. Volunteer or paid local workers have proven quite helpful as they serve as a liaison between villagers and advisors.

Expect Slow Progress

9-49. As civilians begin to see successes from their joint efforts and begin to have new hope, they naturally want a larger hand in their own matters. The advisor may sometimes feel they want to assume more responsibility than they can carry. These evidences of growing pains should be appreciated, for they are a necessary part of becoming able to assume responsibility. The advisor needs to adjust to the growing desires of the people to help themselves.

Transfer Controls Constructively

9-50. Advisors need to help civilians see how they can build the new—what they want—upon the foundations of the old—what they already have. From the beginning of a project, the advisor's team needs to talk with local leaders. Together they envision how to train local personnel, gather financial support, and transfer responsibilities before transferring control of activities. If operating responsibility is transferred too early, some breakage—usually of material things—will likely occur. If the team keeps control too long, the civilians who wanted to take over may grow disillusioned with them for not relinquishing control when they thought it should have been. Working with local leaders, the team determines when to phase out each technical activity.

Expect Little Gratitude from Those Helped

9-51. People who benefit from assistance sometimes feel defensive. In accepting assistance they are, in a sense, admitting their own insufficiency. The self-esteem of a person, community, or a nation as a whole is delicate. The advisor's team should not, therefore, expect thanks. Rather, the team should approach the people in a spirit of fraternity and humility, taking satisfaction in their forward progress. The team should do its job the best it can and accept work well done as is its own reward.

DYNAMICS OF ADVISING FOREIGN SECURITY FORCES

9-52. Advisors work in a complex environment that is heavily influenced by the unified effort of multiple organizations, relationships, compartmentalization, and organizational cultures. Paragraphs 9-53 through 9-57 discusses the probable dynamics that an advisor must take into account.

Advisors and Interdependent Operations

9-53. Advisors commonly operate in complex environments with interdependent operations. Interdependent operations require near-seamless integration of agencies representing the instruments of national power—diplomatic, informational, military, and economic. This same degree of integration must be extended to include external organizations and those agencies representing the instruments of power in partner states. The negotiation and cross-cultural skills that enhance their performance as advisors are also well suited to the complex interrelationships that characterize such operations.

Organizational Relationships

9-54. Often organizational relationships can often be misleading and must be clarified. The actual interrelationships among and within organizations seldom follow a line-and-block diagram. Instead, they are heavily influenced by circumstances, personalities, perceptions, and resources. All relationships and lines of authority are subject to negotiation. Advisors understand that clarifying the roles, functions, and responsibilities is a continuous process and that each statement or action sets a precedent for future interrelationships.

9-55. The advisor must become hypersensitive to the most minor organizational disparities. The apparent similarities—the common language, sociological backgrounds, and citizenship—can be disarming and lead the advisor to overlook very real differences. If ignored, these differences can quickly impede synchronization.

Compartmentalization

9-56. Advisors balance security and synchronization. Overcompartmentalizing must be avoided. Often distrust can be prevented simply by not being overly secretive. Although advisors must conduct a careful, continual risk assessment to avoid compromise, the sharing of critical information is necessary for harmonious operations.

ORGANIZATIONAL CULTURE

9-57. Advisors should approach external organizations as they would foreign counterparts—as unique institutions with distinctive cultures. Organizational differences, biases, and approaches must be researched and—whenever possible—accommodated. Each Service in the Department of Defense has unique perspectives that lead to significant differences in objectives and approaches to various problems. Profound differences in perspective may even be found between the different branches of the Army. If not recognized and accommodated, these characteristics may create distrust, disharmony, and disruption during critical operations.

Chapter 10

Cross-Cultural Influencing and Negotiating

Leveraging foreign counterpart cooperation can prove a difficult and critical task. Although coercion may prompt surrogate forces to perform a specific action, it can never lead to established conduct and legitimacy. Lasting advisors can achieve long-term success only by patiently applying their influence.

Advisors analyze their counterparts and determine their motivations and needs. Basic needs, culture, and individual personalities influence people. Advisors determine what these three things are so the appropriate influence techniques can be applied. Then advisors negotiate. Negotiation is fundamental to problem solving using the study, development, and honing of negotiation skills critical to the advisor.

INFLUENCING

10-1. Advisors need to gain influence with their counterparts, leaders, civilians, coalition members, and others beyond their chain of command. Advisors are self-aware, nonjudgmental, and capable. It is not necessary to accept others' beliefs or values, but advisors need to understand others' ways of thinking and operating. Knowing their own culture and understanding why they believe and value what they do helps advisors find a common cause. A common cause unites a diverse set of decisionmakers with whom advisors interact and try to influence.

10-2. Once advisors gain a minimum amount of influence, they can make things happen through others. They can extend their influence beyond their chain of command. Influence is simply getting others to do what you want.

10-3. Advisors identify the types and principles of influencing. They determine the best techniques with which to influence and review vital considerations. Using measures of effectiveness, they check that their influences have the desired results.

TYPES

10-4. Commitment, compliance, and resistance clarify how advisors understand the impact of their attempts to influence their counterparts. These three types of influencing are determined by how committed the foreign security forces (FSF) are to the task.

Commitment

10-5. Commitment is a willing dedication to a requirement or a cause because one wants to. Commitment-focused influence generally produces longer lasting and broader effects. Whereas compliance only changes immediate behavior, commitment reaches deeper—changing attitudes, beliefs, and behavior. For example, when advisors build desire and responsibility for a task among FSF, they often demonstrate more initiative, personal involvement, and creativity.

Compliance

10-6. Compliance is conforming to a requirement or demand because they feel like they have to avoid punishment or to gain a desired reward. Compliance-focused influence is based primarily on the leader's authority. Giving a direct order to a follower is one approach to obtain compliance during a task.

Compliance is appropriate for short-term, immediate requirements and for situations where little risk can be tolerated. Compliance techniques are also appropriate for leaders to use with others who are relatively unfamiliar with their tasks or unwilling or unable to commit fully to the request. If something needs to be done with little time for delay, and there is not a great need for the target of influence to understand why the request is made, then compliance is an acceptable approach. Compliance-focused influence is not particularly effective when a leader's greatest aim is to create initiative and high esteem with a counterpart.

Resistance

10-7. Resistance is a reaction against change. Resistance generally takes one of four forms—reactance, distrust, scrutiny, and inertia. It is manifested behaviorally (I won't do it), affectively (I don't like it), and cognitively (I don't believe it). Recognizing and lowering potential sources of resistance are key components to gaining compliance and commitment.

MOTIVATIONS

10-8. Different motivations influence behavior. One type of motivation is the desire for something. Another is the fear of something undesirable. As an agent of influence, advisors determine how to increase approach motivations and how to decrease avoidance motivations. For instance, recruiting for a special mission can use special badges as incentive (approach). Advisors can assure Soldiers that the mission is relatively safe by providing extra training and better equipment (lowering risks or avoidance).

PRINCIPLES

10-9. Six influencing principles are relevant to virtually any culture and any group. Understanding and applying these principles can increase an advisor's effectiveness in persuading a group or individual:

- Principle of scarcity.
- Principle of authority.
- Principle of social proof.
- Principle of liking.
- Principle of reciprocity.
- Principle of consistency.

10-10. With the principle of scarcity, people value more of what they can have less. They typically associate greater value with things that are rare, dwindling in availability, or difficult to acquire.

10-11. In the principle of authority, people are more easily persuaded by individuals perceived to be legitimate authorities or experts. They defer to experts who provide shortcuts to decisions requiring specialized information.

10-12. With the principle of social proof, people often look to the behavior of those around them for direction about what choices to make. This action is heightened when those around them are similar in terms of age, education, social standing, and experience.

10-13. In the principle of liking, people prefer to say yes and to comply with the requests of those they like.

10-14. In the principle of reciprocity, if someone grants favors, invited or uninvited, an overpowering need to repay that favor immediately blooms within the recipient of that favor. This human trait transcends all cultures and races.

10-15. With the principle of consistency, the desire for consistency motivates behavior. The drive to look and be consistent is a highly potent weapon of social influence, often causing people to act in ways clearly contrary to their own best interests. People do not like to appear inconsistent to others. It is human nature that people strive to feel good about themselves, which includes behaving per their important values and beliefs. When the behavior is consistent with whom people are and what they value, they feel good. People align with their clear commitments.

TECHNIQUES

10-16. The art of influencing involves knowing what techniques to use based on the situation. Advisors have to know the limits to which can use their influence; their influence is finite. They need to know when to use their influencing skills to achieve the highest payoffs. Their counterparts will only accept so much influencing if they don't completely trust their advisors. Table 10-1 lists positive and negative techniques delineated in FM 6-22. Table 10-2 lists resistance-based influence techniques. These are used when the counterpart resists the advisor's influences.

Table 10-1. Influencing and rapport techniques

Technique	Description	Characteristics	Outcomes
Pressure	Explicit demands, persistent reminders, sometimes appropriate in high stakes, urgent situations	Frequent checking Nagging	Gains compliance Frequent use may cause resentment, undermine morale, and harm relationship
Legitimate Request	Source of authority is used	Source credibility is important	Gains compliance
Exchange	Quid pro quo Reward offered for task completion; punishment threatened for incomplete task	Offer of reward or punishment not tied directly to getting job done, just something that counterpart wants or doesn't want	Gains compliance
Personal Appeal	Request based on friendship or loyalty	Agent of influence highlights talents of counterpart that make target the right person for the job (this is why you are the right person for the job, will you do it for me?)	Gains commitment Builds relationship
Collaboration	Works with target of influence by providing resources of some type, or by removing barriers to task completion	Offer of assistance is directly tied to request	Gains commitment Builds relationship
Rational Persuasion	Using facts, evidence, or logic in a persuasive message	Relies on agent being viewed as expert	Gains commitment Builds relationship
Apprising	Explaining the consequences of a potential action or decision	Decision is good for unit, career, or mission accomplishment	Gains commitment Builds relationship
Inspiration	Arousing strong emotion to build conviction or commitment	Appealing to values of the counterpart	Gains commitment Builds relationship
Participation	Requests for counterpart to be a part of planning, active participation in development of solution	Promotes commitment Possibly more feasible and culturally appropriate course of action	Gains commitment Builds relationship
Relationship Building	Build positive rapport and trust	Provide the foundation for using influence or inspiring	Gains commitment

Table 10-2. Resistance-based techniques

Technique	Description	Characteristics	Outcomes
Sidestep Resistance	Agent defines own role as consultant or advisor, but not commander Pushes the consequence of choice into future Depersonalize task to gain commitment to general idea Focus on end state instead of expended effort	Does not sell, just advises to prevent immediate resistance Puts less focus on immediate impact, more focus on long-term effect Stating "leaders should" instead of "you should" Stress mission accomplishment	Gains compliance Potentially gains commitment if paired with other influence techniques like rational persuasion
Address Resistance Directly	Address reluctance by lowering costs associated with tasks	Provide some of own resources through collaboration or counterargue concerns	Builds commitment
Address Resistance Indirectly	Build confidence, esteem, and self-efficacy for the task	Acknowledge difficulties while showing support for counterpart's abilities	Builds commitment
Distract Resistance	Interfere with the development of counterarguments	Distracting attention	Gains compliance
Disrupt Resistance	Disrupt complacency to bring attention to message	Shake up to motivate	Gains compliance
Consume Resistance	Resistance is energy	Wear them down prior to opportunities to resist. Persist with requests until they concede	Gains compliance
Use Resistance to Promote Change	Reverse psychology	Issuing a challenge	Gains compliance; risky when you don't know target of influence really well

10-17. To succeed with any influence technique, the influence attempt should be perceived as authentic and sincere. Positive influence comes from leaders who do what is right for Army forces, the mission, the team, and each individual Soldier. Negative influence—real and perceived—emanates from leaders who focus more on personal gain and lack self-awareness. False perception may trigger unintended side effects such as resentment of the leader and the deterioration of unit cohesion. With each and every influence attempt, Army leaders should ensure that the influence objective aligns with Army Values, ethics, the Uniform Code of Military Justice, the Warrior Ethos, and the Civilian Creed.

Considerations when Choosing Techniques for Influencing

10-18. Advisors clearly understand what their counterparts value and fear. This information allows advisors to quickly relate the objective of the influence to something that their counterparts treasure. Also, a strong sense of self supports the advisor's recognition of which techniques they can successfully execute. Someone with an intimidating presence may achieve compliance with pressure more quickly than someone without that presence. Understanding the context of the request is important. For instance, using participation in the midst of flying bullets would not make sense. However, a long-term project that impacts many people might be a worthy situation for participation. Advisors also consider—

- The motivation of their counterparts.
- The relationship they have with their counterparts.
- What they do or not have to offer.
- An understanding of the second- and third-order effects their recommendations will have.
- What has worked in the past.
- Possession of an appropriate power based for use of a technique in particular context.
- Consistency of technique with prevailing social norms and role expectations.
- Level of resistance encountered or anticipated.
- Costs of using a technique in relation to likely benefits.

10-19. The cost of a technique refers to the amount of effort, favor, or resources that accompany the technique. A personal appeal that relies on friendship is quite effective; however, counterparts might expect the favor to be returned. If resources or time are plentiful, then the exchange might not seem expensive, but if they are not plentiful, then advisors must choose a different technique. Advisors balance the costs and benefits of each technique.

Moderating Factors

10-20. Sometimes moderating factors diminish the impact of any influence technique. These factors may include cultural misunderstanding, level of trust, or constraints on the counterpart like threats or corruption. Advisors analyze and learn about these factors. Advisors must be flexible when using influencing techniques. When one technique doesn't work, switching strategies or adding techniques may be appropriate.

CONSIDERATIONS

10-21. Seven considerations need to be addressed to influence behavior. These include determining the goal; who to influence; motives; beliefs, values, and attitudes; cultural beliefs, values, and attitudes; and susceptibility. These techniques are designed to help advisors achieve their goals. Just knowing that the counterpart wants something is not enough. Whether it is time off, new weapons, or money, advisors have a limited supply of stuff they can promise; they can only bend so much on rules, regulations, and laws. It is at this point that the advisor enters the realm of negotiations.

Determine the Goal

10-22. Advisors influence others to achieve some purpose. To be successful at exerting influence, advisors have an end state or goal in mind. Sometimes the goal will be very specific. Many goals are less distinct and measurable but are still valid and meaningful.

Determine Who to Influence

10-23. Advisors develop a mental list of those whom they might need to influence. Getting a FSF platoon to take action may be as simple as influencing the commander, but the advisor may have to influence the soldiers and commander of the platoon. When practical, advisors consider many variations of whom to influence. They look for circumstances in the immediate environment that significantly affect their counterparts. Individual or group characteristics to consider can include—

- Gender.
- Religion.
- Age.
- Socioeconomic status.
- Ethnicity.
- Political affiliation.

- Level of education (important in determining how to access and persuade).
- Occupation.

10-24. The challenge is to determine which characteristics impact emotional or behavioral responses and under what circumstances. Ethnicity may significantly affect behavior and emotions regarding voting and politics but may have almost no effect on the decision to enlist in the military. Individual and group characteristics may provide additional insight on motivations.

Determine and Prioritze Motives

10-25. Motives are anything that is required or desired (needs and wants). Motives are the reasons behind the decisions people make. Motives come from an inner desire to meet a need or want. Advisors use the knowledge of what motivates others to influence them. Knowing one's counterpart and others who may be influenced gives advisors insight into influencing their behavior. The desires to fulfill, alleviate, or eliminate a need or want provides the motivation to change behavior.

Basic Motives

10-26. Basic motives involve physiological needs—food, water, and air—and safety needs—security and shelter. When people feel hunger, they are motivated to eat; when people feel pain, they are motivated to reduce the source of the pain. Such basic motives are extremely powerful in driving behavior and overwhelm psychological needs and wants.

Social Motives

10-27. Social motives associate access to other tangibles (money, goods, education, infrastructure, and healthcare) or more complex psychological motives with basic motives. For example, people learn to want money because it can be exchanged for food and other desired goods. Power, achievement, reassurance, escape, justice, and acceptance illustrate types of learned social motives. Examples include:

- Wanting better educational opportunities for their children.
- Wanting better paying jobs.
- Wanting their interests represented by the government.
- Wanting revenge for perceived wrongs.
- Wanting self rule.

Prioritize Motives

10-28. Advisors prioritize motives by immediacy of the need or want. Critical motives are immediate needs. Short-term motives are currently being satisfied or active efforts are being made to satisfy them in the near future. Long-term motives are desired but are not immediately important; satisfaction may be delayed until some point in the future.

Determine Beliefs, Values, and Attitudes

10-29. Although often difficult to derive, beliefs, values, and attitudes can prove quite effective in persuasion. Beliefs are what is thought to be true. People do not believe in something if they think it is not true. People vary from being gullible to skeptical. Values prioritize what one believes. Values can be ethical, doctrinal, social, aesthetic, or economic in nature. Attitudes reflect whether one likes or dislikes something. Likes and dislikes can often be traced back to a person's beliefs and values. This includes loves, hates, frustrations, and fears. Frustrations occur when a want is not met. By examining causes of frustration, the advisor may find another motive to address.

Compare Beliefs, Values, and Attitudes

10-30. The predominant cultural beliefs, values, and attitudes; traditions; and norms may differ greatly from the group the advisor is trying to influence. Many Americans act differently in different cultures. This change in behavior relates directly to their predominant culture.

Determine Susceptibility

10-31. Susceptibility is the likelihood that the individual or group will be open to persuasion. Determining susceptibility aids in prioritizing efforts. If susceptibility is moderate to high, then advisors may not need to focus as much effort as they would if the susceptibility is low. Advisors determine susceptibility by assessing the strength of the underlying source of the motivation and the perception of the desired corresponding behavior. The strength of the motivation depends on the perceived risk and the perceived reward, while the perception of the behavior directly links directly to how consistent the behavior is with the individual or groups' values and beliefs.

Perceived Risks and Rewards

10-32. Risks may include the threat of physical harm, financial loss, or disapproval of peers. Rewards may include financial gain, political power, approval of peers, or increased self-esteem. Risks and rewards can vary from immediate- to long-term. Immediate risks and rewards are stronger than long-term risks and rewards. For example, few people will walk a tightrope (immediate risk), but many people smoke (long-term risk). The higher the perceived rewards and the lower the perceived risks, the more likely the behavior will occur.

Consistency with Values and Beliefs

10-33. Advisors determine if the desired behavior is consistent with values and beliefs. Generally, behavior is consistent with the relative moral or ethical, religious, political, and cultural beliefs, values, and norms. Behavior that is inconsistent or incongruent with values and beliefs often decreases over time. Alternately, if behavior continues to be incongruent with values and beliefs, then individuals are likely to justify this behavior by modifying their values and beliefs. The more consistent a desired behavior change is with the individual's values and beliefs, the more likely the behavior will occur.

10-34. The lower the susceptibility, the less likely influencing behavior will be successful, which does not mean that the person cannot still be influenced. The lower the susceptibility, the more conditions will need to be modified for behavior change to occur.

APPLICATION

10-35. Influencing is a dynamic process. Advisors make a plan before attempting to influence. If one technique fails, they have to be able to shift to other techniques. Advisors do not need to sit down and write a plan before attempting to influence people. However, even with apparently positive results, advisors review their tactics and techniques.

CHECKS

10-36. Advisors check measures of effectiveness, unintended consequences, and second- and third-order effects. Advisors assess if their influencing is achieving their immediate measures of effectiveness. However, with every action some unintended consequences may occur or the activity may not prepare counterparts for more complicated tasks later on.

CROSS-CULTURAL NEGOTIATING

10-37. The principal form of negotiation is likely to be cross-cultural negotiation. Cross-cultural negotiation brings with it a series of unique challenges and techniques. Culture fundamentally affects language and behavior. It also significantly impacts the way people handle conflict.

10-38. Negotiations among members of the same culture can be stressful; negotiating with members of other cultures can be exceedingly difficult. Logically, working with other cultures is a basic skill for the Soldier and an absolute requirement when acting in an advisory role. Understanding the components, perspectives, roles, and outcomes will help minimize the difficulty of cross-cultural negotiation. Chapter 4 of TC 31-73 discusses cross-cultural negotiations.

Influences on Negotiations

10-39. Cultural differences may lead to a conflict between what the parties expect and what their families and communities expect. This has an obvious impact on negotiating behavior. Negotiations are influenced by perceptions, individualism, punctuality and pace, relationship building, and language.

Perceptions

10-40. Most Americans have certain preconceived ideas about people from other cultures. These perceptions may not be factually based, but they exist and influence negotiations. Similarly, foreign negotiators have certain perceptions about Americans. Again, these perceptions may not be based on fact, but they exist. Advisors need to understand how other cultures perceive Americans to appropriately adjust their negotiating style. Successful advisors find ways to capitalize on the positive preconceptions of their counterparts and find ways to neutralize the negative preconceptions.

Individualism

10-41. The United States encourages a largely individualistic culture. As such, the typical American negotiator prefers to enter into negotiations alone or, if required, as part of a very small negotiating team (two or three people, at most). This behavior is largely a by-product of U.S. culture, which focuses on the individual, performance, initiative, and accomplishments. Most other cultures of the world place less emphasis on the individual and more on the group.

Punctuality and Pace

10-42. Americans place a great emphasis on punctuality, viewing it as an indication of an individual's basic ability and commitment. However, other cultures emphasize punctuality differently. Some emphasize punctuality more than Americans. In other cultures, time is considered relative or may not be important at all.

Relationship Building

10-43. Americans are very competitive during negotiations and are inclined to stress short-term results. Interpersonal relationships are of minimal importance to American negotiators. Building long-term relationships only occurs after the successful completion of negotiations.

Language

10-44. U.S. negotiators must be considerate of others' with respect to language. Even if the counterpart is fluent in English, they may expect negotiators to speak their language or use an interpreter. Members of some cultures do not like to speak English as they are very proud of their language. When U.S. negotiator counterparts do not speak English, they will need an interpreter. Counterparts may also expect American negotiators to present detailed written proposals in the local language. Many cultures appreciate foreigners who take the trouble to learn a few simple courtesy words. However, some prefer that foreigners who are not fluent refrain from using more than a few, basic courtesy words.

Elements for Conducting Cross-Cultural Negotiations

10-45. The basic elements common to all negotiations must be applied in ways that allow for the differing cultures of the participants. Elements that must be considered include opening strategies, directness, strengthening behaviors, movement, power, face-saving, formal agreements, and mediation.

Components of Negotiating

10-46. Two opposite types of negotiating exist: distributive and integrative. They reflect a philosophy towards negotiating in general. Distributive negotiating is win-lose while integrative is win-win. Distributive negotiating involves dividing a fixed amount of a thing to the parties involved. This negotiation normally involves people who have never had a relationship before. Integrative negotiating involves joining two parties to achieve something together. Successful negotiations can involve a mix of both strategies.

10-47. When using distributive negotiating, advisors know the point at which counterparts will walk away. Advisors push them almost to that point by manipulation and withholding information. The distributive negotiator believes—

- The disputants are adversaries and the goal is victory.
- Fixed resources are divided so one side receives more than the other after making concessions.
- One person's interests or position opposes the other's interests or position.
- The dominant concern is usually maximizing one's own interests.

10-48. Integrative strategies used by advisors include cooperation, sharing information, and mutual problem solving. The integrative negotiator believes—

- The disputants are joint problem solvers seeking a wise decision.
- An understanding of needs will allow resources to be divided and both sides can win.
- The relationship between the parties is what is critical.
- The dominant concern is to maximize joint outcomes.

PERSPECTIVES AFFECTING NEGOTIATIONS

10-49. Four perspectives impact negotiations: time, environment, self, and nonverbal communication. Advisors consider how each party views these conditions when negotiating.

Time

10-50. Some cultures and individuals see time as a micro view. It is all about here and now. Other cultures and individuals see time in a macro view. Now is a point that bridges the past with the future. For example, imagine a person with the first scheduled event has a deadline to make an agreement now. The next scheduled event is 1500 so that person has two hours to come to an agreement. The second person might agree that they can come to an agreement now; but to the second person, now could be within the next month or two.

Environment

10-51. Some cultures believe they can impact, if not control, their environment. If something should go wrong, somebody is to blame. Every accident is preventable. Other cultures believe the environment impacts the person. Accidents just happen and no one could have done anything about it. One side may be quite ready to commit to something in the future *knowing* that nothing stands in the way, while the other side may be reluctant to talk in certainties of the future *knowing* that factors beyond their control may prevent them from fulfilling their commitments.

Self

10-52. In some cultures, when a person makes a mistake, blame can be isolated to that person. In other cultures, the individual is seen as an element of the group. So a mistake is made not by the individual but rather the whole group. In negotiations, one might accept a personal slight to get the goal achieved, whereas the counterpart represents the group and cannot allow shame to fall on them even though one might accept it personally.

Nonverbal Communication

10-53. Since advisors will often have to work through interpreters, nonverbal communication becomes more important. Some cultures respect people who mean just what they say and maintain their composure. Other cultures respect the eloquence in which something is said or the emotion used to convey it. Where words may be translated improperly, the advisor or counterpart can hear tone and see emotion.

Roles of Negotiation

10-54. When entering a negotiation, advisors must know what their role. The three major roles associated with negotiation are negotiator, mediator, and arbitrator. Each role has its own tactics or techniques and separate functions.

Negotiator

10-55. Advisors most often find themselves in the role of a negotiator. A negotiation is the process by which parties exchange commitments or promises to resolve disagreements and reach a settlement. Advisors prepare for each negotiation so they will not be at a disadvantage during the negotiation. To prepare for negotiations advisors examine their options, discover what motivates the parties, list all positive scenarios, and dispute resolutions.

10-56. Advisors examine their options. They question their goals, the importance of the goals, and the amount of leverage they have. They determine who is in power and what to do if no agreement is reached. Advisors uncover what motivates their counterparts. They delve into their counterparts' underlying needs.

10-57. By listing all the scenarios that satisfy their needs, advisors can maximize the gain for all parties. Advisors can add features to the scenarios that increase the opportunities for all parties to be satisfied. When advisors dispute resolutions, they often find some form of compromise. One of the best outcomes of compromise is when all parties leave the bargaining table satisfied but no single party feels like they have won or lost the argument. A skilled negotiator should strive to achieve a win-win result.

10-58. The negotiator represents one side and has a vested interest in the outcome of the negotiation. It is best to avoid situations where there is a conflict of interests. By blindly pushing the coalition agenda, the advisor could lose the trust of the counterpart. Considerations for a negotiator should include:

- If the foreign security forces ask the advisor to negotiate for them, they expect the advisor, as their negotiator, to have their best interests in mind. The coalition may try to use the advisor to push their agenda and influence rather than negotiate.
- If the foreign counterparts take illegal action, advisors cannot use commitment techniques. They need to use some form of compliant technique.
- If the foreign counterpart is not violating any laws, the advisor should focus on commitment techniques. Success will not be achieved all at once.

Mediator

10-59. Advisors may find themselves in the role of a mediator. The mediator has three general roles to perform: facilitator, formulator, and manipulator. Table 10-3 identifies the roles and tactics advisors use.

Table 10-3. Mediator roles and tactics

Roles	Tactics
Facilitator	• Make contact. • Arrange for interaction. • Clarify the situation. • Supply missing information. • Transmit messages between parties. • Ensure all interests are discussed.
Formulator	• Control the pace, formality, and physical environment of the meetings. • Highlight common interests. • Help parties save face. • Keep process focused on the issues. • Make substitutive proposals. • Suggest concessions parties could make.
Manipulator	• Keep parties at the table. • Change parties expectations. • Highlight costs of nonagreement. • Reward concessions. • Promise resources. • Threaten the withdrawal of resources. • Offer to verify compliance. • Threaten to withdraw mediation.

Arbitrator

10-60. The foreign security force may try to put the advisor in the position of an arbitrator. An arbitration is the process by which the parties to a dispute submit their differences to the judgment of an impartial person or group selected by mutual consent or statutory provision. They want the advisor to decide between two positions. This is a dangerous position for the advisor. When a win-win solution can be found, the advisor can improve rapport with both parties, but when it is a win-lose situation, the advisor could alienate one or both groups.

Expected Outcomes of Negotiation

10-61. The outcome of negotiations sometimes can determine the outcome. Advisors consider the desired outcome to determine which confrontation preference to use. Negotiation can focuses on physical outcomes of relationships. How advisors combine their preferences to their focus affects their perceptions of the outcome.

Confrontation Preferences

10-62. Negotiators will have one of three preferences for confrontation: avoidance, collaboration, or competition. Those that prefer avoidance will consider accommodation as a viable means to resolve an issue. Negotiators that prefer collaboration consider compromise as the preferred outcome. Those negotiators who use competition want to win. Intentionally losing in a negotiation to avoid confrontation seems counterintuitive; however, it tends to protect the current relationship. Collaboration requires a mutual respect of the existing relationship, while competition is generally indifferent. It would be fairly easy to take advantage of a negotiator who prefers to avoid confrontation. At the same time, expecting to be able to compromise with a highly competitive negotiator may become a point of frustration. For the competitor, winning may be more important than the deal itself.

Negotiation Focus

10-63. Defining whether the negotiation is a win-win situation is not as simple as which side gained the most from the negotiation. Advisors often focus on physical outcomes whereas diplomats focus on relationships. Every negotiation varies on which factor is more important. Just as car dealers may accept less profit on individual services, they gain overall if the customer returns.

Perception of the Outcome

10-64. Comparing confrontation preferences leads to nine potential combinations. If advisors can determine a participant's preference—avoidance, collaboration, and competition—and focus, they can determine how the participant might feel based on the outcome of the negotiation. If one participant is a collaborator and one is a competitor, then a win-win situation is possible, but neither side will be completely satisfied with the outcome. The competitors will not have gotten a clear win, and the collaborators will not have gotten what they considered fair. Subsequently, they will both have some minor negative feelings toward the other.

Appendix A
Army Special Operations Forces Imperatives

This appendix outlines the imperatives for Army special operations forces conducting security force assistance and foreign internal defense operations. Table A-1 lists these imperatives.

Table A-1. Army special operations forces imperatives

Imperatives	*General Descriptors*	*Specific Actions*
Understand the Operational Environment	• Knowing the U.S. command relationships (military and interagency) • Knowing the population (who, what, when, where, why, and how) • Knowing the enemy and other antigovernment forces (crime, militias, gangs) • Knowing the higher commander's intent • Knowing that the press is a part of the operational environment (news reports of events reach the U.S. population quickly)	• Define relationship as partnership or augmentation • Conduct analysis using ASCOPE (areas, structures, capabilities, organizations, people, events) • Analyze the enemy order of battle • Protect the population
Recognize Political Implications	• Identifying the political (and legal) restrictions and implications of actions by U.S. and foreign security forces • Identifying political implications of establishing and enforcing rules of engagement policy • Identifying short and long term political implications of security force assistance • Understanding the importance of establishing human rights (critical for tactical and strategic success) • Acknowledging the effects of public opinion on the mission locally, in the United States, and internationally • Understanding that insurgents conduct operations to have political effects among the local populace, external actors, and the U.S. and multinational civilian public	• Include rules of engagement in the planning and briefs of operations • Include local government leadership in operations • Reinforce human rights briefings and training to all foreign security forces • Prepare all U.S. and foreign security forces to talk with the press • Prepare information engagements for all operations • Anticipate and neutralize the methods enemy forces pass propaganda

Imperatives	*General Descriptors*	*Specific Actions*
Facilitate Interagency Activities	• Conducting security force assistance as both a military and an interagency effort; taking advantage of the greater impact of collective efforts • Coordinating the scope and limitations of each agency's influence and programs • (as opposed to a purely military program) • Obtaining the active cooperation of the foreign security forces • Including military and civilian agencies working together for unity of effort	• Include interagency members in planning and training as appropriate • Provide assistance to support limited agency resources, allowing agencies to prioritize their primary roles • Conduct regular coordination meetings with civil agencies to synchronize security efforts • Include host-nation agencies whenever including U.S. agencies
Engage the Threat Discriminately	• Avoiding the use of overwhelming firepower—to prevent alienating the population and turning them away from the host-nation government • Giving training and advice to foreign security forces based on the mission and threat • Targeting insurgents, criminal activities (lawlessness), or subversion—to avoid alienating the populace • Minimizing the use of force in populace and resources control measures—weighed for potential gain versus potential cost • Ensuring that rules of engagement and escalation of force are balanced between force protection and safeguarding civilians	• Take civilians into account when considering weapons systems and techniques for operations • Include civilians in risk assessment during mission planning • Train and rehearse rules of engagement and escalation of force to prepare Soldiers to react in accordance to policy • Publicize reasons for use of populace and resource control (for example, curfews protect civilians); avoid enemy propaganda or rumors dictating the psychological response of the populace • Update rules of engagement and escalation of force periodically • Rules of engagement or escalation of force reflects engaging targets that are visually verified and not simply suppressive fire in populated areas
Consider Long-Term Effects	• Building up the foreign security forces and alleviating the root causes of the current situation over time—not just treating the symptoms • Solving problems in a broader political, military, social and economic context • Making policies, plans, and operations consistent with national and theater priorities and the objectives they support	• Understand local grievances and underlying issues • Interact with all aspects of the population a government to identify root causes • Ensure the commander's intent remains focused on higher goals and policy related to addressing root causes, and protecting the local population • Monitor and update rules of engagement as local government and police gain control of rule of law. Conduct operation that adhere to local laws as possible • Understand legal and political constraints (rules of engagement) to avoid strategic failure while achieving tactical success

Imperatives	*General Descriptors*	*Specific Actions*
Ensure Legitimacy of Special Operations	• Reinforcing and enhancing the legitimacy and credibility of the foreign security forces and government • Gaining trust and credibility to ensure advisors and U.S. forces obtain legitimacy with the local population • Obtaining legitimacy in theater, in the United States, and internationally to ensure ongoing support	• Conduct joint and multinational patrols and operations. • Ensure information engagement is common in all operations and contributes to maintaining credibility • Engage the media with the host nation to inform as well as to justify actions • Focus on what the population considers the root causes, not the U.S. perspective • Develop long-term solutions to restore or improve basic services with host-nation leadership
Anticipate and Control Psychological Effects	• Considering the psychological effects of all operations (such as populace and resources control measures, combat operations, and civic action) • Avoiding negative psychological effects that can negate tactical victories • Appreciating the power of people's perceptions versus reality	• All operations are developed, analyzed and executed with information engagement • Continue to gauge the psychological effects with the local population, local government, and security forces as well as the enemy (propaganda and interrogation)
Apply Capabilities Indirectly	• Teaching host-nation forces to plan instead of giving them tasks • Supporting U.S. efforts enhancing the legitimacy and credibility of the host-nation government and foreign security forces • Balancing control and pressure on foreign security forces to ensure they conduct successful tactical operations and accomplish missions • Fulfilling primary role to advise, train, and help foreign security forces take the lead and stabilize the host nation • Influencing a foreign security force without commanding it; not accepting a lack of effort or a high level of corruption	• Conduct training of foreign security forces only to teach basics • Transition to mentor or advisor relationship during planning and operations to encourage the host-nation to take the lead • Use discretion when advising counterparts as not to embarrass foreign security forces leadership • Use host-nation methods whenever possible

Imperatives	General Descriptors	Specific Actions
Develop Multiple Options	• Considering the complex and irregular environment • Planning to use the full range of capabilities • Anticipating that an operational environment may dictate a change of rules of engagement or type of operations • Ensuring foreign security forces are capable of shifting from one option to another before and during mission execution	• Continue to refine and train on battle drills and contingencies • Develop an understanding of skills and abilities of the members of the unit and community • Elicit the input of foreign security forces on how to address situations to widen the options beyond conventional techniques • Include government agencies, special operations forces, nongovernmental organizations, local police, and the host-nation government to develop additional methods or options for solving problems
Ensure Long-Term Sustainment	• Avoiding advising and training foreign security forces in techniques and procedures beyond their capabilities to sustain • Modifying tactics, techniques, and procedures; training; operations; and sustainment to fit the culture, education, and technological capability of the foreign security forces • Developing host-nation sustainment before handing over economic, social, political, military, and security initiatives and projects • Avoiding host-nation dependency on U.S.-funded programs	• Understand and utilize the equipment and manpower present in the foreign security forces unit's inventory • Assess the capabilities of foreign security forces for leadership, training, sustainment (logistics, administration, and equipment), and ongoing development of a professional security force • Recognize the programs that are durable, consistent, and sustainable by the host nation • Do not begin programs beyond the economic or technological capacity of the host nation
Provide Sufficient Intelligence	• Ensuring detailed, near real-time, all-source tactical intelligence products support all activities and programs • Ensuring detailed and comprehensive intelligence on all aspects of the operational environment and its internal dynamics support operations • Ensuring human intelligence, urban recon, and internal security elements assess internal threats, warn the government, and take action to penetrate and neutralize instability • Considering aspects of the society not directly related to the tactical combat situation when assessing the enemy threat • Using information from intelligence assets to advise, train, and help foreign security forces counterparts and ease interagency efforts	• Ensure host nation forces are conducting operations using accurate, real-time intelligence • Establish priority of effort when identifying intelligence requirements • Assist foreign security forces to prioritize and focus counterintelligence on the commander's critical information requirement to receive timely and specific intelligence • Assess other organizations that impede the rule of law and government functions (such as gangs, criminals, smugglers, militias, mercenaries, drug lords, and narcoterrorists)

Imperatives	*General Descriptors*	*Specific Actions*
Balance Security and Synchronization	• Maintaining unity of effort and avoiding compartmentalizing key security forces from planning and intelligence contingencies even when security concerns exist • Avoiding either excessive or insufficient security • Anticipating reasons foreign security forces may not provide intelligence, such as lack of internal security or lack of U.S. intelligence support	• Encourage combined planning, preparation, and operations with foreign security forces' military, police, and governmental units • Resolve conflicting demands on mission planning and execution • Employ measures to mitigate risk of operations security instead of dismissing the host-nation government, police, or military needed for the overall effort • Strive to gain the trust of the local population to gain the most human intelligence and support from the foreign security forces

Appendix B
Legal Considerations

GENERAL LEGAL BASIS

B-1. Law and policy govern the actions of U.S. forces in all military operations, including security force assistance. For U.S. forces to conduct operations, a legal basis must exist. This legal basis profoundly influences many aspects of an operation. It affects the rules of engagement, how U.S. forces organize and train foreign forces, the authority to spend funds to benefit the host nation, and the authority of U.S. forces to detain and interrogate. Under the Constitution of the United States, the President is the Commander in Chief of U.S. forces. Therefore, orders issued by the President or the Secretary of Defense to a combatant commander provide the starting point in determining the legal basis. Laws are legislation passed by Congress and signed into law by the President and treaties to which the United States is party. Policies are executive orders, departmental directives and regulations, and other authoritative statements issued by government officials. This appendix summarizes some of the laws and policies that bear upon U.S. military operations in support of security force assistance. This summary does not replace a consultation with the unit's supporting staff judge advocate.

LEGAL AUTHORITY FOR SECURITY FORCE ASSISTANCE

B-2. If the Secretary of State requests and the Secretary of Defense approves, U.S. forces can participate in security force assistance. The request and approval can go through standing statutory authorities in Title 22, U.S. Code. Title 22 contains the Foreign Assistance Act of 1961, the Arms Export Control Act of 1976, and other laws. It authorizes security assistance, developmental assistance, security force assistance and other forms of bilateral aid.

B-3. The request and approval for security force assistance might also occur under various provisions in Title 10, U.S. Code. Title 10 authorizes certain types of military-to-military contacts, exchanges, exercises, and limited forms of humanitarian and civic assistance in coordination with the U.S. ambassador to the host nation. In such situations, U.S. forces may be granted status as administrative and technical personnel based on a status-of-forces agreement or an exchange of letters with the host nation. This cooperation and assistance is limited to liaison, contacts, training, equipping, and providing defense articles and services. It does not include direct involvement in operations. Assistance to police by U.S. forces is permitted but, generally, Department of Defense (DOD) does not serve as the lead government department. Without receiving a deployment or execution order from the President or Secretary of Defense, U.S. forces may be authorized to make only limited contributions during operations that involve security assistance.

EXISTING UNITED STATES LAW

B-4. The Constitution divides the power to wage war between the executive and legislative branches of government. Under Article I, Congress holds the power to declare war; to raise and support armies; to provide and maintain a navy; and to make all laws necessary and proper for carrying out those responsibilities. Balancing that legislative empowerment, Article II vests the executive power in the President and makes the President the Commander in Chief of the Armed Forces. This bifurcation of the war powers created an area in which the coordinate political branches of government exercise concurrent authority over decisions relating to the use of Armed Forces overseas as an instrument of U.S. foreign policy.

B-5. The Supremacy Clause of the Constitution (Article VI) states, in part, that all treaties made by the United States are the supreme law of the land. Therefore, ratified treaties, such as the United Nations (U.N.) Charter (see paragraphs B-7 through B-8) and the Geneva Conventions, create legal obligations on U.S. forces regarding their ability to perform various types of missions and functions.

B-6. The War Powers Resolution of 1973 ensures the collective judgment of both the executive and legislative branches to deploy U.S. forces by requiring consultation of and reports to Congress, in any of the following circumstances:

- Introduction of troops into actual hostilities.
- Introduction of troops, equipped for combat, into a foreign country.
- Greatly enlarging the number of troops, equipped for combat, in a foreign country.

B-7. The President is required to make such reports within 48 hours of the triggering event, detailing—

- The circumstances necessitating introduction or enlargement of troops.
- The constitutional or legislative authority upon which he bases his action.
- The estimated scope and duration of the deployment or combat action.

B-8. Since the War Powers Resolution was passed over the veto of President Nixon, no President has either conceded its constitutionality or complied fully with its mandates.

International Law and Treaties

B-9. The U.N. Charter became effective on 24 October 1945 after being ratified by the United States and a majority of other signatories. The U.N. Charter mandates that all member states resolve their international disputes peacefully. It requires that member states refrain in their international relations from the threat or use of force. The U.N. Charter also provides that all nations have the right to use self-defense to combat acts of aggression against them until such time as the Security Council takes action.

B-10. The U.N. Charter provides the essential framework of authority for the use of force, effectively defining the foundations for a modern *jus ad bellum.* Inherent in its principles are the requirements for necessity, proportionality, and timeliness. Necessity involves considering—

- The exhaustion or ineffectiveness of peaceful means of resolution.
- The nature of coercion applied by the aggressor State.
- The objectives of each party.
- The likelihood of effective community intervention.

B-11. Proportionality refers to limiting force in magnitude, scope, and duration to that which is reasonably necessary to counter a threat or attack. Timeliness refers to situations in which the delay of a response to an attack or the threat of attack attenuates the immediacy of the threat and the necessity to use force in self-defense.

B-12. U.S. forces obey the law of war. The law of war is a body of international treaties and customs, recognized by the United States as binding. It regulates the conduct of hostilities and protects noncombatants. The main law-of-war protections come from The Hague and Geneva Conventions. They apply at the tactical and operational levels and are summarized in ten rules shown in table B-1.

Table B-1. Summary of the law-of-war rules

• Fight only enemy combatants.
• Do not harm enemies who surrender; disarm them and turn them over to the chain of command.
• Do not kill or torture detainees.
• Collect and care for the wounded, whether friend or foe.
• Do not attack medical personnel, facilities, or equipment.
• Destroy no more than the mission requires.
• Treat all civilians humanely.
• Do not steal; respect private property and possessions.
• Do one's best to prevent violations of the law of war.
• Report all violations of the law of war to superiors.

B-13. During security force assistance operations, commanders must be aware of Common Article 3 of the Geneva Conventions and the status of insurgents under the laws of the host nation. One article contained in all four of the Geneva Conventions—Common Article 3—is specifically intended to apply to internal armed conflicts:

> *In the case of armed conflict not of an international character occurring in the territory of one of the High Contracting Parties, each Party to the conflict shall be bound to apply, as a minimum, the following provisions:*
>
> *(1) Persons taking no active part in the hostilities, including members of armed forces who have laid down their arms and those placed "hors de combat" by sickness, wounds, detention, or any other cause, shall in all circumstances be treated humanely, without any adverse distinction founded on race, colour, religion or faith, sex, birth or wealth, or any other similar criteria. To this end, the following acts are and shall remain prohibited at any time and in any place whatsoever with respect to the above-mentioned persons: (a) Violence to life and person, in particular murder of all kinds, mutilation, cruel treatment and torture; (b) Taking of hostages; (c) Outrages upon personal dignity, in particular humiliating and degrading treatment; (d) The passing of sentences and the carrying out of executions without previous judgment pronounced by a regularly constituted court, affording all the judicial guarantees which are recognized as indispensable by civilized peoples.*
>
> *(2) The wounded and sick shall be collected and cared for. An impartial humanitarian body, such as the International Committee of the Red Cross, may offer its services to the Parties to the conflict. The Parties to the conflict should further endeavor to bring into force, by means of special agreements, all or part of the other provisions of the present Convention. The application of the preceding provisions shall not affect the legal status of the Parties to the Conflict.*

HOST-NATION LAW AND STATUS-OF-FORCES AGREEMENTS

B-14. After considering the type of baseline protections represented by fundamental human rights law, the military leader must be advised in regard to the other bodies of law that leader should integrate into planning and execution. This includes consideration of host-nation law. Because of the nature of most noninternational armed conflict missions, commanders and staffs must understand the technical and pragmatic significance of host-nation law within the area of operations. Although in theory, understanding the application of host-nation law during military operations is perhaps the simplest component, in practice it is perhaps the most difficult.

B-15. Judge advocates must recognize the difference between understanding the technical applicability of host-nation law and that law's application to control the conduct of U.S. forces during operations. In short, the significance of host-nation law declines in proportion to the movement of the operation toward violent conflict. Judge advocates should understand that U.S. forces enter other nations with a legal status that exists anywhere along a notional legal spectrum.

B-16. Judge advocates and planners must be familiar with any status-of-forces agreements or status-of-mission agreements that may be applicable. In any given country, there may be agreements short of status-of-forces agreements, such as diplomatic notes, on point. Relevant international documents affecting military operations may be difficult to locate. Several sources are available for locating applicable international agreements governing the status of U.S. forces or affecting military operations. Department of State (DOS) publications, such as *Treaties in Force,* contain unclassified international agreements. Both the relevant geographic combatant command's legal office and the defense attaché or military assistance group at the embassy normally have access to host-nation or international agreements impacting a military operation.

B-17. Status-of-forces and other forms of agreements frequently exist. They are essentially contractual agreements or treaties between two or more nations that establish the legal status of military personnel in foreign countries. Topics that are usually covered in a status-of-forces agreement include criminal and civil jurisdiction, taxation, and claims for damages and injuries. In the absence of an agreement or some other arrangement with the host nation, DOD personnel in foreign countries may be subject to its laws. Commanders ensure that all personnel understand status of U.S. forces in the area of operations and are trained accordingly.

LEGAL CONSTRAINTS ON THE SECURITY FORCE ASSISTANCE MISSIONS

B-18. U.S. law and regulation play a key role in establishing the parameters by which military forces may conduct security force assistance missions. These parameters tend to constitute constraints on the activities of military units. They range from the rules of engagement in combat situations to the ability to spend government funds in furtherance of a training or support mission.

GENERAL PROHIBITION ON ASSISTANCE TO POLICE

B-19. Historically, DOD is not the lead government department for assisting foreign governments. DOS is the lead when U.S. forces provide security assistance—military training, equipment, and defense articles and services—to host-nation governments. The Foreign Assistance Act of 1961 specifically prohibits assistance to foreign police forces except within specific exceptions and under a Presidential directive. When providing assistance to training, DOS provides the lead role in police assistance through its Bureau of International Narcotics and Law Enforcement Affairs. The President, however, may delegate this role to other agencies. This was done in 2004, when the President signed a decision directive granting the commander, United States Central Command, authority to train and equip Iraqi police.

TRAINING AND EQUIPPING FOREIGN SECURITY FORCES

B-20. All training and equipping of foreign security forces are specifically authorized. U.S. laws require Congress to authorize expenditures for training and equipping foreign forces. The laws of the United States also require DOS to verify that the host nation receiving the assistance is not in violation of human rights. The Secretary of Defense may authorize deployed U.S. forces to train or advise host-nation security forces as part of the mission. In this case, DOD personnel, operations, and maintenance appropriations provide an incidental benefit to those security forces. Numerous other programs to assist foreign forces are paid for with funds appropriated by Congress for that purpose. Consultation with a staff judge advocate or legal advisor early in the planning ensures the availability of funds for missions to train and equip foreign forces.

RULES OF ENGAGEMENT

B-21. *Rules of engagement* are directives issued by competent military authority that delineate the circumstances and limitations under which United States forces will initiate and/or continue combat engagement with other forces encountered (JP 1-02). Often these directives are specific to the operation. If there are no operation-specific rules of engagement, U.S. forces apply standing rules of engagement. When operating with a multinational force, commanders must coordinate the rules of engagement thoroughly.

FISCAL LAW CONSIDERATIONS

B-22. In security force assistance missions, like all operations, commanders require specific authority to expend funds normally found in the annual DOD appropriations act. As a general rule, operation and maintenance funds may not be used for security force assistance missions. In recent years, Congress appropriated additional funds to commanders for conducting more complex stability operations that were not typically covered by operation and maintenance funds. Examples include the Commander's Emergency Response Program (CERP), the Iraq Relief and Reconstruction Fund, Iraq Freedom Fund, and Commander's Humanitarian Relief and Reconstruction.

B-23. DOS has the primary responsibility, authority, and funding to conduct foreign assistance on behalf of the U.S. government. Foreign assistance encompasses any and all assistance to a foreign nation, including security assistance (assistance to the internal police forces and military forces of the foreign nation), development assistance (assistance to the foreign government in projects that will assist the development of the foreign economy or their political institutions), and humanitarian assistance (direct assistance to the population of a foreign nation). The legal authority for DOS to conduct foreign assistance is found in the Foreign Assistance Act of 1961.

B-24. There are two exceptions to the general rule requiring the use of Title 22 funds for foreign assistance. The first exception is for interoperability, safety, and familiarization training. DOD may fund the training (as opposed to goods and services) of foreign militaries with operation and maintenance dollars only when the purpose of the training is to enhance interoperability, familiarization, and safety training. Operation and maintenance funds may not be used for security force assistance training. This exception applies only to interoperability training. The second exception is congressional appropriation or authorization to conduct security foreign assistance. This exception allows DOD to fund foreign assistance operations if Congress has provided a specific appropriation or authorization to execute the mission.

Constraints on Funding

B-25. The Foreign Operations Appropriation Act (an annual DOS appropriations act) was enacted in 1997. Known as the Leahy Amendment, it prohibits the U.S. government from providing funds to a unit of the security forces of a foreign country if DOS has credible evidence that the unit has committed gross violations of human rights. The Leahy Amendment contains additional constraints on government funding of security force assistance missions. The law restricts funding until the Secretary of State determines and reports that the government of such country is taking effective measures to bring the responsible members of the security forces unit to justice.

B-26. Congress specifically appropriates funds for foreign assistance. The United States Agency for International Development (USAID) expends such funds under the legal authorities in Title 22. Provisions of Title 10 may also authorize amounts of money for these purposes. Standing funding authorities are narrowly defined and generally require advance coordination within DOD and DOS.

B-27. Effective foreign forces need training and equipment. U.S. laws require Congress to authorize such expenditures. U.S. laws also require DOS to verify that the host nation receiving the assistance is not in violation of human rights.

Funding Authorization for Security Force Assistance

B-28. All training and equipping of foreign security forces must be specifically authorized. Military and civilian personnel, operations, and maintenance appropriations should typically provide only an incidental benefit to those security forces. All other weapons, training, equipment, logistic support, supplies, and services provided to foreign forces must be paid for with funds appropriated by Congress for that purpose. Examples include the Iraq Security Forces Fund and the Afghan Security Forces Fund. Moreover, the President must give specific authority to DOD for its role in training and equipping foreign security forces. In May of 2004, the President signed National Security Presidential Directive (NSPD) 36 that made the commander, U.S. Central Command, under policy guidance from the chief of mission, responsible for coordinating all U.S. government efforts to organize, train, and equip Iraqi Security Forces, including police. Absent such a directive, DOD lacks authority to take the lead in assisting a host nation to train and equip its security forces.

Key Security Assistance and Foreign Assistance Funding Programs

B-29. Congress has appropriated funds to allow DOS to conduct its foreign assistance mission, including the following activities:

- Foreign Military Financing Program.
- International Military Education and Training Program.
- Economic Support Fund.
- Peacekeeping operations.
- Antiterrorism assistance.
- Global humanitarian demining.

- Refugee assistance.
- Personnel details.

B-30. DOS administers the following programs but no funds are appropriated to sustain them:

- Foreign Military Sales Program.
- Foreign Military Lease Program.
- Economy Act Security Assistance.
- U.S. Government Commodities and Services Program.
- Direct Commercial Sales Program.

B-31. Additional special support created by law for the foreign assistance mission includes—

- Excess defense articles.
- Presidential drawdowns.

B-32. DOS directly, or indirectly through USAID, finances numerous development assistance activities to address the following needs:

- Rule of law.
- Governance.
- Agriculture and nutrition.
- Population control.
- Health.
- Education.
- Energy.
- Environment improvement.

B-33. CERP (see paragraph B-19) is an example of a targeted humanitarian assistance fund program. CERP's primary purpose is to enable military commanders to respond to urgent humanitarian relief and reconstruction requirements within their fields of interest by carrying out programs that will immediately assist the host-nation populace. CERP was originally funded with seized Iraqi assets, but Congress later appropriated U.S. funds for the purpose. CERP is not applicable to missions outside Iraq and Afghanistan. Future missions, though, may have similar funding sources established to facilitate a humanitarian assistance mission. Commanders consult with the servicing judge advocate to determine the availability of funds.

Appendix C
Information Engagement

INFORMATION SUPERIORITY

C-1. Information shapes a commander's operational environment at every level. Twenty-first century conflicts occur in operational environments with instant communication. Because information systems are everywhere and exposure to news and opinion media is pervasive, an individual's words and actions can have immediate operational and strategic implications. Gaining and maintaining the trust of key publics has never been more important. Army forces conduct operations in the public eye and must maintain public confidence by being consistent in word and deed.

C-2. *Information superiority* is the operational advantage derived from the ability to collect, process, and disseminate an uninterrupted flow of information while exploiting or denying an adversary's ability to do the same (JP 3-13). Advisors build partner capacity in this mission area. Advisors ensure that host-nation counterparts integrate information superiority into operations as carefully as fires, maneuver, protection, and sustainment. Army information tasks contribute to information superiority. Information tasks are used to shape the operational environment (see FM 3-0). The information tasks are—

- Information engagement.
- Command and control warfare.
- Information protection.
- Operations security.
- Military deception.

C-3. Advisors prioritize the five information tasks based on the host nation's capabilities and strengths. This appendix discusses some information engagement capabilities that advisors can expect to employ while conducting security force assistance, as well as considerations when working with the media. For a further discussion of the other information tasks, see FM 3-0.

INFORMATION ENGAGEMENT CAPABILITIES

C-4. *Information engagement* is the integrated employment of public affairs to inform U.S. and friendly audiences; psychological operations, combat camera, and U.S. government strategic communications and defense support to public diplomacy, and other means necessary to influence foreign audiences; and leader and Soldier engagements to support both efforts (FM 3-0).

C-5. In general, advisors focus on information engagement because land operations occur among civilian populations. Advisors can have a significant impact by focusing on this task. Advisors work with host-nation counterparts to communicate information to the populace, build their trust and confidence in friendly forces, promote popular support for friendly operations, and positively influence public perceptions and behavior. Advisors integrate information engagement throughout the operations process.

C-6. Advisors ensure host-nation counterparts focus their information engagement activities on achieving desired effects locally. However, land operations always occur in a regional and even global context. Therefore, local information engagement activities must support and complement higher headquarters information engagement activities, host-nation and U.S. government strategic communications guidance (when available), and broader host-nation and U.S. government policy where applicable.

C-7. Three information engagement capabilities deserve particular attention as the advisor can directly influence them. These are public affairs, combat camera, and leader and Soldier engagements.

PUBLIC AFFAIRS

C-8. First, public affairs can have a profound and lasting effect on public perceptions of operational success or failure. This effect may have strategic consequences, even when related to tactical operations. Therefore, advisors ensure that public affairs efforts are integrated at tactical, operational, and strategic levels. (See paragraphs C-10 through C-20 for more information to help advisors working directly with the media.)

COMBAT CAMERA

C-9. Second, combat camera is used to document a wide range of tactical and operational successes and to counter enemy propaganda. If there is no dedicated host-nation combat-camera capability, the advisor recommends that a host-nation soldier be designated, equipped, and trained for this function. If that is not possible, the advisor should obtain a digital camera and, if necessary, photograph relevant host-nation activities personally. Positive photographs can have a tremendous benefit, and digital photos are easily taken and transmitted.

LEADER AND SOLDIER ENGAGEMENTS

C-10. Third, leader and Soldier engagements are a fundamental and highly effective method to influence people. They involve face-to-face interaction by leaders and Soldiers to influence the perceptions of the local populace. Military leaders engage key communicators and civilian leaders (often referred to as key leaders) through carefully planned meetings. Examples of key leaders include local tribal elders, elected or appointed officials, military and police commanders, and nongovernmental organization leaders.

WORKING WITH THE MEDIA

C-11. The appropriate public affairs officer normally manages media contacts. However, this is not always possible. A deployed advisor team seldom has a public affairs officer representative.

C-12. Occasions may arise when the host-nation press wants to question advisors. Even if an advisor team lacks a public affairs officer representative, advisors must avoid refusing to speak to the media. Apparent refusal to speak with accredited members of the media, for whatever reason, may inadvertently create strong negative impressions with strategic implications. Therefore, advisor teams designate a media spokesperson and keep this individual well informed. This individual coordinates media contacts for the advisor team. This individual ensures that official, published public affairs officer guidance and command policies regarding media relations, including anonymity and operations security, are observed. The following paragraphs provide general media relations guidance for advisor teams.

ENHANCING RELATIONSHIPS WITH THE MEDIA

C-13. Advisors strive to convey a positive impression and enhance relationships between the media and the military. If deemed acceptable within the public affairs guidance and command policies, the advisor team's designated media spokesperson—

- Maintains a list of reliable reporters and editors that cover operations in the team's operational area and keeps them informed of significant activities.
- Answers media inquiries promptly, accurately, and courteously.
- Finds out reporters' deadlines and uses them to the friendly force's advantage.
- Stresses the human aspects of a story that readers, viewers, or listeners can identify with, including the negative impact of opposing operations on the people.
- Points out how advisors and host-nation counterparts are working to address the needs of the unfortunate.
- Encourages the media to observe constructive work the host nation and advisors are doing and to interview them about their efforts.
- Identifies a location in advance for the press to take photos and videos.

CORRECTING PUBLIC MISCONCEPTIONS

C-14. If information is withheld, journalists often fall back on speculation. Such speculation is usually inaccurate. However, it is often near enough to the truth that it is easily accepted by large sections of the public and possibly by established governments. In addition, partisan sources may find it advantageous to leak part of a story to the press to build up public support for their position. On occasion, such activities can grow into fully orchestrated press campaigns.

C-15. With modern satellites and communications technology, media are able to distribute reports and photographs extremely quickly. Incidents, sometimes fabricated or slanted toward a partisan viewpoint, may be aired in living rooms across the globe the same day.

C-16. However, advisors and host-nation counterparts frequently can correct public misconceptions and provide accuracy, adequate context, and proper characterization promptly. They can turn a potential liability into a major advantage by remaining prepared to coordinate and communicate with the media. They can take advantage of the media's powerful and far-reaching ability to communicate with critical audiences.

C-17. If the media report inaccurate or damaging information, advisors use caution in deciding on a course of action. Advisors (usually through the team's designated media spokesperson) inform their chain of command. Advisors consider the following questions:

- Is it important enough to correct or is it a detail that—in the long term—is not really important?
- How damaging is the charge, criticism, or error? Will a correction simply give greater visibility to an unfavorable point of view?
- Is a correction worth a restatement of the entire problem to new audiences who did not read or see it the first time?
- Is it possible to target only the audience originally exposed to the story?
- Can significant gain result from pointing out an error?

C-18. When incorrect or misleading information is released to the public through host-nation government or media sources, the advisor team determines a course of action appropriate to the situation. It must include a plan to ensure that relevant host-nation government officials and media recognize the inaccuracy of the information. This may prevent future incidents. Operational and combat experience since 2001 shows that in the struggle for information superiority, friendly forces must be candid and honest. Additionally, they must not allow lies disseminated by enemies to stand unchallenged, particularly in areas of great international interest.

INTERVIEWS WITH ADVISOR TEAM MEMBERS

C-19. Advisor teams sometimes arrange for media interviews with selected advisor team members. Properly conducted interviews can support the mission and even improve morale. When preparing for an interview, the team's designated spokesperson—

- Provides all team members with a single and simple theme to convey to the press, sometimes in the form of a mission narrative (a succinct account of the mission as a whole, explained consistently with the host nation's cultural view).
- Conducts rehearsals in front of video cameras, using mock interviews to practice communication of the theme.
- Selects interviewees based on their understanding of the theme covered, not their comfort level in front of a reporter. (Shyness is a normal trait that may add a human touch to a sensitive situation.)

C-20. If an advisor team member does not know the answer to a reporter's question, the individual should attempt to find out and get back to the reporter or refer the reporter to an appropriate source. This may establish the team as a helpful information source and develop a relationship for future balanced coverage.

C-21. Communicating with the media is the most difficult aspect of information engagement. Anything advisors say or do publicly can impact the mission. Therefore, advisors ensure consistency with national policy, professionalism, and clear communication. Advisors must know and follow the policies of the geographic combatant commander, the American Embassy, and the appropriate public affairs officer regarding media interviews. Additionally, advisors follow the specific guidelines shown in table C-1.

Table C-1. Specific guidelines for media interviews

• Maintain a friendly, yet professional attitude.
• Remain positive, greet interviewers, and welcome questions. (This is difficult under stressful or tragic conditions; however, a calm, mature appearance earns respect from audiences.)
• Use language that is clear and easy to understand.
• Avoid military jargon or terminology that others can misinterpret.
• Never enter an interview unprepared.
• Prepare notes and study them carefully; use facts to back up statements.
• Ensure the facts supporting their position are up to date and come to mind easily.
• Anticipate possible questions and think about various responses.
• Rehearse before interviews on location if possible and, if time permits, role-play tough questions.
• Learn and use reporters' names.
• Research interviewers' organizations, including—
    ▪ The organizations' views or views they defend.
    ▪ Their underlying motivations.
    ▪ How members of the organizations previously conducted interviews.
• Keep the interests of the local nationals or other beneficiaries of the mission at the forefront.
• Avoid talking from the point of view of U.S. interests.
• Finish answering one question before answering another if two or more questions are asked at the same time.
• Avoid speculation and give only factual information that can be verified.
• Use caution in quoting statistics as such data may be easily disputed or reinterpreted.
• Avoid repeating a question that contains incorrect information or inflammatory language (this may result in a misquotation).
• Strive to tell the truth.
• Avoid saying *no comment*. (It gives the appearance the advisor has something to hide.)
• If unsure of something, admit it rather than mistakenly lie.
• If questioned about another agency's activities, refer to that agency for comment.
• Do not speak for other organizations.
• Avoid exaggeration or claims that cannot be backed up.
• Refuse all off-the-record discussions.
• Expect any statement they make to be quoted publicly.
• Avoid criticizing other groups or organizations, and avoid committing information fratricide.
• Remind reporters that photographing recognizable dead Soldiers, charts, maps, supply depots, or electronic warfare assets is off limits.
• Remind reporters that their personal security is not a primary military concern.
• Remain objective and calm.
• Demonstrate conduct befitting a representative the U.S. government
• Never give personal opinions and beliefs.
• Avoid reacting emotionally when interviewers appear skeptical or hostile.
• Maintain focus on the mission.

Glossary

SECTION I – ACRONYMS AND ABBREVIATIONS

ABCA	American, British, Canadian, Australian, and New Zealand Armies Program
AR	Army regulation
BCT	brigade combat team
CCIR	commander's critical information requirement
CERP	Commander's Emergency Response Program
COA	course of action
COIST	company intelligence support team
DA	Department of the Army
DOD	Department of Defense
DODD	Department of Defense directive
DOS	Department of State
DOTMLPF	doctrine, organization, training, materiel, leadership and education, personnel, facilities
FM	field manual
FSF	foreign security forces
IDAD	internal defense and development
ISR	intelligence, surveillance, reconnaissance
JP	joint publication
METT-TC	mission, enemy, terrain and weather, troops and support available, time available, civil considerations (mission variables)
MTT	military transition team
NSPD	national security Presidential directive
PIR	priority intelligence requirement
S-2	intelligence staff officer
S-3	operations staff officer
S-4	logistics staff officer
SFA	security force assistance
TC	training circular
UN	United Nations
US	United States
USAID	United States Agency for International Development

SECTION II – TERMS

foreign internal defense

(joint) Participation by civilian and military agencies of a government in any of the action programs taken by another government to free and protect its society from subversion, lawlessness, and insurgency. (JP 3-05)

foreign military sales

(joint) That portion of United States security assistance authorized by the Foreign Assistance Act of 1961, as amended, and the Arms Export Control Act of 1976, as amended. This assistance differs from the Military Assistance Program and the International Military Education and Training Program in that the recipient provides reimbursement for defense articles and services transferred. (JP 1-02)

***foreign security forces**

Forces, including but not limited to military, paramilitary, police, and intelligence forces; border police, coast guard, and customs officials; and prison guards and correctional personnel, that provide security for a host nation and its relevant population or support a regional security organization's mission.

host nation

(joint) A nation that receives the forces and/or supplies of allied nations, coalition partners, and/or NATO organizations to be located on, to operate in, or to transit through its territory. (JP 3-57)

humanitarian and civic assistance

(joint) Assistance to the local populace provided by predominantly US forces in conjunction with military operations and exercises. This assistance is specifically authorized by Title 10, United States Code, section 401, and funded under separate authorities. Assistance provided under these provisions is limited to (1) medical, dental, and veterinary care provided in rural areas of a country; (2) construction of rudimentary surface transportation systems; (3) well drilling and construction of basic sanitation facilities; and (4) rudimentary construction and repair of public facilities. Assistance must fulfill unit training requirements that incidentally create humanitarian benefit to the local populace. (JP 3-57)

internal defense and development

(joint) The full range of measures taken by a nation to promote its growth and to protect itself from subversion, lawlessness, and insurgency. It focuses on building viable institutions (political, economic, social, and military) that respond to the needs of society. (JP 3-07.1)

international military education and training

(joint) Formal or informal instruction provided to foreign military students, units, and forces on a nonreimbursable (grant) basis by offices or employees of the United States, contract technicians, and contractors. Instruction may include correspondence courses; technical, educational, or informational publications; and media of all kinds. (JP 1-02)

security assistance

(joint) Group of programs authorized by the Foreign Assistance Act of 1961, as amended, and the Arms Export Control Act of 1976, as amended, or other related statutes by which the United States provides defense articles, military training, and other defense-related services, by grant, loan, credit, or cash sales in furtherance of national policies and objectives. (JP 3-57)

security force assistance

The unified action to generate, employ, and sustain local, host-nation, or regional security forces in support of legitimate authority. (FM 3-07)

security sector reform

The set of policies, plans, programs, and activities that a government undertakes to improve the way it provides safety, security, and justice. (FM 3-07)

References

Field manuals and selected joint publications are listed by new number followed by old number.

REQUIRED PUBLICATIONS

These documents must be available to intended users of this publication.

FM 1-02 (101-5-1). *Operational Terms and Graphics.* 21 September 2004.

JP 1-02. *Department of Defense Dictionary of Military and Associated Terms.* 12 April 2001.

RELATED PUBLICATIONS

These documents contain relevant supplemental information.

JOINT AND DEPARTMENT OF DEFENSE PUBLICATIONS

Most joint publications are available online: <http://www.dtic.mil/doctrine/jpcapstonepubs.htm.>

DOD 5105.38-M. *Security Assistance Management Manual.* 3 October 2003. <http://www.dsca.osd.mil/samm/>

DODD 3000.05. *Military Support for Stability, Security, Transition, and Reconstruction (SSTR) Operations.* 28 November 2005. <http://www.dtic.mil/whs/directives/corres/html/300005.htm>

JP 3-0. *Joint Operations.* 17 September 2006.

JP 3-05. *Doctrine for Joint Special Operations.* 17 December 2003.

JP 3-07.1. *Joint Tactics, Techniques, and Procedures for Foreign Internal Defense (FID).* 30 April 2004.

JP 3-07.4. *Joint Counterdrug Operations.* 13 June 2007.

JP 3-57. *Civil-Military Operations.* 8 July 2008.

ARMY PUBLICATIONS

Most Army doctrinal publications are available online: <http://www.army.mil/usapa/doctrine/Active_FM.html>. Army regulations are produced only in electronic media. Most are available online: < http://www.army.mil/usapa/epubs/index.html>.

AR 11-31. *Army International Security Cooperation Policy.* 24 October 2007.

AR 12-1. *Security Assistance, International Logistics, Training, and Technical Assistance Support Policy and Responsibilities.* 24 January 2000.

FM 3-0. *Operations.* 27 February 2008.

FM 3-05.137. *Army Special Operations Forces Foreign Internal Defense.* 30 June 2008.

FM 3-05.202 (FM 31-20-3). *Special Forces Foreign Internal Defense Operations.* 2 February 2007.

FM 3-07. *Stability Operations.* 6 October 2008.

FM 3-24. *Counterinsurgency.* 15 December 2006.

FM 3-34.400 (FM 5-104). *General Engineering.* 9 December 2008.

FM 4-0 (100-10). *Combat Service Support.* 29 August 2003.

FM 4-02 (FM 8-10). *Force Health Protection in a Global Environment.* 13 February 2003.

FM 6-0. *Mission Command: Command and Control of Army Forces.* 11 August 2003.

FM 6-22 (22-100). *Army Leadership.* 12 October 2006.

FM 7-0. *Training for Full Spectrum Operations.* 12 December 2008.

FM 7-15. *The Army Universal Task List.* 27 February 2009.

TC 31-73. *Special Forces Advisor Guide.* 2 July 2008.

OTHER PUBLICATIONS

National Defense Strategy of 2008. Washington, DC: U.S. Government Printing Office, 2008. <www.defenselink.mil/news/2008 National Defense Strategy.pdf>

National Military Strategy of the United States of America. Washington, DC: U.S. Government Printing Office, 2004. <www.defenselink.mil/news/Mar2005/d20050318nms.pdf>

National Security Strategy of 2006. Washington, DC: U.S. Government Printing Office, 2006. <www.marforres.usmc.mil/docs/nss2006.pdf>

Quadrennial Defense Review of 2006. Washington, DC: U.S. Government Printing Office, 2006. <www.defenselink.mil/qdr/report/Report20060203.pdf>

Armed Forces. Title 10, U.S. Code. <http://www.loc.gov/law/help/guide/federal/uscode.php>

Arms Export Control Act of 1976. <http://www.loc.gov/law/help/guide.php>

The United States Constitution. <http://www.house.gov/house/Constitution/Constitution.html>

Foreign Relations and Intercourse. Title 22, U.S. Code. <http://www.loc.gov/law/help/guide/federal/uscode.php>

Foreign Operations Appropriation Act of 1997. <http://www.loc.gov/law/help/guide.php>

Foreign Assistance Act of 1961. <http://www.loc.gov/law/help/guide.php>

The Geneva Conventions of 1949. <http://www.loc.gov/rr/frd/Military_Law/MLR_site-map.html>

National Security Presidential Directive 36. "United States Government Operations in Iraq." 11 May 2004.

Treaties in Force 2007. <http://www.state.gov/s/l/treaty/treaties/2007>

Charter of the United Nations. <http://www.un.org/aboutun/charter/>

War Powers Act of 1973. <http://www.thecre.com/fedlaw/legal22/warpow.htm.>

OUTSIDE RECOMMENDED READING

von Steuben, Friedrick Wilhelm. *Regulations for the Order and Discipline of the Troops of the United States*. Philadelphia: Styner and Cist, 1779.

PRESCRIBED FORMS

None

REFERENCED FORMS

DA Form 2028. Recommended Changes to Publications and Blank Forms

Index

Entries are by paragraph number.

Entries are by paragraph number.

Entries are by paragraph number.

Entries are by paragraph number.

Made in the USA
Monee, IL
05 November 2026

67811387R00275